Modern Public Information Technology Systems:

Issues and Challenges

G. David Garson
North Carolina State University, USA

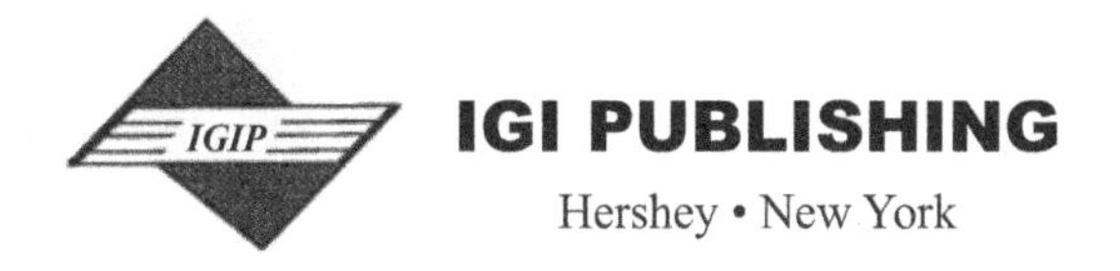

Acquisition Editor:	Kristin Klinger
Senior Managing Editor:	Jennifer Neidig
Managing Editor:	Sara Reed
Assistant Managing Editor:	Sharon Berger
Development Editor:	Kristin Roth
Copy Editor:	Shanelle Ramelb
Typesetter:	Sharon Berger
Cover Design:	Lisa Tosheff
Printed at:	Yurchak Printing Inc.

Published in the United States of America by
IGI Publishing (an imprint of IGI Global)
701 E. Chocolate Avenue
Hershey PA 17033
Tel: 717-533-8845
Fax: 717-533-8661
E-mail: cust@idea-group.com
Web site: http://www.idea-group.com

and in the United Kingdom by
IGI Publishing (an imprint of IGI Global)
3 Henrietta Street
Covent Garden
London WC2E 8LU
Tel: 44 20 7240 0856
Fax: 44 20 7379 0609
Web site: http://www.eurospanonline.com

Library of Congress Cataloging-in-Publication Data

Modern public information technology systems : issues and challenges / G. David Garson, editor.
p. cm.
Summary: "The nature of governance is rapidly changing, due to new technologies which expand public sector capabilities. This book examines the most important dimensions of managing IT in the public sector and explores the impact of IT on governmental accountability and distribution of power, the implications of privatization as an IT business model, and the global governance of IT"--Provided by publisher.
Includes bibliographical references and index.
ISBN 978-1-59904-051-6 (hardcover) -- ISBN 978-1-59904-053-0 (ebook)
1. Internet in public administration. 2. Information technology--Government policy. 3. Public administration--Information technology. 4. Electronic government information. I. Garson, G. David.
JF1525.A8M63 2007
352.3'802854678--dc22

2007007221

British Cataloguing in Publication Data
A Cataloguing in Publication record for this book is available from the British Library.

Modern Public Information Technology Systems:

Issues and Challenges

Table of Contents

Foreword

As public administration enters a difficult new era characterized by budgetary austerity, the privatization of public functions, and public demands for greater accountability, many have looked to information and communications technology as a critical component of public-service adaptation to the 21st century. Much of this response has been left to technically trained individuals who have not had the background afforded by public-administration education for managerial positions in the public sector. Public administration as a discipline, reflected in the essays contained in this book, brings a more holistic viewpoint to bear on central problems of e-government and the implementation of public information technology. Widespread and spectacular failures of information-technology projects, often driven by technocratic ignorance of human and political factors, which the public-administration literature emphasizes, provide an immediate and compelling reason to go beyond technical reasoning to a broader, more systemic understanding of the challenges involved in this domain. If this work can provide some small contribution to the field, it is simply that a government that works must be based on principles long understood in public administration and adapted to the context of modern public information systems given by authors in this anthology.

Preface

The readings that compose this volume are intended to constitute a survey of many of the most important dimensions of managing information technology in the public sector. This work updates and replaces earlier volumes: *Public Information Technology: Policy and Management Issues* (Idea Group, 2003) and *Information Technology and Computer Applications in Public Administration: Issues and Trends* (Idea Group Press, 1999). All contributions to the present volume have been substantially revised and updated, and five entirely new chapters replace outdated topics. Essays in Part I address general policy and administrative issues in this arena while those in Part II are more applied and address information technology skills needed by public managers. Taken together, it is hoped that a contribution is made by these essays toward the knowledge and competencies needed by graduate students of public administration and by practitioners new to this field.

Section I: Managing Information Technology in the Public Sector

In "Lip Service? How PA Journals and Textbooks View Information Technology," Alana Northrop points out the continued need for a reader on information technology by reviewing the early importance given to computing education by MPA programs and practitioners. Her chapter surveys current textbooks' and general public-administration journals' treatment of information technology, finding scant attention given to the topic. Northrop's chapter concludes by briefly discussing a range of issues with which public administrators should be conversant if they are to successfully utilize information technology in the delivery of public-sector services.

In "The Evolution of Federal Information Technology Management Literature: Does IT Finally Matter?" Stephen Holden finds federal agencies rely extensively on information technology to perform basic missions. His chapter analyzes federal IT management literature over time and compares federal IT-management literature to a normative model of management maturity focusing on the strategic objectives for IT and related management approaches. Public administration's minimal contribution to federal IT management literature raises profound questions of whether federal agencies are performing commensurate with public expectations as the theory and practice of IT management may be moving into a new, post-information-age era.

In "Politics, Accountability, and Information Management," Bruce Rocheleau provides examples of the politics of managing information in public. He shows how, within the organization, politics is involved in structuring decision making, struggles over purchases of hardware and software, interdepartmental sharing of information, and the flow of communications such as e-mail among employees. The chapter analyzes examples of each of these internal aspects of politics. Rocheleau also discusses evidence concerning whether political appointees or career administrators are more effective as information managers. Externally, this chapter discusses how information management has been used to attempt to achieve greater political accountability through e-reporting and examples of cases where purchasing problems spill over into the realm of external politics such as through attempts to privatize governmental information management function. The attempts to use governmental Web sites as mechanisms to achieve e-governance and greater citizen participation in the political process also make it impossible for information managers to insulate themselves against politics, Rocheleau argues.

In "Reconciling Information Privacy and Information Access in a Globalized Technology Society," George Duncan and Stephen Roehrig find that in reconciling information privacy and information access, agencies must address a host of difficult problems. These include providing access to information while protecting confidentiality, coping with health information databases, and ensuring consistency with international standards. The policies of agencies must interpret and respect the ethical imperatives of democratic accountability, constitutional empowerment, individual autonomy, and information justice. In managing confidentiality and data-access functions, agencies utilize techniques for disclosure limitation through restricted data and administrative procedures and through restricted access.

Chris Demchak and Kurt Fenstermacher, in "Privacy-Sensitive Tracking of Behavior with Public Information Systems: Moving Beyond Names in a Globalizing Mass Society," explore the roles of names and name equivalents in social tracking and control. This is particularly timely and urgent as increased interest in biometrics offers an insidious expansion of unique identifiers of highly personal data. Demchak and Fenstermacher review the extent of privacy-sensitive databases accumulating today in U.S. legacy federal systems and propose an alternative that reduces the likelihood of new security policies violating privacy. The authors conclude with a proposed conceptual change to focusing the social-order mission on the behavior of individuals rather than their identities (behavior identity knowledge model [BIK]).

Next, in "E-Government: An Overview," Shannon Schelin surveys the rapid growth of e-government. Her essay offers an overview of the historical premises, theoretical constructs, and associated typologies of e-government. These typologies posit a framework for understanding e-government, its potential benefits, and its related challenges. In the subsequent essay, "E-Participation Models," Suzanne Beaumaster discusses participation as the cornerstone of governance, now rendered more diverse through the possibilities offered by technology. She asks what kind of participation should we be fostering and what do we hope to gain through participative processes? Her chapter provides a definition and discussion of three e-participation models: information exchange, general discourse, and deliberative models of electronic participation in the governance process.

Picking up on the e-participation theme, Christopher Reddick in "E-Government and Creating a Citizen-Centric Government: A Study of Federal Government CIOs" examines the relationship between e-government and the creation of a more citizen-centric government. Using a conceptual framework, the author shows a possible relationship among management, resources, security, and privacy issues that would lead to creating a more citizen-centric government. Reddick explores the opinions of chief information officers (CIOs) on e-government issues and effectiveness. He finds that CIOs who have higher management capacity and project management skills are associated with creating a more citizen-centric federal government.

Rounding out Section I, in "The Federal Docket Management System and the Prospect for Digital Democracy in U.S. Rulemaking," Stuart Shulman traces how an interagency group led by the Environmental Protection Agency (EPA) worked to establish a centralized docket system for all U.S. federal rulemaking agencies. The resulting Federal Docket Management System (FDMS) reflects technical, administrative, financial, and political challenges and lessons. Actual progress, Shulman notes, has come with mixed results falling short of original visions. Shulman concludes that in the great American tradition of incrementalism, the FDMS represents a small step toward a number of worthy but perennially elusive goals linked to the prospect for digital democracy.

Section II:
Computer Applications in Public Administration

In this chapter on computer applications ("IT Innovation in Local Government: Theory, Issues, and Strategies"), Charles Hinnant and John O'Looney consider the social and technical factors that lead to technological innovation in general and e-government in particular for U.S. local governments. They discuss local governments' motivations to innovate and their technology characteristics, available resources, and stakeholder support, as well as other factors. Hinnant and O'Looney make the case that local governments should seek to formally assess the need to adopt e-government technologies, develop new funding strategies, and develop a mix of in-house and contracted IT services.

In "Information Technology as a Facilitator of Results-Based Management," James Swiss discusses IT in relation to results-based management, which encourages planning and target setting to make the organization more proactive, an emphasis on outcomes to make the organization better focused on its mission, quick performance feedback to make the organization more responsive, and continuous process improvements to make the organization better able to serve its clients. These changes are possible only with supporting information technology. This chapter discusses ways that IT can support the new management model if properly designed. Swiss discusses design considerations by which top public and nonprofit managers must determine what information would best guide upcoming major decisions and must also decide how they wish to balance system integration vs. costs, disintermediation efficiencies vs. client guidance, internal information accessibility vs. security, and frontline worker empowerment vs. organizational uniformity.

In "Managing IT Employee Retention: Challenges for State Governments," Deborah Armstrong, Margaret Reid, Myria Allen, and Cynthia Riemenschneider review the literature on factors that may reduce the voluntary turnover of public-sector IT professionals. Examples are presented that illustrate what states have been doing to improve their ability to retain their technology workforces. The authors conclude that while public-sector IT employees may not weigh the financial rewards associated with their jobs as heavily as private-sector IT employees might, workplace and job characteristics are important in ways public managers must recognize.

In "Computer Tools for Public-Sector Management," Carl Grafton and Anne Permaloff survey basic elements of the public-sector office, including word processing, spreadsheet, statistics, and database management programs. Web authoring software, presentation software, graphics, project planning and management software, decision analysis, and geographic information systems (GISs) are also surveyed as part of the public-office software suite.

In "Computers and Social Survey Research for Public Administration," Michael Vasu, Ellen Storey Vasu, and Ali Ozturk discuss electronic tools for citizen surveys and the integration of computing into survey research and focus groups used in research and practice in public administration. Their chapter reviews uses of computers in computer-assisted information collection (CASIC), computer-assisted telephone interviewing (CATI), computer-assisted personal interviewing (CAPI), and transferring survey research methods onto the Web. A second portion of the chapter gives special attention to continuous audience response technology (CART), using as an example a citizen survey focused on economic growth issues.

In "Geographic Information System Applications in the Public Sector," Douglas Carr and T. R. Carr trace how geographic information systems emerged in the 1970s to become a significant decision-making tool. Their chapter discusses various GIS applications and highlights issues that public managers should consider when evaluating implementation of a geographic information system to achieve its effective use in the public sector.

Charles Prysby and Nicole Prysby, in "You Have Mail, but Who is Reading It? Issues of E-Mail in the Public Workplace," discuss the increasing use of electronic mail in the workplace. This increase has been accompanied by important legal questions for public organizations. These questions fall into two basic categories: (1) issues of employee privacy regarding e-mail messages and (2) public access to e-mail under applicable freedom-of-information legislation. Privacy concerns have generated demands for greater protection of employee confidentiality, and some states have responded with legislation that covers e-mail in the workplace. Government organizations must treat at least some of their e-mail as part of the public record, making it open to public access, but this also can lead to conflict between public administrators, who may feel that much of their e-mail represents thoughts that were not intended for public disclosure, and external groups, such as the press, who feel that all such information belongs in the public domain. Given the uncertainty and confusion that frequently exist regarding these legal questions, it is essential that public organizations develop and publicize an e-mail policy that both clarifies what privacy expectations employees should have regarding their e-mail and specifies what record-keeping requirements for e-mail should be followed to appropriately retain public records.

In an essay titled "World Wide Web Site Design and Use in U.S. Local Government Public Management," Carmine Scavo and Jody Baumgartner explore the promise and reality of Web applications for U.S. local government. Four types of Web utilizations are analyzed—bulletin-board applications, promotion applications, service-delivery applications, and citizen-input applications—based on a survey of 145 municipal and county government Web sites originally conducted in 1998 and replicated in 2002 and again in 2006. The authors conclude that local governments have made progress in incorporating many of the features of the Web but have a long way to go in realizing its full promise.

Finally, in "An Information Technology Research Agenda for Public Administration," G. David Garson outlines research questions that frame the dimensions of a research agenda for the study of information technology in public administration. The dimensions selected as being the most theoretically important include the issues of the impact of information technology on governmental accountability, the impact of information technology on the distribution of power, the global governance of information technology, information resource equity and the digital divide, the implications of privatization as an IT business model, the impact of IT on organizational culture, the impact of IT on discretion, centralization and decentralization, restructuring the role of remote work, the implementation success factors, the regulation of social vices mediated by IT, and other regulatory issues.

G. David Garson

North Carolina State University, USA

Section I

Managing Information Technology in the Public Sector

Chapter I

Lip Service? How PA Journals and Textbooks View Information Technology

Alana Northrop, California State University, Fullertan, USA

Abstract

This chapter first points out the continued need for a reader on information technology by reviewing the early importance given to computing education by MPA programs and practitioners. Next, the chapter surveys current textbooks' and general public-administration journals' treatment of the topic. Three highly respected public-administration journals and three textbooks are reviewed. The journals are found to typically give little attention to the topic of computing, whether as a main focus or as merely a mention in articles. The textbooks also barely mention computing. In addition, there was no consistent rubric or chapter topic under which computing is discussed. The continued and vital need for a reader on information technology and computer applications in public administration is apparent. Finally, the chapter concludes by briefly discussing a range of issues that public administrators should be conversant with if they are to successfully utilize computer applications in the delivery of public-sector services.

Introduction

In 1985, a special computing education committee recommended to the National Association of Schools of Public Affairs and Administration (NASPAA) that a sixth skill, computing, be added to the original five skills that must be taught in an MPA program. This recommendation applied to the accreditation of schools starting in 1988. Now over 20 years have passed since the original recommendation. Let us turn to evaluate the progress that has been made.

Computing Education in MPA Programs

There have been two published studies that surveyed MPA programs and assessed the level of computing education. Cleary (1990) mailed out questionnaires to 215 public affairs and public administration (PA) master's programs affiliated with the National Association of Schools of Public Affairs and Administration in 1989. Of the 80% returned, about one out of four reported that they had a course dealing with information systems and computer skills. The respondents were quick to note that the information systems and computer-skills areas needed more attention in the future. Yet, 1989 was a long time ago, especially when it comes to the massive changes in the computer field.

Brudney, Hy, and Waugh (1993) did a little more recent survey of MPA programs. Close to 90% of the programs said they use computers in their instruction. Over half of the institutions offer a course in computers, yet only 30% had made computing a requirement. The study also suggested that computing skills need to be taught beyond the typically taught statistical applications.

Without an absolutely current survey of programs, one can only surmise, though pretty safely, that computer use in MPA courses has greatly expanded. Word processing, spreadsheets, graphics, e-mail, the Internet, geographic information system (GIS), and online classes are now part and parcel of MPA programs and assumed student skills.

What PA Practitioners Advise in Computing Education

Four studies surveyed public managers. Lan and Cayer (1994) surveyed administrators in one state. The recommendations were that MPA programs need course work in computer literacy, specifically knowledge of applications and hands-on skills. The respondents said they use information technology (unfortunately this includes phone and fax) an average of 56% of their day. The respondents also said that they were involved with the management of the information system, so management issues as well as computer skills are important for PA students.

Crewson and Fisher (1997) surveyed 371 city administrators in the United States. In terms of importance for public administrators in the future, 37% of the sample rated computer skills as most important, with 57% giving such skills moderate importance. Similar ratings were given by the sample to knowing about computers.

An earlier study (Poister & Streib, 1989) of 451 municipal managers indicates the extensive diffusion of management information systems in the 1980s. Other indications of computer use can be obliquely inferred from usage of such management tools as revenue forecasting and performance monitoring.

A 1988 study of 46 technologically advanced cities was intended to predict the common state of computerization in U.S. cities in the late 1990s (Kraemer & Northrop, 1989). That study indicated that no city department or staff role was spared from the diffusion of computers. In fact, 84% of managers respondents and 85% of staff respondents indicated that their work involved major interaction with computers.

More recent studies (Moon, 2002; Norris & Moon, 2005) indicate the absolute spread of IT to city and county governance through Web sites and their evolving nature.

In essence, the word from public managers is that the use of computers has become essential to daily municipal business.

Computing Education in Public Administration Journals and Textbooks

We know there is a need for computing education in MPA programs, as practitioners in both the 1985 recommendation and 1989 update pointed out (Kraemer & Northrop, 1989; NASPAA Ad Hoc Committee on Computers in Public Management Education, 1986). Yes, schools say they have integrated computing into their curriculum (Brudney et al., 1993), but how has the field of public administration pushed computing education in print?

One way to answer that question is to review research and textbooks in public administration. This third approach to looking at computing education is based on the theory of triangulation. Triangulation means using different data sources trained on the same problem, in this case, computing education. Triangulation not only involves using data from different sources but also from different perspectives. In this section, the sources are general public-administration journals and textbooks. The different perspective is the belief that one can learn about computing education not just from what university program directors say they teach, but also from looking at the published sources of information commonly available to public-administration academics and professionals.

Professional Public Administration Articles

The journals were selected based on Forester and Watson's (1994) survey of all editors and editorial board members of 36 journals who mention public administration in general or

Table 1.

	All Respondents	Minus Board Members
Public Administration Review	8.34	8.19
Administration & Society	5.36	5.17
The American Review of Public Administration	3.85	3.40
Journal of Public Administration Theory	3.20	2.78
Public Administration Quarterly	2.88	2.45

Note: There is a very clear drop-off in ratings for journals rated lower.

public administration topics, such as personnel and finance, in their mission statements. The study used a 10-point scale: 10 representing the best journal in the field according to the respondents, and a 0 indicating that no respondent rated the journal in the top 10. The top 5 general public-administration journals, whether you include or exclude the board members of those journals in the rankings, can be seen in Table 1.

In deciding which journals to evaluate, the quality of the journal was considered as well as the requirement that the journal be recognized as one that dealt with the field of public administration in general. The latter requirement was based on the recommendation of the 1985 NASPAA committee that the computing topic be integrated into all courses vs. segregated into one or a part of one course. Thus, the computing topic should be relevant to all academics and practitioners interested in public administration, not just those in a particular specialized area. The quality issue obviously speaks to the dissemination of information as well as the importance of computing as demonstrated by its acceptability as a topic in esteemed publishing outlets.

Clearly, *Public Administration Review* (PAR) and *Administration & Society* (A&S) stand out as the top general public administration journals and, in fact, as the top public-administration journals, period (Forester & Watson, 1994). We also felt that *The American Review of Public Administration* (ARPA) should be selected. Although it is closer in ratings to other lower ranked journals than it is to the two leaders, it stands out in its ratings' gap from the lower journals more than it is similar to them in ratings differences.

Table 2 shows how often computing appeared as a topic in the three selected journals over the last 10 years. There is no trend but instead a turning point. Articles that mention computing or have computing as the main focus are rare, with a notable increase in 2002 but slipping downward by 2004. It should be noted that PAR, the main journal outlet for IT issues, had a special issue on 9/11 and terrorism in 2002 that often mentioned IT issues. It should also be noted that articles that mention computing might involve as little as a one-sentence mention in the whole article.

In sum, while the academic field and the world of government practice increasingly recognize the importance of computing, the research world in terms of top-quality journals really does not.

Table 2. Appearance of computing topic in public administration journals

	1996	1997	1998	1999	2000	2001	2002	2003	2004	2005
Administration & Society										
Number of articles that mention computing	0	0	0	0	0	0	0	0	2	2
Number of articles in which main focus is computing	0	1	1	1	1	1	2	0	2	1
American Review of Public Administration										
Number of articles that mention computing	0	1	0	1	1	0	0	4	3	1
Number of articles in which main focus is computing	0	0	1	0	0	3	2	2	1	2
Public Administration Review										
Number of articles that mention computing	2	4	2	0	0	1	10	10	2	5
Number of articles in which main focus is computing	2	0	1	0	1	2	4	3	3	1
Totals										
Number of articles that mention computing	2	5	2	1	1	1	10	14	7	8
Number of articles in which main focus is computing	2	1	3	1	2	6	8	5	6	4
Number of articles that mention computing or in which main focus is computing	4	6	5	2	3	7	18	19	13	12

Table 3. Appearance of computing topic in public administration textbooks

Textbook name and author	**Number of pages mentioning computers or information systems and percentage***	
Public Administration Concepts and Cases by Stillman (2005)	0	**Chapter in which pages appear**
Managing the Public Sector by Starling (2005)	10 (2)	Chapter 12, "Information Management"
Public Administration the New Century by Greene (2005)	2 (.5)	Chapter 1, "Understanding Public Administration"

*Note: *Percentages (in parentheses) are calculated using number of textbook pages as bases (i.e., excluding appendixes, references, and indexes).*

Public Administration Textbooks

Professional public-administration journal articles are a common outlet for academics and professionals to keep up on the latest research and trends in the field. Articles can be assigned in class or incorporated into lectures. Another common outlet on what is essential to the public administration field is textbooks. While one can avoid keeping up by not reading all journal articles, it is hard not to read the book assigned in class on both the professor's and student's sides. So if one were seeking to learn about the essential topics in the field of public administration, what would you learn by reading the textbooks?

In this instance, we looked at how often computers or information systems were mentioned in current public-administration textbooks. Three texts were chosen based on their most current printing date of 2005. Table 2 indicates the remarkable lack of attention that these textbooks give to computing. Similar to the three general public-administration journals studied earlier, computing is just not a textbook topic of major importance.

An additional concern, besides the amount of attention given to computing in these texts, is how it is treated. As Table 3 indicates, computing does not have its own chapter except loosely in Starling's textbook. Computing as a topic also appears to not have any consistent rubric under which it is treated. Such inconsistent treatment suggests that computing has not been integrated into all areas of public administration and, in fact, has not even found a home in one area.

Need for a Book on Computing for Public Administrators

Without a doubt computing has permeated the practice of public administration at all levels of government in the United States. NASPAA has recognized this by requiring all accredited MPA programs to include in their programs information management, including computer literacy and applications. Yet in spite of the importance the work world and NASPAA has put on computing education, the two tables in this chapter show that textbooks and general public-administration journals barely treat computing as a topic worthy of mention. Consequently, there is the strong sense that we all say computing is important, but it is more lip service than actual service. If it is truly accepted as important, computing should be a common research topic in our leading journals, a common topic in our textbooks, and thus a topic on which we are working hard to build a common body of knowledge. This is not true today, 21 years since we as a profession formally recognized computing's importance. How can there be a common theme or treatment to computing education if the textbooks and respected journals offer minuscule help or encouragement? A major way to begin correcting this dismal state is the present publication of an edited book on information technology and computer applications. This author also refers the reader to the articles that mention computers, listed at the end of this chapter. Articles going back to 1985 from the three journals are listed there.

Management Issues

A master's in public administration signifies the recipient has the skills to manage people and tasks in an environment of both internal and external political demands and responsibilities. What are some key issues about which an MPA graduate should be conversant when it comes to managing in a computerized environment? First and foremost is the fact that the computerization of a task does not necessarily lead to payoffs and more than likely will underachieve compared to expectations. It is important to understand what factors affect payoffs and then address how to deal with them. The following section briefly points out factors that have been shown to influence the usefulness of computer applications.

Quality of Data

An absolute condition for achieving payoffs from computerization is that the data must be accurate. A system to control input errors and to change data must be instituted. In contrast, the length of time to get information from a computerized task does not need to be made as short as possible. Data that can be quickly retrieved are very nice, but data that many users think take too long to be retrieved will still be used if they are considered useful.

Training

Managers must devote more resources and ongoing thought to training. Based on an informal survey of over 450 public employees, this was one of the top two lessons that was a constant theme from department to department, application to application, and employee to employee (Northrop, 2002).

One obviously needs a training program to teach users how to use new applications. Another related consideration is having a way to train new hires in ongoing applications. In addition, an initial training program should not be considered the end of training. Follow-up training programs or help sessions need to be routinized; training should be considered an ongoing process.

Then there is the issue of who should do the training. There is no clear answer whether in-house trainers or external trainers are best or whether professional trainers or just experienced employees are best (Northrop, Kraemer, Dunkle, & King, 1994). Professional trainers bring experience in conveying application knowledge, but sometimes an employee who uses the program is better at answering questions about how to use the program within the context of the work product.

Resources

Computerized applications may be able to provide all sorts of new and valuable information to decision makers. However, the information is only useful if the organization has the resources to take full advantage of the new information. To illustrate, a computerized manpower application system can outline where and how many police officers one needs to deploy at a certain time of day across the city. If the police department has that many officers available, all is well and good. If not, whether due to limits on force size or just scheduling variations, the computerized manpower information will not help much in the fight against crime.

What One Automates

When one is automating a task or upgrading a task system, the success of the present system to do the work needs to be considered. Often an organization just automates the way they presently do a task. If, for example, the present way one tracks the names of people who should be paying child support only finds and receives payment from 20% of the list, then the automated system will likely not do better. Therefore, an evaluation of the extent to which policy goals are currently being met should be required before a task is automated or upgraded.

Who to Involve in Adoption Decisions

Oddly enough, some organizations never consult the very employees who will use the new computer application to see what they need and to get their input on the weaknesses of the old application. Employees from all levels who will use the application should be asked for their input, from line personnel to managers.

Purchasing a Customized System

Experience at the federal, state, and local level point out that purchasing an information system is rife with risks. Millions have been spent on systems that were delivered years after being promised or that never worked. Choosing a company and writing a contract should be done with much care. Consider purchasing a customized system as similar to choosing and working with a contractor to remodel one's house. Payments should not be made until each stage is approved. A final payment should be held back until the developer fixes the little problems in the system. Expect to pay more if the organization changes its specifications of the system but builds into the contract a cost. A bonus can be offered if the system is finished before the agreed-upon date, and penalties can be assessed if the system is delivered after the expected date. Above all, ask for references and check them. Send a representative to other organizations that have contracted with the system-development company. Do not just call. Remember that line staff may have a different perspective than a manager; all input is relevant.

Management Support

This is the second most frequent lesson passed on by over 450 public employees (Northrop, 2002). Management must be supportive of the computer application. Staff personnel have been known to just not use the application because management has given the signal indirectly that they do not see the usefulness of the application. One way that management can effectively show support is to actively use the application or the generated reports. If staff members have a question about the application, management should be able to answer the question even if this only means referring the staff to someone else who can help. Management must show they care about staff knowing and effectively using the application. It is up to management to sell the usefulness of the application to employees.

Security

Security has been considered a major management concern since the early computerization security of data. If data are accessible and changeable by inappropriate personnel, major legal issues involving rights compromise the usefulness of the computerized database. Depending on the department or agency, security issues vary. For example, police field reports once entered should not be able to be changed at will by any patrol officer. Incident reports must be protected from being expunged or altered to protect the integrity of the court case from bribes or favoritism. Personnel files need to be more widely accessible to change to update job titles, benefits, and addresses. However, the extent of access to personnel information must also be limited. The security needed for some national computer databases, such as that for social security, is monumental.

Conclusion

A PA graduate needs hands-on skills in computer applications. Said graduate also needs to be conversant with the issues involved in successfully managing information technology and computer applications.

Professional journals do offer useful articles on information technology, even if they are typically few in number. However, general PA textbooks are not a source for building one's knowledge in this important area. Thus, edited books, like the present one, are critical to provide a common grounding in computer education for PA graduates.

References

Brudney, J., Hy, R. J., & Waugh, W. L. (1993). Building microcomputing skills in public administration graduate education: An assessment of MPA programs. *Administration & Society, 25*, 183-203.

Cleary, R. (1990). What do MPA programs look like? Do they do what is needed? *Public Administration Review, 50*, 663-673.

Crewson, P. E., & Fisher, B. S. (1997). Growing older and wiser: The changing skill requirement of city administrators. *Public Administration Review, 53*, 380-386.

Forester, J. P., & Watson, S. S. (1994). An assessment of public administration journals: The perspective of editors on editorial board members. *Public Administration Review, 54*, 474-482.

Kraemer, K. L., & Northrop, A. (1989). Curriculum recommendations for public management education in computing: An update. *Public Administration Review, 49*, 447-453.

Lan, Z., & Cayer, J. (1994). The challenges of teaching information technology use and management in a time of information revolution. *The American Review of Public Administration, 24*, 207-222.

Moon, M. J. (2002). The evolution of e-government among municipalities: Rhetoric or reality? *Public Administration Review, 64*, 515-528.

National Association of Schools of Public Affairs and Administration (NASPAA) Ad Hoc Committee on Computers in Public Management Education. (1986). Curriculum recommendations for public management education in computing. *Public Administration Review, 46*, 595-602.

Norris, D. G., & Moon, M. J. (2005). Advancing e-government at the grassroots: Tortoise or hare? *Public Administration Review, 65*, 64-75.

Northrop, A. (2002). Lessons for managing information technology in the public sector. *Social Science Computer Review, 20*, 194-205.

Northrop, A., Kraemer, K. L., Dunkle, D. E., & King, J. L. (1994). Management policy for greater computer benefits: Friendly software, computer literacy, or formal training. *Social Science Computer Review, 12*, 383-404.

Poister, T., & Streib, G. (1989). Management tools in municipal government: Trends over the past decade. *Public Administration Review, 49*, 240-248.

Further Reading

Journal Articles That Mention Computers

Allcorn, S. (1997). Parallel virtual organizations. *Administration & Society, 29*(4), 412-439.

Bajjaly, S. T. (1998). Strategic information systems planning in the public sector. *The American Review of Public Administration, 28*, 75-86.

Barth, T. J., & Arnold, E. (1999). Artificial intelligence and administrative discretion. *The American Review of Public Administration, 29*, 332-352.

Bolton, M. J., & Stolcis, G. B. (2003). Ties that do not bind: Musings on the specious relevance of academic research. *Public Administration Review, 63*, 626-630.

Botner, S. B. (1985). The use of budgeting management tool by state government. *Public Administration Review, 45*, 616-619.

Bovens, M., & Zouridis, S. (2002). From street-level to system-level bureaucracies: How information and communication technology is transforming administrative discretion and. *Public Administration Review, 62*, 174-184.

Bozeman, B., & Pandey, S. K. (2004). Public management decision making: Effects of decision content. *Public Administration Review, 64*, 553-565.

Brainard, L. A., & Siplon, P. D. (2002). Cyberspace challenges to mainstream nonprofit health organizations. *Administration & Society, 34*, 141-175.

Bretschneider, S. (1990). Management information systems in public and private organizations: An empirical test. *Public Administration Review, 50*, 536-545.

Brown, M. M., & Brudney, J. L. (1998a). A "smarter, better, faster, and cheaper" government: Contracting and geographic information systems. *Public Administration Review, 58*, 335-346.

Brown, M. M., & Brudney, J. L. (1998b). Public sector information technology initiatives. *Administration & Society, 30*, 421-443.

Brown, M. M., & Brudney, J. L. (2003). Learning organizations in public sector? A study of police agencies employing information and technology to advance knowledge. *Public Administration Review, 63*, 30-43.

Brudney, J. L., Hy, R. J., & Waugh, W. L., Jr. (1993). Building microcomputing skills in public administration graduate education: An assessment of MPA programs. *Administration & Society, 25*, 183-203.

Brudney, J. L., & Selden, S. C. (1995). The adoption of innovation by smaller local governments: The case of computer technology. *The American Review of Public Administration, 25*, 71-86.

Brudney, J. L., & Wright, D. S. (2002). Revisiting administrative reform in the American states: The status of reinventing government during the 1990s. *Public Administration Review, 62*, 353-361.

Buyers, K. M., & Palmer, D. R. (1989). The microelectronics and computer technology corporation: An assessment from market and public policy perspectives. *Administration & Society, 21*, 101-127.

Carroll, J. D., & Lynn, D. B. (1996). The future of federal reinvention: Congressional perspectives. *Public Administration Review, 56*, 299-304.

Cats-Baril, W., & Thompson, R. (1995). Managing information technology projects in the public sector. *Public Administration Review, 55*, 559-566.

Caudle, S. L. (1990). Managing information resources in state government. *Public Administration Review, 50*, 515-524.

Cleary, R. (1990). What do MPA programs look like? Do they do what is needed? *Public Administration Review, 50*, 663-673.

Cleveland, H. (1985). The twilight of hierarchy: Speculations on the global information society. *Public Administration Review, 45*, 185-195.

Comfort, L. K. (2002). Rethinking security: Organizational fragility in extreme events. *Public Administration Review, 62*, 98-107.

Considine, M., & Lewis, J. M. (2003). Bureaucracy, network, or enterprise? Comparing models of governance in Australia, Britain, the Netherlands, and New Zealand. *Public Administration Review, 63*, 131.

Corder, K. (2001). Acquiring new technology. *Administration & Society, 33*, 194-216.

Crewson, P. E., & Fisher, B. S. (1997). Growing older and wiser: The changing skill requirements of city administrators. *Public Administration Review, 57*, 380-386.

Czarniawska, B. (2002). Remembering while forgetting: The role of automorphism in city management in Warsaw. *Public Administration Review, 62*, 163.

Danziger, J., & Kraemer, K. L. (1985). Computerized data-based systems and productivity among professional workers: The case of detectives. *Public Administration Review, 45*, 196-209.

Danziger, J., Kraemer, K. L., Dunkle, D., & King, J. L. (1993). Enhancing the quality of computing service: Technology, structure, and people. *Public Administration Review, 53*, 161-169.

Donley, M. B., & Pollard, N. A. (2002). Homeland security: The difference between a vision and a wish. *Public Administration Review, 62*, 138-144.

Dumont, G., & Candler, G. (2005). Virtual jungles: Survival, accountability and governance in online communities. *The American Review of Public Administration, 35*, 287-299.

Durant, R. F. (2002). Whither environmental security in the post-September 11th era? Assessing the legal, organizational, and policy challenges for the national security state. *Public Administration Review, 62*, 115-123.

Edmiston, K. D. (2003). State and local e-government: Prospects and challenges. *The American Review of Public Administration, 33*, 20-45.

Fraumann, E. (1997). Economic espionage: Security missions redefined. *Public Administration Review, 57*, 303-308.

Frederickson, H. G., & LaPorte, T. R. (2002). Airport security, high reliability, and the problem of rationality. *Public Administration Review, 62*, 33-43.

Gianakis, G. A., & McCue, C. P. (1997). Administrative innovation among Ohio local government finance officers. *The American Review of Public Administration, 27*(3), 270-286.

Globerman, S., & Vining, A. R. (1996). A framework for evaluating the government contracting out decision with an application to information technology. *Public Administration Review, 56*, 577-586.

Gooden, V. (1998). Contracting and negotiation: Effective practices of successful human service contract managers. *Public Administration Review, 58*, 499-510.

Goodsell, C. T. (2003). The concept of public space and its democratic manifestations. *The American Review of Public Administration, 33*, 361-383.

Grizzle, G. A. (1985). Essential skills for financial management: Are MPA students acquiring the necessary competencies? *Public Administration Review, 45*, 840-844.

Haines, D. W. (2003). Better tools, better workers: Toward a lateral alignment of technology, policy, labor and management. *The American Review of Public Administration, 33*, 449-478.

Haque, A. (2001). GIS, public service, and the issue of democratic governance. *Public Administration Review, 61*, 259-266.

Haque, M. S. (2002). Government responses to terrorism: Critical views of their impacts on people and public administration. *Public Administration Review, 62*, 170-180.

Hendrick, R. (1994). An information infrastructure for innovative management of government. *Public Administration Review, 54*, 543-550.

Ho, A. T. (2001). Information technology planning and the Y2K problem in local governments. *The American Review of Public Administration, 31*, 158-181.

Ho, A. T. (2002). Reinventing local governments and the e-government initiative. *Public Administration Review, 62*, 434-444.

Ho, A. T., & Ni, A. Y. (2004). Explaining the adoption of e-government features: A case study of Iowa county treasurers' office. *The American Review of Public Administration, 34*, 164-180.

Holliday, I., & Wong, L. (2003). Social policy under one country, two systems: Institutional dynamics in China and Hong Kong since 1997. *Public Administration Review, 63*, 269-282.

Hou, Y., Moynihan, D. P., & Ingraham, P. W. (2003). Capacity, management, and performance exploring the links. *The American Review of Public Administration, 33*, 295-315.

Jreisat, J. E. (2005). Comparative public administration is back in, prudently. *Public Administration, 65*, 231-242.

Kakabadse, A., Kakabadse, N. K., & Kouzmin, A. (2003). Reinventing the democratic governance project through information technology? A growing agenda for debate. *Public Administration Review, 63*, 44-60.

Kellogg, W. A., & Mathur, A. (2003). Environmental justice and information technologies: Overcoming the information-access paradox in urban communities. *Public Administration Review, 63*, 573-585.

Kettl, D. F. (2003). Contingent coordination: Practical and theoretical puzzles for homeland security. *The American Review of Public Administration, 33*, 253-277.

Kim, P., Halligan, J., Namshin, C., Oh, C. H., & Eikenberry, A. M. (2005). Toward participatory and transparent governance: Report on the Sixth Global Forum on Reinventing Government. *Public Administration Review, 65*, 646-654.

Kim, S. (2005). Factors affecting state government information technology employee turnover intentions. *The American Review of Public Administration, 35*, 137-156.

Koppell, J. G. S. (2005). Pathologies of accountability: ICANN and the challenge of "multiple accountabilities disorder." *Public Administration Review, 65*, 94-108.

Kouzim, A. (2000). Mapping institutional impacts of lean communication in lean agencies. *Administration & Society, 32*, 29-70.

Kraemer, K. L., Guraxani, V., & King, J. L. (1992). Economic development, government policy, and the diffusion of computing in Asia-Pacific countries. *Public Administration Review, 52*, 146-154.

Kraemer, K. L., & King, J. L. (1987). Computers and the constitution: A helpful, harmful or harmless relationship? *Public Administration Review, 47*, 93-105.

Kraemer, K. L., & Northrop, A. (1989). Curriculum recommendations for public management education in computing: An update. *Public Administration Review, 49*, 447-453.

Kulchitsky, R. D. (2004). Computerization, knowledge and information technology initiatives in Jordan. *Administration & Society, 36*, 3-37.

Lan, Z., & Cayer, J. (1994). The challenges of teaching information technology use and management in a time of information revolution. *The American Review of Public Administration, 24*, 207-222.

Landry, R., Lamari, M., & Amara, N. (2003). The extent and determinants of the utilization of university research in government agencies. *Public Administration Review, 63*, 94-108.

Landsbergen, D., Jr. (2001). Realizing the promise: Government information systems and the fourth generation of information technology. *Public Administration Review, 61*, 206-220.

Lane, L. M., Wolf, J. F., & Woodard, C. (2003). Reassessing the human resource crisis in the public service: 1987-2002. *The American Review of Public Administration, 33*, 123-145.

LaPorte, T. M., Demchak, C. C., & Dejong, M. (2002). Democracy and bureaucracy in the age of the Web: Empirical findings and theoretical speculations. *Administration & Society, 34*, 411-446.

Lee, R. D., Jr. (1997). A quarter century of state budgeting practices. *Public Administration Review, 57*, 133-140.

Levin, M. A. (1991). The information-seeking behavior of local government officials. *The American Review of Public Administration, 21*, 271-286.

Lewis, D. L. (1991). Turning rust into gold: Planned facility management. *Public Administration Review, 51*, 494-502.

Lonti, Z. (2005). How much decentralization? Managerial autonomy in the Canadian public service. *The American Review of Public Administration, 35*.

Lu, H., & Facer, R. L., II. (2004). Budget change in Georgia counties: Examining patterns and practices. *The American Review of Public Administration, 34*, 67-93.

Mahler, J., & Regan, P. M. (2002). Learning to govern online: Federal agency Internet use. *American Review of Public Administration, 32*, 326-349.

Manring, N. J. (2005). The politics of accountability in national forest planning. *Administration & Society, 37*, 57-88.

Menzel, D. (1998). www.ethics.gov: Issues and challenges facing public managers. *Public Administration Review, 58*, 445-453.

Merget, A. E. (2003). Times of turbulence. *Public Administration Review, 63*, 390-395.

Milam, D. M. (2003). Practitioner in the classroom: Bringing local government experience into the public administration curriculum. *Public Administration Review, 63*, 364-369.

Miller, H. T. (2005). Some evidence of a pluralistic discipline: A narrative analysis of public administration symposia. *Public Administration Review, 65*, 728-738.

Moon, M. J. (2002). The evolution of e-government among municipalities: Rhetoric or reality? *Public Administration Review, 62*, 424-433.

Moynihan, D. P. (2004). Building secure elections: E-voting, security, and systems theory. *Public Administration Review, 64*, 515-528.

Moynihan, D. P., & Ingraham, P. W. (2004). Integrative leadership in the public sector: A model of performance-information use. *Administration & Society, 36*, 427-453.

Nedovic-Budic, Z., & Godschalk, D. R. (1996). Human factors in adoption of geographic information systems (GIS): A local government case study. *Public Administration Review, 57*, 554-567.

Nelson, L. (2002). Protecting the common good: Technology, objectivity, and privacy. *Public Administration Review, 62*, 69-73.

Nelson, L. (2004). Privacy and technology: Reconsidering a crucial public policy debate in the post-September 11 era. *Public Administration Review, 64*, 259-269.

Newcomer, K., & Grob, G. (2004). Federal offices of the inspector general: Thriving on chaos? *The American Review of Public Administration, 34*, 235-251.

Newcomer, K. E., & Caudle, S. (1991). Evaluating public sector information systems: More than meets the eye. *Public Administration Review, 51*, 377-384.

Norris, D. G., & Kraemer, K. L. (1996). Mainframe and PC computing in American cities: Myths and realities. *Public Administration Review, 56*, 568-576.

Norris, D. G., & Moon, M. J. (2005). Advancing e-government at the grassroots: Tortoise or hare? *Public Administration Review, 65*, 64-75.

Northrop, A., Kraemer, K. L., Dunkle, D., & King, J. L. (1990). Payoffs from computerization: Lessons over time. *Public Administration Review, 50*, 505-514.

Nunn, S. (2001). Police information technology: Assessing the effects of computerization on urban police functions. *Public Administration Review, 61*, 221-235.

Pandey, S. K., & Welch, E. W. (2005). Beyond stereotypes: Multistage model of managerial perceptions of red tape. *Administration & Society, 37*, 542-575.

Peled, A. (2001). Do computers cut red tape? *The American Review of Public Administration, 31*, 414-436.

Poister, T. H., & Streib, G. (1989). Management tools in municipal governments: Trends over the past decade. *Public Administration Review, 49*, 240-248.

Regan, P. M. (1986). Privacy, government information, and technology. *Public Administration Review, 46*, 629-634.

Relyea, H. C. (1986). Access to government information in the information age. *Public Administration Review, 46*, 635-639.

Roberts, N. (1997). Public deliberation: An alternative approach to crafting policy and setting direction. *Public Administration Review, 57*, 124-132.

Roberts, N. (2004). Public deliberation in an age of direct citizen participation. *The American Review of Public Administration, 34*, 315-353.

Rocheleau, B. (1992). Information management in the public sector: Taming the computer for public managers. *Public Administration Review, 52*, 398-400.

Rocheleau, B. (1994). The software selection process in local governments. *The American Review of Public Administration, 24*, 317-330.

Rocheleau, B. (2000). Prescriptions for public-sector information management. *The American Review of Public Administration, 30*, 414-436.

Rocheleau, B., & Wu, L. (2002). Public versus private information systems: Do they differ in important ways? A review and empirical test. *The American Review of Public Administration, 32*, 379-397.

Romzek, B. S., & Johnston, J. M. (2005). State social services contracting: Exploring the determinants of effective contract accountability. *Public Administration Review, 65*, 436-449.

Rubenstein, R., Schowartz, A. E., & Stiefel, L. (2003). Better than raw: A guide to measuring organizational performance with adjusted performance measures. *Public Administration Review, 63*, 607-615.

Scherpereel, J. A. (2004). Renewing the socialist past or moving toward the European administrative space? Inside Czech and Slovak ministries. *Administration & Society, 36*, 553 -593.

Slack, J. (1990). Information, training, and assistance needs of municipal governments. *Public Administration Review, 50*, 450-457.

Smith, D. C. (2001). Strategic planning for municipal information systems. *The American Review of Public Administration, 31*, 139-157.

Stanley, J. W., & Weare, C. (2004). The effects of Internet use on public participation: Evidence from an agency online discussion forum. *Administration & Society, 36*, 503-527.

Stewart, D. W., Sprinthall, N. A., & Kem, J. D. (2002). Moral reasoning in the context of reform: A study of Russian officials. *Public Administration Review, 62*, 282-297.

Stewart, J., & Kringas, P. (2003). Change management-strategy and values in six agencies from the Australian public service. *Public Administration Review, 63*, 675-688.

Thomas, J. C., & Streib, G. (2005). E-democracy, e-commerce and e-research: Examining the electronic tie between citizens and governments. *Administration & Society, 37*, 503-527.

Thompson, F. J. (2002). Reinvention in the states: Ripple or tide? *Public Administration Review, 62*, 362-367.

Tjoregas, C. (2000). Lessons from the "Y2K and You" campaign for the local government community. *Public Administration Review, 60*, 84-89.

Van Blijswijk, J. A. M., Van Breuklen, R. C. J., Franklin, A. L., Raadschelders, J. C. N., & Slump, P. (2004). Beyond ethical codes: The management of integrity in the Netherlands tax and customs administration. *Public Administration Review, 64*, 718-727.

Ventura, S. J. (1995). The use of geographic information systems in local government. *Public Administration Review, 55*, 61-467.

Walker, D. M. (2002). 9/11: The implications for public-sector management. *Public Administration Review, 62*, 94-97.

Weare, C., Musso, J. A., & Hale, M. L. (1999). Electronic democracy and the diffusion of municipal Web pages in California. *Administration & Society, 31*, 3-28.

Werlin, H. H. (2003). Poor nations, rich nations: A theory of governance. *Public Administration Review, 63*, 329-342.

West, D. M. (2004). E-government and the transformation of service delivery and citizen attitudes. *Public Administration Review, 64*, 15-27.

Williams, F. P., III, McShane, M., & Seehnst, D. (1994). Barriers to effective performance review: The seduction of raw data. *Public Administration Review, 54*, 537-542.

Chapter II

The Evolution of Federal Information Technology Management Literature: Does IT Finally Matter?

Stephen H. Holden, SRA Touchstone Consulting Group, USA

Abstract

Federal agencies rely extensively on information technology to perform basic missions. Arguably, public administration should be driving the theory, policy, and practice for managing these increasingly important resources. While there has been some maturation in the literature for managing IT in federal agencies in the last several years, academics from the field of information systems and practitioners have contributed more recently to the theory and practice of IT management at the federal level than public administration. This chapter analyzes federal IT-management literature over time and compares federal IT-management literature to a normative model of management maturity focusing on the strategic objectives for IT and related management approaches. Public administration's minimal contribution to federal IT-management literature raises profound questions about whether federal agencies are performing commensurate with public expectations as the theory and practice of IT management may be moving into a new, post-information-age era.

Introduction

Given the growing importance of effective IT management to the basic functioning of most public programs, the sophistication of the policy, theory, and practice in this area should be evolving quickly. Unfortunately, that is not so (Fountain, 2001; Holden, 1996; Holden & Hernon, 1996). As a result, it is quite possible that the current generation of public-administration scholars and practitioners may be ill equipped to face the challenges of the information age in which we find ourselves trying to govern.

A mere gap in IT-management theory might not be fatal, but in reality, the implications for the practice of public administration, and therefore governance, are quite grim. Press accounts of the interoperability challenges first responders faced during the World Trade Center attacks on 9/11 offer just one example of how integral public-sector IT has become to the safety and economic well-being of the country. While the billions of dollars currently spent by the federal government on IT make up an insignificant portion of the budget, IT underpins almost the whole budget directly or indirectly. Just ponder the implications to the government's cash flow if the Internal Revenue Service (IRS) could not collect taxes or the Social Security Administration could not post employee earnings.

This chapter compares the federal IT-management literature with a normative model of management maturity, examining the strategic objectives for IT and the related management approaches. The academic disciplines that contribute to an understanding of the management of IT in the federal government include business administration, state and local government management, information sciences, and public administration. Although the analysis of the literature does include government publications, it does not discuss the pertinent public law or government-wide policy (see instead Beachboard & McClure, 1996; Holden, 1994; Plocher, 1996).

Like public administration more generally, IT management draws on several different sources. Unlike other management topics in public administration, though, the literature covering IT management lacks breadth and maturity. Even more alarming, Kraemer and Dedrick's (1997) review of the public-administration literature for managing IT found that research on public-sector computing was declining when federal agencies were relying more heavily on IT. The following quotation summarizes the state of the literature at the turn of the 21st century:

A century from now, social and policy scientists will look back with amusement and no small amount of condescension at the glacial pace with which social scientists moved to consider fundamental changes in information processing and their implications. (Fountain, 2001, p. 10)

While the chapter documents some progress in the field in the last several years since a similar review was published (Holden, 2003), it also points to continuing shortcomings. It is particularly troubling that disciplines besides public administration are responsible for the few recent developments in federal IT-management literature. Compared against the management maturity model presented below, there is clearly much work left to do.

A Model for Information Technology Management Maturity

This analysis of federal IT-management literature proceeds in an order that reflects the maturation of the strategic objectives for IT and the attendant management philosophy over the last 40 years. As a normative standard, this analysis adapts a model of maturation of theory and practice that Donald Marchand (1985) first used for the field of information management (Holden, 1994). He identified four stages of evolution for information management encompassing the 20th century.

This chapter adapts his model as an organizing principle in several ways. First, since this analysis deals with the management of IT, it does not include his Stage 1, which addresses the physical control of information before automation in the 1950s. Second, Marchand discusses the evolution of information management for five distinguishing characteristics: (1) precipitating forces, (2) strategic objective, (3) basic technologies, (4) management approaches, and (5) organizational status. This chapter stresses how strategic objectives for and management approaches to IT have evolved, placing little emphasis on the other three characteristics. Third, since Marchand covers information management, which is broader than IT management, the names for the stages differ. Finally, the stages used to organize this chapter reflect the stage of development of the literature, despite the date of its publication. In contrast, Marchand demarcated his stages by time periods, assuming that all activities within a specified period of time conformed to the same stage of development in management.

The adaptation of Marchand's model is quite consistent with stage or phase models used in the IS literature. For instance, one of the classic texts (Applegate, McFarlan, & McKenney, 1999) for what business schools call management information systems (MISs) discusses three major eras of computing in organizations. The eras cited in this MIS text mirror the three stages used to organize this chapter in terms of the prevailing technology and level of management attention in each era or stage. More recently, Peppard and Ward (2004) used what they call a widely accepted notion of these three eras of computing in organizations as the organizing frame to assert that there is a new era emerging in IT management in both theory and practice. These two examples lend support to the credence of this chapter's adaptation of Marchand's work and also point to the emergence of a new phase or stage that should be informing both the theory and practice of federal IT management as discussed in the concluding section of the chapter.

Stage 1: Management Information Systems

This stage, which Marchand (1985) labeled "Management of Automated Technology," spanned the 1960s through the mid-1970s. Then, business and public-sector professionals often used the term MIS to describe IT. The management of technology dominated this era at the expense of the management of information. Because the management of technology was limited to the data center, there was little concern for relating those resources to other facets of management in organizations.

The management approaches for IT of this era reflected the isolation of the IT professional from broader functional and executive oversight. Organizations used IT to automate back-room operations, with the primary strategic objective being to improve the efficiency of clerical activities (Zuboff, 1988). Personnel in the data-processing function took primary responsibility for the management of these resources and focused on the development of applications and systems. Typically, these systems consisted of applications run by data-processing professionals in centralized processing facilities that consisted of mainframe computers. Users had few direct contacts with these systems other than to fill out punch cards, do manual key entry, and receive printouts from the data-processing department. As a result, line functions in the organization rarely controlled their own computing resources (Ackoff, 1967).

The early MIS literature dealt almost exclusively with private-sector applications. It was not until much later that public administration adapted MIS literature to public-sector theory and practice. In 1986, Bozeman and Bretschneider articulated a case for a separate body of literature to address the unique information needs of public organizations. They proposed that this body of theory and practice fall under the heading of public management information systems (PMISs).

To support this argument, they asserted that MIS literature ignores variables external to the organization, such as the political environment and the annual appropriation process. The political control of public organizations, which entail uncertain and variable goals, means that public- and private-sector methods for establishing IT performance indicators differ dramatically. Though private and public organizations can acquire the same hardware and software, their different organizational environments require unique system design techniques. While this notion of PMIS brought attention to the use of IT in the public sector, it had limited applicability as a management approach. Instead of viewing IT as a strategic resource for public-sector organizations, this perspective examined the development of one application at a time.

Some public-administration research on IT has examined the availability of IT at the state and local government levels. For instance, the International City Management Association (ICMA, 1986) surveyed local and county governments' use of computers several times. Kraemer, King, Dunkle, and Lane (1989) at the University of California have published several works resulting from the Urban Information Systems (URBIS) research project. These initial efforts addressed the use of IT in local governments, placing much less emphasis on the management of those resources or their strategic importance.

Building on the URBIS study, Norris and Kraemer (1996) did an analysis of the results of ICMA's 1993 survey of local government computer use. While based on more recent survey data, this study looked more at the adoption and use of IT at the local level more so than how those resources were managed. The authors did examine the degree of computing centralization and found a statistically significant relationship between population and the kind of technology deployed. Large cities were more likely to have deployed central systems (such as mainframes and minicomputers) and smaller communities were more likely to have deployed only personal computers.

Some theoretical work in evaluating information systems has surfaced nonetheless, which applies generally to public-sector organizations. Newcomer and Caudle (1991) provide some insights into how and why the evaluation of information systems in the public sector should go beyond mere return-on-investment criteria. In particular, they offer a framework for evaluation that includes qualitative and quantitative measures and recognizes the multiple uses of most public information systems. Although this framework does not help agency decision makers choose between competing projects, it, nonetheless, broadened the theoretical base for evaluating individual systems projects.

Stevens and McGowan's work (1985) presents an overview of information-systems management for public administrators from the local to the federal level. Their book adapts contingency-based organization theory to explain how public organizations must process information effectively to respond to their environment. For the most part, the book focuses on managing single applications and makes illustrative points through the discussion of three case studies. Although it does cover a variety of topics, including management, policy, and technology, it does not provide an organization-wide or strategic view of managing IT.

Despite some maturation of the literature during the MIS stage, theory was apparently not meeting the needs of practice. There were two notable contributions to the literature during this period. First, an emerging view recognized that IT required broader management attention. Second, management approaches matured to recognize that private-sector IT management might not suffice for public-sector managers. Even in the private sector, the MIS philosophy fell prey to vocal criticism as the developers rarely interpreted user needs accurately, and even if they did, it took so long to write the programs to run the systems that the original user requirements changed. After corporate managers spent millions of dollars to buy and subsequently upgrade MISs that did not provide the expected results, this criticism became more widespread.

Frustration found a voice in John Dearden (1972), who wrote that MIS would never meet managers' expectations or needs. He asserted the following:

The notion that a company can and ought to have an expert (or a group of experts) create for it a single, completely integrated supersystem—an MIS—to help govern every aspect of its activity is absurd. (p. 92)

After fixating on the technology and the life cycle, evidently organizations still did not build the information systems they needed. A new, broader perspective for managing IT emerged as an alternative to MIS in the early 1980s.

Stage 2: Information Resources Management

Ironically, the private sector did not lead the next phase of IT-management theory. Stage 2 signaled the replacement of MIS with a theory of information resources management (IRM). The federal government ushered in this new state of IT management with the passage of the Paperwork Reduction Act (PRA) (*Paperwork Reduction Act of 1980*, 1980). The PRA articulated a need for the federal government to manage information and IT as a resource, much like financial and human resources (Caudle, 1987). Though the PRA created requirements for managing both information and IT in the federal government, Marchand (1985) uses the term IRM to refer to a philosophy for managing IT more generally, as does this chapter.

IRM, as the Stage 2 perspective for managing IT, reflected change in both the technical and the external environments of organizations from the mid-1970s to the mid-1980s. Information technology began to move out of the data center with the arrival of minicomputers and the introduction of microcomputers. With this decentralization of computing power, management approaches had to shift from the data center to include user organizations. Strategic objectives for IT also changed as line organizations began to realize that they could use IT for more than just backroom functions. As a result, program offices often acquired their own IT when data-processing organizations could not keep pace with their needs. These changes, combined with the increasing level of spending on IT in organizations, brought these resources to the attention of a broader range of managers (Marchand, 1985).

In response to the disillusionment toward the management perspective focused on single applications mentioned earlier (i.e., MIS), private-sector organizations and business schools began to think more about how to tie together disparate systems to form an organization-wide perspective. Often, data-processing organizations call such perspectives architectures. This view represented a departure from the earlier philosophy of managing IT because it assumed that systems existed outside the physical and management control of a central data center.

McFarlan and McKenny (1983) also took an organization-wide view of managing IT, which they called the "IS function." Their work distinguished itself in recognizing that organizations, especially large ones, often manage many information-systems projects at once. As a result, it dealt with issues arising from coordinating several systems projects, going beyond exerting life-cycle management control over individual projects.

McFarlan and McKenny (1983) acknowledged that cultural (what they call environmental) factors often determine the effectiveness of control mechanisms for managing information systems. These factors include: (a) the penetration of information systems in the working environment, (b) the level of maturity of information-systems development efforts, and (c) the planning style of the organization. More specifically, McFarlan and McKenny asserted that the effectiveness of information-systems planning depends on the perceived importance and status of the systems manager, the physical proximity of the systems group and the general management team, the corporate management style, and organizational size and complexity.

International Business Machines (IBM, 1984) developed a systems planning methodology called business systems planning (BSP) that offered a framework for managing a collection of information systems across organizations. Specifically, BSP provided organizations, public or private, with a method for creating an information-systems plan.

Experience has shown that BSP can be applied to all institutions in the public sector and all industries in the private sector because the requirements for developing information systems are similar no matter the business served or the products and services provided (IBM, 1984).

Even thought BSP does not address the planning or use of individual systems, it does create a framework for linking systems planning to the broader purposes of the organization.

IBM's stated intent in creating BSP was to develop a methodology for creating a plan that would overcome the historic weaknesses of information-systems implementation, generally attributed to a lack of planning. In part, IBM built on its own attempts to deal with the plethora of systems that each of its own functions had developed over time. The keys to this new method, as IBM espoused them, were using top-down planning and analysis of organizational processes, relying on bottom-up implementation, translating organizational objectives into information-systems requirements, and using a structured methodology (IBM, 1984).

Consistent with IBM's view that BSP methods applied to public- and private-sector organizations, collections of articles on IRM issues appeared during this stage of the literature. For instance, Rabin and Jackowski (1988) edited a volume that touches on a variety of IRM issues, including information-systems management, data administration, and applications such as decision-support systems and databases. While sections of the volume dealt with public-sector IRM (Marchand & Kresslein, 1988), the balance of the work was so general that applying it to any particular kind of organization is difficult. Specifically, it devoted little attention to federal government.

In response to a request from the House Committee on Government Operations, the General Accounting Office's (GAO, 1992b) Division of Information Management and Technology (IMTEC) compiled a report that highlighted some problems it consistently found in federal agencies' management of information resources. It identified 11 themes across 132 agency-specific reports that spanned the period between October 1, 1988, and May 31, 1991. Of those 11 themes, 8 dealt with IT management and 2 addressed agency-wide management and control of information resources, including IT and data management. GAO did not set out to find root causes to these broadly defined problems, but did identify several plausible explanations in the report. For management mechanisms such as life-cycle management, project evaluation, and the coordination of information resources, GAO found that agencies lacked sufficient policy controls or did not effectively use the controls that existed.

Based on its experiences reviewing agency-specific IT projects, IMTEC developed what it described as a generic framework for developing systems architectures. In the preface to the report describing this framework, IMTEC expressed hope that the framework would help to address a prevalent problem of agencies trying to manage IT projects, which is the lack of planning and analysis of alternatives. While this report provided a very high-level view of the steps an agency should consider in developing a particular systems project, it did not address how to provide cohesiveness to information-systems efforts across the agency (GAO, 1992a).

To date, only two academic studies have addressed the implementation of the PRA by federal agencies. Caudle (1987) conducted the first study, documented in a report for the National Academy of Public Administration (NAPA). This report included the results of interviews with federal IRM officials in cabinet-level agencies, with selected subagency organizational units, and with the central oversight agencies. Caudle used these interviews to learn how agencies had organized to meet the mandates of the PRA and to assess whether the principles contained in the act had begun to pervade agency attitudes and behavior. In addition, her study presented agency views on the Office of Management and Budget's (OMB) and General Services Administrations (GSA) policy-making mechanisms for overseeing the achievement of the PRA and the Brooks Act.

Caudle's (1987) work provided groundwork for subsequent public-administration advances in IT management. In particular, she found that agency staff identified the strategic planning and budget processes as the primary control mechanisms for IRM. This attitude manifested itself in the respondents viewing IT management as guiding a project through the acquisition approval process at GSA and the budget approval process at OMB. In the conclusion to the report, Caudle recommended future research into control mechanisms for information resources. Such mechanisms, she noted, might differ from those used for financial and human resources. Additionally, her work serves as the foundation for GAO's strategic information management study discussed later.

The second study, conducted by Levitan and Dineen (1986), also relied on interviews with selected federal IRM officials. This research differed from Caudle's (1987) report for NAPA in that Levitan and Dineen first established a model of what they called "integrative IRM" and used that as a basis for assessing the state of the art of integration of federal IRM in 1985. In the authors' view, managing IT is complex because such management issues transcend normal organizational boundaries. Specifically, Levitan and Dineen cited strategic planning, the implementation of technology, and interaction with the organizational constituencies as examples of integrative issues for federal IRM.

Having constructed this model of the integrative nature of federal IRM, the authors conducted interviews with representatives from several federal agencies and one bureau of a federal agency. They posed questions about whether managing information resources in the organization extended beyond information systems to include information management (e.g., functions such as dissemination and records management) and whether notions of IRM extended to the program offices.

The findings confirmed several of those cited by Caudle (1987) in her study published by NAPA. For the most part, agencies had not fulfilled the mandate of the PRA by integrating the management of information systems and information content. The federal IRM offices represented in the interviews cited the management of information systems as their major concern overwhelmingly. This reflected the information-systems background of those staff who had risen to management positions in IRM. Agencies relied on task forces to achieve the integration of IRM issues across normal organizational boundaries. Levitan and Dineen (1986) found no consistent patterns of whether agencies had succeeded in integrating IRM as a management discipline in program operations.

Bishop, Doty, and McClure (1989) prepared a compendium of views on federal IRM from the two aforementioned studies, academic literature, and GAO, OMB, and GSA publications that went beyond just the carrying out of the PRA. Their paper presented a matrix of various observers' critiques on the status of IRM in the federal government. Comparing these critiques, they identified strong agreement on several points such as (a) insufficient integration of IRM at the agency level, (b) insufficient integration of IRM with the agency mission and program management, and (c) a need for better planning. Only one index dealt with the management of IT, and the respondents split evenly on the issue of whether IRM suffered from an overemphasis on technology.

Several other contributions from information sciences bring more insight into the state of IRM policy development and to some extent implementation. Beachboard and McClure (1996) point out some inherent inconsistencies among the various federal oversight bodies and policy instruments. To help deal with weaknesses they found in the IRM policy they cite, Beachboard and McClure advocate changes in IRM policy, federal agency commitment to IRM as a management discipline, and further empirical research into the lack of effectiveness of IRM policy and practice.

A similar publication on federal IRM finds that neither policy nor practice seems able to bring federal agencies into the information age, thereby enabling wide-scale electronic service delivery (Bertot, McClure, Ryan, & Beachboard, 1996). Without much empirical grounding, the authors offer a mix of policy, organizational, and investment choices that they assert will enhance the effectiveness of federal IRM policy and practice. These contributions add a certain richness to an understanding of federal IRM policy, but despite some hints of loftier strategic objectives for IT (i.e., electronic service delivery), they offer little in the way of concrete management processes to make that vision a reality.

Consistent with the maturing view of IRM, the research on local governments' use of IT began to explore the benefits from investments in automation. An analysis of survey data collected from 46 U.S. cities found that payoffs accrued in fiscal control, cost avoidance, and improved service-delivery mechanisms. This study also found that it took longer than anticipated to realize some benefits and that expected payoffs in better information for management and planning had yet to appear (Northrop, 1990).

Other research has included IT management as part of an examination of IRM in state governments. Caudle (1989) produced a nationwide study on the maturity of IRM in state governments. This study used data collected from a survey of 2,200 program managers and information-systems directors to characterize their views on the state of the art across several IRM activities, including IT usage and acquisition. It did not, however, discuss management techniques for IT or the strategic role of IT for stage governments.

At the county level, Fletcher and Bertot (1993) examine the extent and impact of the centralization of information services. The article is a based on a national survey of the largest 450 counties in the United States, supplemented with case studies of 13 counties, ranging from very small to very large. The survey gathered data on the extent of user-agency vs. central IT department equipment ownership, IT operation, network configuration, software-development strategies, and the operation of office systems. The analysis of these data reflected the shift of the role of the central IT organization from controller of all IT resources to provider of resources. The authors called this a collaborative model between central IT and user organizations and asserted that these data support a finding that county governments were succeeding at IRM while federal agencies were not at that time.

A consistent theme emerges from the literature of Stage 2 covering IRM. The application of theory and practice from this era has been very uneven. The federal government, in particular, still relies extensively on centralized, mainframe information processing and a similarly centralized management approach, which does not support the sharing and integration envisioned by IRM proponents. A lack of agreement on the meaning and relevance of IRM, as manifest in the publications advocating continued changes to basic federal law and government-wide policy for IRM, continues to thwart the adoption of the core management principles. This may be in part because researchers outside public administration, most notably information sciences, have done most of the research on IRM policy.

There is still work being done, though, to solidify the role of information and IT in public organizations in the tradition of PMIS started in the mid to late 1980s. Rocheleau (2006) has published a text devoted solely to the management of IT in the public sector. It begins with a discussion of how the management of IT differs between the public and private sectors. The bulk of the book, though, outlines key management issues and techniques for IT management that public administrators should know. This is text is one of the few examples that focuses exclusively on public IT management and includes discussions of some of the political and ethical issues that are unique to the public sector.

The strategic focus for IT remains largely the same: efficiency. Although efficiency of government service delivery to the public, as compared with the efficiency of just backroom functions, represents some progress, the ultimate goal remains lower costs. These notions of IRM as a management philosophy and strategic focus for IT remain mostly inwardly focused to the organization, except for preliminary discussions of electronic service delivery. Despite, or perhaps because of, the limited success of IRM, a new era for managing IT has begun to emerge.

Stage 3: Managing Information Technology in an Information Age

While many organizations strive to make the transition from Stage 1 (MIS) to Stage 2 (IRM), further developments in managing IT are forming the outline of a new stage. Marchand (1985) calls this stage "knowledge management" to highlight the shift in emphasis from the physical management of technology and information to the management of information content. Although there is now considerable interest in the area of knowledge management as a field of study and practices with the information-systems field, Marchand's use of the term has a narrower meaning than we now generally find. He contends this stage of development began in the mid-1980s and would continue through the 1990s. While the management of information content is emerging as a new skill, the successful application of IT makes it possible to maximize the benefits of information content (Marchand). The IS literature typically refers to this stage as strategic information systems, where the benefits of managing information content translates into competitive advantage in the marketplace (Applegate et al., 1999).

At the outset, though, a more precise definition of the term information age is in order. Information age has crept into the day-to-day language used in most organizations and the popular press. It is generally used to describe a future state where information will be quickly and universally accessible in electronic form. Dizard (1985, 1989) described his view of an information age, but implied that our society has not yet arrived there. The mature phase, which he believed would arrive in the 1990s, would entail the mass availability of IT and technology-based information services to consumers in their homes. This third stage would be made possible by the existence and use of a broad network of networks for sharing information between individuals, not only large organizations and businesses. Many of these new services would be made available, not through the personal computer, but with existing appliances in many homes, for example, the television and the telephone. To a great extent, this third stage of the information age is becoming a reality quicker than even Dizard might have envisioned. For that reason, this part of the chapter uses that description of the term information age to help frame the most mature management philosophy for IT.

As an indication of how powerful the notion has become, the Clinton administration described an information-age initiative promoting new uses for IT in federal agencies (Office of Science and Technology Policy, 1993). This report brought a new strategic focus to the uses of IT for federal agencies as a service-delivery vehicle, requiring new management approaches and supporting policies. The Clinton administration envisioned the federal government will use IT to make agencies more responsive and service oriented. Instead of using automation to make government operations more efficient for internal processing, agencies would use IT to reach out to the public to provide timelier and higher quality service.

Comparing differences between an information-age philosophy and IRM points to the maturation of management approaches for federal IT management. One difference is a more robust understanding in the information age of how important the technological infrastructure is. Emerging and maturing technologies such as graphical user interfaces, client-server computing, and workstations create a foundation for more substantial user involvement in information management. Additionally, this reality requires recognition that technology cannot be managed effectively from the mainframe data center to the exclusion of management of the desktop and everything between.

Several themes emerge from Strassman (1990), extending the IRM view of managing information systems from an organization-wide perspective. Strassman created a 137-item policy checklist for managing IT based on his work in the private sector prior to becoming the CIO (chief information officer) for the Department of Defense. He stresses that the policies for managing information systems should place responsibility in the hands of the users, not the IRM specialists. These policies include: (a) using economic analysis that promotes trading off between information resources and other resources and (b) providing mechanisms that charge users for information and information systems. He also introduces the notion that organizations should examine and redesign work processes before automation to ensure that automation does not speed up archaic and unneeded functions. The unifying theme presented by his list of policies is to ensure that investments in information systems add measurable value to the core missions of the organization.

Without identifying it as such, Strassman's (1990) approach of aligning information resources to organizational missions falls under the rubric of strategic management. Strategic management represents an effort by private- and public-sector managers to make strategic planning more useful in meeting the short-run needs of organizations. Much like Strassman, business authors writing about strategic management talk about aligning corporate resources and missions to take advantage of opportunities in the market. Strategic management emphasized looking at resource-allocation processes and support systems to see how they added value to a business (Hax & Majluf, 1984). For use in the public sector, public-administration authors have adapted these models for differences in civil-service traditions in political environments (Nutt & Backoff, 1992). Some public-sector models of strategic management explicitly include and highlight the role of IT in strategic management processes (Adler, McDonald, & MacDonald, 1992; DeLisi, 1990).

An analysis of a national survey of public managers' views of IT issues provided further insight into differences and similarities between public- and private-sector management in this area. Caudle, Gorr, and Newcomer (1991) compared the results of a 1988 survey of managers in federal, state, and local agencies to similar studies of private-sector organizations. The comparison of the public-sector survey to the previous survey of private-sector managers found that 7 of 10 top issues are common between the two groups. The authors found some differences in the relative ranking of the issues, leading them to conclude that the public sector was still grappling with some issues that private-sector organizations had been able to deal with more quickly. For instance, public-sector respondents ranked issues like the integration of technologies and end-user computing as high-priority issues where their importance had been declining over time for private-sector managers. They did find that both groups rated the issues of aligning IT with agency goals and IT planning near or at the top of their respective lists, indicating that a strategic orientation was beginning to take hold in the public sector.

GAO (1994) has taken a more preventive approach to dealing with federal agencies' problems with managing IT and grounded it in the literature of strategic management. In doing so, GAO not only presents a new management approach for IT grounded in systematic research, but it also brings the new strategic objective of mission performance into discussions of why the successful management of IT is so important. Within the federal IRM community, GAO's work in this area often goes by the label of its best-practices study (Caudle, 1996).

Part of what separates this GAO effort from typical GAO reports is the rigor of the analysis supporting GAO's recommendations for change. GAO went to some lengths to find case-study examples of effective IT-management practices in both the private and public sectors and included federal agencies beyond state governments. Throughout the process of deriving the key best practices from the case-study research, the GAO team consulted with federal IRM executives and oversight organizations such as Congress, OMB, and GSA to help ensure that the practices were applicable in the federal environment. As part of the case-study analysis, the research also went to great lengths to point to empirical proof of improved mission performance resulting from or at least relating to the application of the best practices.

While not explicitly linking the research or the findings to a particular body of literature, it becomes apparent from viewing the list of practices that GAO has created a strategic management framework specifically for managing IT. The practices GAO advocated are (a) directing IRM changes, (b) integrating IRM decision making in a strategic management process, (c) linking mission goals and IRM outcomes through performance management, (d) guiding IRM project strategy and follow-up through an investment philosophy, (e) using business-process innovation to drive IRM strategies, and (f) building IRM and line partnership through leadership and technical skills (Caudle, 1996). Although some of the language and the term IRM pervade the best-practices report, GAO's work has clearly brought both federal IT theory and practice to another level of management maturity.

Compared with the model of management maturity posited earlier, GAO's research, the resulting report, and the associated tool kits contribute new viewpoints to the strategic objectives and management approaches for IT. The strategic objective for GAO's best-practices work is unambiguously clear with the discussion of both strategic planning and mission performance, grounded in related program effectiveness measures. In effect, this implies a life-cycle management approach to linking IT plans, investments, and results. The mission focus also becomes apparent as the management approach seeks high alignment of IT investments with strategic plans and greater partnership between IRM and program staff. An investment management philosophy, grounded in the budget process, gives the whole management approach "teeth" and helps to guard against the tendency for more traditional IT plans to lie unused.

One academic study examined federal agency implementation of GAO's strategic information management (SIM) framework to assess how the practices from leading organizations, mostly from outside the federal government, were being implemented. Westerback (2000) analyzed data gathered from 20 case studies identified by the federal CIO council and the Industry Advisory Council in a report of federal IT-management best practices. By this time, most of the SIM framework had been codified into public law and government-wide policy (see *Information Technology and Management Reform Act of 1996*, 1996; *Paperwork Reduction Act of 1995*, 1995), and supporting government-wide policy (i.e., OMB, 1994).

The study therefore examined whether the case-study agencies were following the new laws and related policy and whether the practices were as relevant to federal agencies as GAO asserted. The analysis of the case-study data found strong support in practice for 3 of GAO's 11 practices and support for the remaining 8. In addition to finding support in practice for the GAO SIM framework, the study identified another practice (modified acquisition practices) that the study organizations found key to success but was not included in GAO work. This study helps to ground this important work, initiated by a practitioner organization, in the public-administration literature and also provide some much-needed external evaluation.

The Kennedy School of Government (Mechling & Applegate, 2000) published a series of papers titled *Eight Imperatives for Leaders in a Networked World.* The papers included an overview document and a guideline paper devoted to each of the eight imperatives. The intended audience for the paper series was government leaders, although the findings and suggested practices are grounded in the literature of both public policy and business administration. The topics for the papers were (a) a focus on how IT can reshape work and public-sector strategies, (b) how IT can be used for strategic innovation, (c) the utilization of best practices in implementing IT initiatives, (d) improving budgeting and financing for promising IT initiatives, (e) protecting privacy and security, (f) the formation of IT-related partnerships to stimulate economic development, (g) using IT to promote equal opportunity and healthy communities, and (h) preparing for digital democracy. Though not particularly deep in explanation, these guidelines provide a good overview of issues government leaders face in leveraging the full potential of investments in IT in the public sector.

One recent addition to the literature covers an IT management project that is quite unique to the federal government: megaprojects. Bozeman (2003) uses the IRS Tax Systems Modernization (TSM) as an extended case study to explore the challenges of deploying desperately needed IT and business modernization on a scale not often seen. Bozeman analyzes the TSM case using the literatures of organizational culture and risk aversion. While some of these lessons are unique to the IRS, they may well prove instructive to federal agencies with risk-adverse cultures seeking major systems and business-process improvements.

There is also some state-level research that addresses the strategic role of IT in the public sector.

Although the primary focus of the research is on strategic planning, Holley, Dufner, and Reed (2002) nonetheless build on the GAO studies of strategic information management to assess the extent to which state governments are managing information resources more strategically. The authors analyzed data from a Government Performance Project (GPP) survey gathered in 2000. For this study, the authors analyzed a subset of the GPP data that examined the processes and end results of strategic information-systems planning in state governments. The authors concluded that much like at the federal level, middle managers instead of top-level career and political appointees were involved in planning for IT implementation and as a result, the resulting plans were more tactical and inwardly focused than they were strategic and targeted at improving mission performance.

Garson (2006) addresses the policy, management, and implementation issues required for the success of a virtual state. One of the first chapters provides a brief history of public-sector information policy that chronicles the evolution of federal IT policy since the 1940s. Garson labels the most recent phase the virtual state, which places this work in the stage of IT management for the information age. The text points out how changing technology and public expectations have necessitated changes in public-sector policy, management, and implementation of IT.

One final article rounds out the most recent stage of managing IT in an information age and addresses the issue of whether public-sector investments in IT have and will yield the expected benefits of organizational transformation. Danziger and Andersen (2002) analyzed empirical research published in journal articles between 1987 and 2000 to understand the impacts of IT on public administration and the public sector. They organized their findings into four broad groupings of impact and 22 specific impacts. While the capabilities grouping focused on traditional impact measures like efficiency and effectiveness, the other groupings examined impacts that might be considered more political, such as how the interactions among governments, citizens, and stakeholders might change and whether value distributions might change as a result of public-sector IT. The authors concluded that while positive impacts included improved efficiency and productivity, negative impacts of public-sector IT included changes in public-employee work environments and citizens' interactions with government. On the larger issue of whether IT had transformed the public sector, the authors concluded that the impacts were substantial, but were not fundamental transformations.

Stage 4: Conclusion—Emerging New Models of IT Management

It is quite apparent how integral IT has become to administering federal programs specifically and all public programs more generally. Relegating this management domain to "computer people" is no longer sufficient. Since governance is an information-rich endeavor, public administrators need to understand how to manage the infrastructure that collects, processes, stores, and disseminates information about and for public-program delivery.

Unfortunately, most advancement in federal IT-management theory, policy, and practice continues to come from outside public administration. In particular, the information-systems field leads public administration in identifying the strategic role of IT in supporting organizational missions. Beyond that, there is evidence that researchers in IS are articulating a new era of computing beyond what they called strategic information systems and this chapter refers to as managing IT in an information age.

Carr's (2003) polemic, "IT Doesn't Matter," in the *Harvard Business Review* (HBR) asked the question of whether investing in IT for strategic advantage was still advisable. A subsequent debate in the editorial pages of HBR, the trade press, and even the popular press probed Carr's article title, data, analyses, and conclusions. I used this article to help generate debate in my undergraduate and graduate management information systems courses over the last three years, and my students and I concluded that the title was a bit disingenuous, but still quite effective for provoking debate. We also felt that Carr overreached with an analogy that asserted that IT had become a commodity like electricity and therefore was no longer a source of competitive advantage and, as a result, should be managed by minimizing risk and cost. What Carr said in his article, which is sometimes overlooked in the hyperbole over the title, is that while the underlying technology may no longer be the source of competitive advantage it was once, what really matters now is how organizations deploy and manage IT.

At about the same time, Peppard and Ward (2004) argued that organizations should be moving beyond identifying the "killer application" that will provide strategic advantage. Instead, the authors argue that organizations should be developing and sustaining an "information systems capability" that can be a source of continuous value. They assert this will be a fourth era of computing in organizations that recognizes the whole organization's performance will depend on IS capability instead of just a strategy of identifying a few select strategic applications. The resulting management focus should shift, in their view, to developing a full set of IS competencies, and the success of this approach should be measured by the performance of the organization as a whole.

It is not clear (and it may not matter) whether IT management is entering a fourth era. What seems clearer is that public-administration literature for managing IT still lags behind practice and theory from other disciplines. Given the growing popularity and spending on electronic government, it raises issues of whether public organizations lack a sufficient management infrastructure for building new Internet-based forms of service delivery. The depth and rigor of public-administration research in budgeting, personnel, and organization theory compared to IT management points to a need to reexamine the core management competencies for public administration. Peppard and Ward (2004) argue this for the private sector, and it likewise makes sense that the management of IT should be viewed as a competency as important as the management of human and financial resources within public administration.

Public administration should provide a vision for federal agencies as they move from a technology base and management philosophy grounded in concepts of management information systems and information resources management to a more mature notion of an information age. This more mature notion should recognize that as IT becomes even more widely available and used by the public, citizens will demand changes in the way they interact with their governments. Increasingly, the public will ask, if not demand, to do business with the federal government electronically. As a result, it is imperative that IT management theory, policy, and practice mature more quickly to enable the electronic service delivery the public will demand and current state-of-the-art technology enables. It remains to be seen whether public administration will contribute to this maturation any more than it has in the past.

References

Ackoff, R. (1967). Management misinformation systems. *Management Science, 14*, B147-B156.

Adler, P. S., McDonald, W., & MacDonald, F. (1992). Strategic management of technical functions. *Sloan Management Review, 33*, 19-37.

Applegate, L. M., McFarlan, W. F., & McKenney, J. L. (1999). *Corporate information systems management: Text and cases* (5th ed.). New York: Irwin McGraw-Hill.

Beachboard, J. C., & McClure, C. R. (1996). Managing federal information technology: Conflicting policies and competing philosophies. *Government Information Quarterly, 13*(1), 15-34.

Bertot, J. C., McClure, C. R., Ryan, J., & Beachboard, J. C. (1996). Federal information resources management: Integrating information management and technology. In P. Hernon, C. R. McClure, & H. C. Relyea (Eds.), *Federal information policies in the 1990's: Views and perspectives* (pp. 105-136). Norwood, NJ: Ablex.

Bishop, A., Doty, P., & McClure, C. R. (1989). *Federal information resources management (IRM): A policy review and assessment.* Paper presented at the 52nd Annual Meeting of the American Society for Information Science.

Bozeman, B. (2003). Risk, reform and organizational culture: The case of IRS tax systems modernization. *International Public Management Journal, 6*(2), 117-143.

Bozeman, B., & Bretschneider, S. (1986). Public management information systems: Theory and prescription. *Public Administration Review, 46*, 475-487.

Carr, N. G. (2003). IT doesn't matter. *Harvard Business Review, 81*(5), 5-12.

Caudle, S. L. (1987). *Federal information resources management: Bridging vision and action.* Washington, DC: National Academy of Public Administration.

Caudle, S. L. (1989). *Managing information resources: New directions in state government.* Syracuse, NY: Syracuse University School of Information Studies.

Caudle, S. L. (1996). Strategic information resources management: Fundamental practices. *Government Information Quarterly, 13*(1), 83-97.

Caudle, S. L., Gorr, W. L., & Newcomer, K. E. (1991). Key information systems management issues for the public sector. *MIS Quarterly*, 171-188.

Danziger, J. N., & Andersen, K. M. (2002). The impacts of information technology on public administration: An analysis of empirical research from the "golden age" of transformation. *International Journal of Public Administration, 25*(5), 591-627.

Dearden, J. (1972). MIS is a mirage. *Harvard Business Review, 50*, 90-99.

DeLisi, P. S. (1990). Lessons from the steel axe: Culture, technology, and organizational change. *Sloan Management Review, 32*, 83-93.

Dizard, W. P., Jr. (1985). *The coming information age: An overview of technology, economics, and politics* (1st ed.). White Plains, NY: Longman.

Dizard, W. P., Jr. (1989). *The coming information age: An overview of technology, economics, and politics* (3rd ed.). White Plains, NY: Longman.

Federal Property and Administrative Services Act (Brooks Act, 40 U.S.C. Sections 759, 487). (1962).

Fletcher, P. D., & Bertot, J. (1993). The central information services function in county government. *Government Information Quarterly, 11*(1), 73-88.

Fountain, J. E. (2001). *Building the virtual state: Information technology and institutional change.* Washington, DC: Brookings Institution Press.

Garson, G. D. (2006). *Public information technology and e-governance.* Sudbury, MA: Jones and Bartlett Publishers.

General Accounting Office. (1992a). *Strategic information planning: Framework for designing and developing systems architectures* (GAO/IMTEC-92-51). Washington, DC: Author.

General Accounting Office. (1992b). *Summary of federal agencies' information resources management problems* (GAO/IMTEC-92-13FS). Washington, DC: Author.

General Accounting Office. (1994). *Improving mission performance through strategic information management and technology* (GAO/AIMD-94-115). Washington, DC: Author.

Hax, A. C., & Majluf, N. S. (1984). *Strategic management: An integrative perspective.* Englewood Cliffs, NJ: Prentice-Hall.

Holden, S. H. (1994). *Managing information technology in the federal government: New policies for an information age.* Unpublished doctoral dissertation, Virginia Polytechnic and State University, Blacksburg.

Holden, S. H. (1996). Managing information technology in the federal government: Assessing the development and application of agency-wide policies. *Government Information Quarterly, 13*(1), 65-82.

Holden, S. H. (2003). The evolution of information technology management at the federal level: Implications for public administration. In G. D. Garson (Ed.), *Public information technology: Policy and management issues* (pp. 53-73). Hershey, PA: Idea Group Publishing.

Holden, S. H., & Hernon, P. (1996). An executive branch perspective on managing information resources. In P. Hernon, C. R. McClure, & H. C. Relyea (Eds.), *Federal information policies in the 1990's: Views and perspectives* (pp. 96-100). Norwood, NJ: Ablex Publishing Corporation.

Holley, L. M., Dufner, D., & Reed, B. J. (2002). Got SISP? Strategic information systems planning in state governments. *Public Performance & Management Review, 25*(4), 398-412.

Information Technology and Management Reform Act of 1996 (Clinger-Cohen Act, 40 U.S.C. Chapter 25). (1996).

International Business Machines. (1984). *Business systems planning guide* (GE20-0527-4). Atlanta, GA: Author.

International City Management Association (Ed.). (1986). *Local government yearbook: 1986.* Washington, DC: Author.

Kraemer, K. L, & Dedrick, J. (1997). Computing and public organizations. *Journal of Public Administration and Theory, 7*(1), 89-112.

Kraemer, K., & King, J. L. (2006). Information technology and administrative reform: Will e-government be different? *International Journal of Electronic Government Research, 2*(1), 1-20.

Kraemer, K. L., King, J. L., Dunkle, D. E., & Lane, J. P. (1989). *Managing information systems: Change and control in organizational computing.* San Francisco: Jossey-Bass.

Levitan, K. B., & Dineen, J. (1986). Integrative aspects of information resources management (IRM). *Information Management Review, 1*(4), 61-67.

Marchand, D. A. (1985). Information management: Strategies and tools in transition. *Information Management Review, 1*(1), 27-34.

Marchand, D. A., & Kresslein, J. C. (1988). Information resources management and the public administrator. In J. Rabin & E. M. Jackowski (Eds.), *Handbook of information resources management* (pp. 395-456). New York: Marcel Dekker.

McFarlan, F. W., & McKenny, J. L. (1983). *Corporate information systems management: The issues facing senior executives.* Homewood, IL: Richard D. Irwin, Inc.

Mechling, J., & Applegate, L. (2000). *Eight imperatives for leaders in a networked world: Guidelines for the 2000 election and beyond.* Boston: Harvard University. Retrieved from *http://www.ksg.harvard.edu/stratcom/hpg/*

Newcomer, K. E., & Caudle, S. L. (1991). Evaluating public sector information systems: More than meets the eye. *Public Administration Review, 51*, 377-384.

Norris, D. F., & Kraemer, K. L. (1996). Mainframe and PC computing in American cities: Myths and realities. *Public Administration Review, 56*, 568-576.

Northrop, A. (1990). Payoffs from computerization: Lessons over time. *Public Administration Review, 50*, 505-514.

Nutt, P. C., & Backoff, R. W. (1992). *Strategic management of public and third sector organizations: A handbook for leaders.* San Francisco: Jossey-Bass.

Office of Management and Budget. (1994). *Management of federal information resources: Transmittal 2, OMB Circular No. A-130* (Federal Register Vol. 59, pp. 37906-37928). Washington, DC: U.S. Government Printing Office.

Office of Science and Technology Policy. (1993). *Technology for America's economic growth: A new direction to build economic strength.* Washington, DC: Author.

Paperwork Reduction Act of 1980 (44 U.S.C. 3501 et seq). (1980).

Paperwork Reduction Act of 1995 (44 U.S.C. 3501 et seq). (1995).

Peppard, J., & Ward, J. (2004). Beyond strategic information systems: Towards an IS capability. *Journal of Strategic Information Systems,* 13 (2), 167-194.

Plocher, D. (1996). The Paperwork Reduction Act of 1995: A second chance for information resources management. *Government Information Quarterly, 13*(1), 35-50.

Rabin, J., & Jackowski, E. M. (Eds.). (1988). *Handbook of information resource management* (Vol. 31). New York: Marcel Dekker.

Rocheleau, B. (2006). *Public management information systems.* Hershey, PA: Idea Group Publishing.

Stevens, J. M., & McGowan, R. P. (1985). *Information systems and public management.* New York: Praeger Publishing.

Strassman, P. A. (1990). *The business value of computers: An executive's guide.* New Caanan, CT: Information Economics Press.

Westerback, L. K. (2000). Toward best practices for strategic information technology management. *Government Information Quarterly, 17*(1), 27-42.

Zuboff, S. (1988). *In the age of the smart machine: The future of work and power.* New York: Basic Books.

Chapter III

Politics, Accountability, and Information Management

Bruce Rocheleau, Northern Illinois University, USA

Abstract

This chapter provides examples of the politics of managing information in public organizations by studying both its internal and external aspects. Within the organization, politics is involved in structuring decision making, struggles over purchases of hardware and software, interdepartmental sharing of information, and the flow of communications such as e-mail among employees. The chapter analyzes examples of each of these internal aspects of politics. The chapter also discusses evidence concerning whether political appointees or career administrators are more effective as information managers. Externally, the chapter discusses how information management has been used to attempt to achieve greater political accountability through e-reporting and examples of cases where purchasing problems spill over into the realm of external politics such as through attempts to privatize governmental information management function. Certain topics such as municipal broadband systems and information management disasters are highly likely to involve information managers in politics. The attempts to use governmental Web sites as mechanisms to achieve e-governance and greater citizen participation in the political process also make it impossible for information managers to insulate themselves against politics.

Introduction

The message of this chapter is that information management has always been political and will become increasingly political due to several important trends that are occurring. First of all, information technology has become a central aspect of organizations, so more people care about it. This high interest can lead to struggles over strategic and operational issues. Second, there are emerging issues that push technology into areas that are potentially fraught with politics. For example, many local governments are interested in establishing governmentally supported broadband and wireless areas and these efforts have already resulted in major political battles with more likely to come. Also, information management is viewed as a method of obtaining increased citizen participation in the political process through various electronic mechanisms such as governmentally supported online e-governance mechanisms such as online rule-making dockets, public Listservs, public blogs, and other forms of computer-mediated communication (CMC). Each of these mechanisms has the potential to achieve positive goals, but they are also fraught with potential for generating political conflict. The underlying premise of this chapter is that information is power and consequently information management is inherently political. Information asymmetries give an advantage of one actor over others (Bellamy, 2000). Maintaining control over information can allow individuals, departments, and organizations to control how successful they appear to others and thus may protect autonomy, job security, and funding. Therefore, in order to provide effective leadership for IT, the generalist and head IT manager will need to actively engage themselves in both internal and external politics. An excellent case illustrating the importance of political issues in managing IT occurred in California. The California Department of Information Technology (DOIT) was eliminated in June of 2002 (Peterson, 2002). The department had been created in 1995 in order to solve the problem of several disastrous contracts in the IT area including a Department of Motor Vehicles (DMV) project that cost over $50 million but never functioned as planned (Peterson). Peterson cites accounts from observers to support the argument that a major reason for the failure was due to the other major agencies that viewed the new department as a threat to their power and lobbied to reduce the authority of the agency in the legislation creating it. In particular, the opponents lobbied to deny the new DOIT control over operations in the legislation creating DOIT. Those with interests opposed to the new DOIT included existing departments that had major authority in the IT field and/or those with large data centers. The opposition was successful so that the legislation limited DOIT's role mainly to authority over the budget. Consequently, the DOIT did not have control over data centers and was not able to achieve one of its major goals to centralize and consolidate these data centers (Peterson). This lack of operational authority limited its ability to influence other departments as Peterson summarizes:

Without controlling data centers or California's telecommunications network, DOIT simply had no juice, some sources argued. Because DOIT didn't add value to other state agencies, it couldn't exert any leverage on those agencies. DOIT could present ideas, but it couldn't make any real contribution to making those ideas happen. In other words, with the Department of Finance controlling IT budget processes, the Department of General Services controlling IT procurement and the state data centers handling computing needs, what was the DOIT's responsibility?

Also, according to observers, the head of DOIT was not allowed to sit in on cabinet meetings and there were reported cases of other departments doing "end arounds" concerning the formal requirement for DOIT to approve all major new projects. Another symbol of the weakness of the DOIT was that the governor appointed a new head of e-government who was independent of the DOIT, again lending credence to the perception that the DOIT lacked respect and power. The precipitating event in the death of DOIT was the quick approval by DOIT of a controversial project with the Oracle Corporation that resulted in an investigation and the resignation of several of the state's top IT officials. The California case illustrates how IT can become enmeshed in both internal and external political issues that I will analyze in this chapter.

In some cases such as those above, politics appears to refer to actions that tend to be viewed by outside observers as narrow-minded and self-serving. However, it is important to note that I use the term politics in a nonjudgmental manner. Politics can be about money and the "mobilization of bias" as Schattschneider (1983) described it as different forces struggle to prevail. However, politics can also be thought of as the attempt to mobilize resources to achieve public objectives and thus a necessary part of implementing any major project. I agree with Dickerson (2002, p. 52) that politics need not be a "lot of nasty back-stabbing and infighting" but is most often about "working and negotiating with others ... to get things done." It can be as simple as practicing good communication skills to keep others informed.

Although information management involves many technical issues, it is important to understand that it involves major political challenges. A large portion of governmental information managers come from technical backgrounds such as computer science and business. They usually have excellent technical skills and they can quickly rise to leadership positions such as chief information officer (CIO). However, decision making concerning the management of information technology requires more than technical knowledge as Towns (2004) notes: "There's increasing talk that CIO's don't need to be technologists because the position's nature is changing. Project management skills and people skills now mean more to a CIO than IT skills, the argument goes ..."

The most important critical success factors involve organizational and political skills that the technologically skilled often lack; however, these skills can be learned. In this chapter, I identify some of the key political problems that are likely to be faced. Many technical staff dislike politics and try to avoid dealing with political dilemmas. Molta (1999, p. 23) says, "engineers and programmers frequently appear oblivious to the strategic issues that keep management awake at night." He goes on to state that managing information technology is "the most politicized issue of the modern organization" and that technical staff "need to get in the game." Refraining from politics will lead to more serious problems and result in ineffective management of information technology.

Before the days of the World Wide Web and electronic government, managers and user departments often deferred computing decisions to technical staff because information management was not central to the organization (Lucas, 1984); generalist managers had little knowledge to contest decisions made by technicians. Now, since the information system has become a central concern, user departments often have their own technical staff, and generalist managers may become "technical junkies" (Molta, 1999, p. 24) who keep abreast of technological trends. The result is that information management is a much more prominent issue and the potential to become the source of disagreement. Consequently, technical skills themselves are not sufficient to be effective for information managers to achieve their goals. A study (Overton, Frolick, & Wilkes, 1996) of the implementation of executive information systems found that political concerns were perceived as the biggest obstacles to success. Many people felt threatened by the installation of such systems for a variety of reasons including fears of loss of their jobs and increased "executive scrutiny" (Overton et al., p. 50). Feldman (2002) also sees politics as one of the biggest challenges for technical managers. Feldman (2002, p. 46) observed that technicians often adopt a "bunker mentality" on important technology decisions and fail to take effective steps to achieve their goals. Peled (2000) also argues that information technology leaders need to bolster their politicking skills to boost the rate of their success. A survey (Anonymous, 1994) of over 500 British managers revealed that the majority believe that information flows were constrained by politics and that individuals use information politics for their own advancement. In short, the effective use of political skills is an important component of effective information management.

Over the past decade, computing has become much more important in governmental as well as private organizations. Major decisions about computing have always involved politics. Detailed studies of cities in the 1970s (Laudon, 1974) and 1980s (Danziger, Dutton, Kling, & Kraemer, 1982) demonstrate several cases involving computing and politics. However, computing was less central to public organizations then. Most employees had minimal contact with computing but now routinely employ computer technology in many of their day-to-day practices. They use e-mail, the Internet, and a variety of computer modules to accomplish their tasks. Computing is a central part of their jobs and they care about technology. The development of e-government and the Internet has greatly broadened the end users of governmental information systems so that now they include citizens and groups such as contractors. Information technology now is employed to provide greater forms of accountability to the public such as using computers to derive sophisticated assessments of performance and posting performance measures on the Internet. Many elected chief executives such as governors wanted to be associated with information technology. Coursey (in press) points out, for example, that Jeb Bush wants to be known as an "e-governor." As a result, information technology decisions are more complex and subject to the influence of external politics.

I analytically differentiate the politics of information management into two major categories: (a) internal organizational politics concerning issues involving organizational members and (b) external politics concerning how the governmental organization relates to its councils or boards, other organizations, external groups, and general citizenry. However, these two forms of politics frequently overlap and influence each other as they did in the California case. I will not be providing many prescriptions to managers on how to behave politically, not only because research about politics is sketchy but also due to the fact that politics is highly contextual. The course of action that should be pursued depends on the complex interplay of political resources of the actors involved, ethical concerns, and legal issues, as well as economic and technical factors. My purpose is to sensitize information managers and generalist administrators to major political issues that are likely to affect IT decisions. I outline examples of both successful and unsuccessful strategies that managers have employed to deal with information politics.

Sources

This chapter makes use of my own experiences as well as drawing on literature concerning public information management. I found very little formal research in recent years that explicitly focuses on the politics of internal information management. This aspect has not changed since I wrote earlier versions of this chapter. There is a rapidly growing literature on the use of external issues such as the use of the Internet to spur political involvement among citizens. However, my experience shows that internal political issues continue to be pervasive and important for IT managers; most of the major and rigorous academic studies concerning the politics of computing remain those from the mainframe era (e.g., Danziger, 1977; Dutton & Kraemer, 1985). Before IT became pervasive, these early studies by the "Irvine group" demonstrated that political and social factors generally affected how technology was structured and used (e.g., Northrop, Kraemer, Dunkle, & King, 1990). Due to its lack of coverage in traditional academic journals, I make use of periodicals such as computer magazines and newspaper articles concerning computing. Also, many of my examples are based on 25 years of experience in the IT field and also my study of public organizations at the municipal, state, and federal levels.

Information Systems and Internal Organizational Politics

Although almost any decision about computing can become embroiled in politics, my experience is that the most prominent and important political issues involve questions of control and power over the following kinds of decisions.

1. **Information management structures:** How should information management be structured? Where should control over information be placed in the organizational structure? A related issue concerns what kind of background is most effective preparation for technology department heads—technological or political?
2. **Hardware and software acquisitions:** What should be the nature of the process? How centralized should it be? Who should be involved? Should outsourcing be used? Should open-source or other preferences be mandated?
3. **Information management, information sharing, and interdepartmental relations:** What process should be used to determine information sharing and exchange? How will computing influence and be influenced by other aspects of interdepartmental and interorganizational IT issues? How can obstacles to sharing information be overcome?
4. **Managing personnel and communication flows:** How does computing influence employee relations and communication flows? What, if any, rules and procedures should be established? How do e-mail and other forms of CMC influence communication and organizational politics? How does information technology influence the careers of organizational members?

Although each of these issues has technical aspects, nontechnical issues such as concerns about autonomy and power often prevail. Below, I outline how each of these decision areas involves important political aspects. Many information managers prefer to avoid these political aspects. Likewise, generalist managers such as city managers have often ignored direct involvement in these decisions due to their lack of expertise concerning computers. As a consequence, it has been my experience that persons other than information system or generalist managers often dominate these decisions. Consequently, these issues are often decided without adequate attention from those with the most expertise or broadest perspective.

Information Management Structures

There is no consensus as to the best method of organizing computing. How centralized or decentralized should computing be? There are advantages and disadvantages to the centralization of computing: One major review of the centralization-decentralization debate (King, 1983) concluded that political factors are paramount in decisions on how to structure computing. Business organizations have encountered the same dilemma. Markus (1983) has described how business departments have resisted efforts toward the integration of information systems. Davenport, Eccles, and Prusak (1992) found several reasons behind information politics, including the following: (a) Units that share information fully may lose their reason for existence, and (b) weak divisions may be reluctant to share information when they are sensitive about their performance. Overall, in recent years, there has been an emphasis on the centralization of authority as organizations move toward an enterprise-wide approach in which databases are centrally organized and standards govern the hardware and software systems of organizations.

Where should computing be placed in the organizational structure of a public organization such as a city? Should it be a separate line department, a subunit of another department (e.g., budgeting and finance), or a staff unit to the manager or mayor? Should mayors or managers require that the head of computing report directly to them, or should they place a staff member in charge? Except for small organizations, the centrality of today's computing to all departments would suggest that information management should be in a separate unit and not be structurally placed under another department such as budget and finance. There has been a strong movement at the federal and state levels to establish a CIO to deal with problems of technological issues that cut cross-departmental boundaries and to head efforts at building corporate-wide e-government and intranet systems (Fabris, 1998). However, managers, both computing and generalist, need to think carefully about the implications of these different arrangements and which structure is most likely to meet the needs of the organization. Decisions concerning the structure will be based on a number of factors including the degree of interest of the generalist managers in computing as well as what goals managers have for IT.

Molta (1999, p. 23) defines politics as the "allocation of resources within an organization." Some departments will want to control their own computing as much as possible through hiring their own IT staff. Eiring (1999, p. 17) defines politics as the "art of negotiation, compromise, and satisfaction," and urges information management staff to form strategic alliances that are beneficial to the IT cause and to "nurture them as one would a good lasting friendship" (p. 19). Feldman (2002, p. 46) warns that "when departments are doing their own thing—namely hiring their own IT staff—a central IT department is in political trouble." He goes on to point out that hiring a staff implies unhappiness with the services of the central IT unit. However, in many organizations, it has been common and perhaps necessary for line departments to have staff dedicated to IT. For example, police departments often have their own dedicated systems and staff because of the early development of police information systems, the centrality of computer searches to their function, and the need to have secure and easy exchange with other police departments. When non-IT-department IT staff exist, one political issue is how they should relate to the central IT staff. Feldman (p. 48) argues that the smart strategy is for the central IT staff to offer "to exchange information and support" with non-IT-department staff. By taking these steps, Feldman argues that they can at least establish "dotted line relationships." Feldman argues further that these non-IT-department technological staff are often isolated and appreciate the support from the central IT department. Anderson, Bikson, Lewis, Moini, and Strauss (2003) found examples of dotted-line relationships in states like Pennsylvania where the formal authority of the state IT head over state agencies was weak: "Additionally, all agency CIO's have a 'dotted line' relationship to the state CIO even though they formally report to their own agency heads; they meet quarterly with him" (p. 23). The only alternative is to try to control all computing from the central IT department, but this strategy can either work very well or turn out to be a disaster (Feldman).

A major rationale for the existence of a CIO (as opposed to a traditional data-processing manager) is that he or she will not be restricted to technical issues but act as a change agent, politician, proactivist, and integrator as well (Pitkin, 1993). The federal government (Koskinen, 1996; Pastore, 1995) has firmly established the use of CIOs in order to improve information management. Will it have a positive impact? Should the CIO model be followed in municipalities and other public organizations? Merely assigning the CIO title does not ensure that these functions will be performed. For example, a study (Pitkin) of CIOs in universities found that, despite their title, CIOs did not view themselves as executives and often do not perform nontechnical roles. Without a push from a CIO, public organizations may fail to make good use of information technology. For example, one study found that police regularly used their database systems for reports to external agencies but rarely for internal management purposes (Rocheleau, 1993). CIOs and centralized information structures help to fix responsibility and that can mean that they themselves become targets of unhappiness with technological decisions. There are many cases in which CIOs in the private sector have not been viewed favorably by their fellow managers (Freedman, 1994), and CIOs in both the public and private sector are blamed for failures (Cone, 1996; Newcombe, 1995).

With the growth and importance of IT, organizations of moderate to larger size now tend to have an independent IT department (Gurwitt, 1996) because IT is now viewed as infrastructure serving everyone and should not be under control of any single department. Also, because there is now a strong acceptance of the need to have as much standardization of hardware and software as possible, the IT unit is often vested with final approval over major purchasing decisions. However, the California case shows that this centralizing trend is not inevitable or necessarily linear. According to Peterson (2002), the structure recommended to replace the State Department of Information Technology was to decentralize, with the authority of the extinguished department being reassigned to the finance and general-services departments. By way of contrast, a study by Rand Institute researchers (Anderson et al., 2003) recommended a more powerful and centralized department to replace California's deposed DOIT. However, it is interesting to note that the Rand Institute researchers studied the governance structures of four other states and found that some of the states (e.g., Illinois) had IT structures that were weak in formal authority but nevertheless worked effectively due to the fact that the IT leadership worked through brokering relationships. Likewise, according to the Rand study, Pennsylvania's system does not vest strong authority in the state CIO, but it depends on successful dotted-line relationships. Their conclusion is that there is no one best way to organize and that CIOs who have weak formal authority can use their negotiating and brokering skills to be successful. They also argue that management style is important. Successful state IT managers have a style that is "participative, collaborative," and emphasizes positive "carrots" rather than "sticks" in seeking change (Anderson et al., p. ix). In the California case, the politics of IT involved the legislature and also key vendors, so IT leaders have to practice their communication and political skills on key external as well as internal constituencies. Recently, the head of the National Association of State Chief Information Officers (NASCIO) reported that several state CIOs had told him that they preferred not to report directly to the governor because they wanted to focus on operational issues and avoid partisan politics (Towns, 2005).

In my experience, there is a wide variety in the amount of attention devoted by generalist managers to structural issues concerning computing. In one city, a city manager was very much focused on information management and devoted great attention to decisions made concerning computing by the city, in effect acting as the municipality's CIO. His focus on information technology enabled him to establish a positive reputation for innovation that helped to secure his next job. When this manager left for another city, he was replaced by another manager who was not especially interested in computing. Devoting great attention to computing can be both productive and counterproductive. The first manager who was heavily involved in computing decisions became embroiled in severe struggles with his new board and organization over IT issues that contributed to his resignation from his new job.

Although there is no single right way of organizing computing, each manager needs to ensure that the structure will provide relevant, timely, and reliable information. Kraemer and King (1976) argue that public executives spend too much of their time on decisions concerning the purchase of equipment and too little on other important information management issues that have less visibility but are equally important. Kraemer and King emphasize the need for generalist managers to take personal responsibility for computing and to be engaged in the following decisions: how to structure computing, the purposes to be served by computing, and implementation issues such as the goals of computing and the structures used to achieve them.

Some experts (e.g., Severin, 1993) argue that the chief executive officer (CEO) of an organization should also be the true CIO. The former mayor of Indianapolis, Stephen Goldsmith, is an example of a CEO who took charge of the information technology function and instituted a number of important policy changes such as privatizing many IT functions as well as encouraging e-mail from any employees directly to himself (Poulos, 1998). A related issue concerns the question of to what extent the head of IT needs to have a technical background. John Kost (1996) was appointed to be CIO of the state of Michigan by Governor Engler and instituted several policy goals of Engler's such as the consolidation of state data centers, the establishment of statewide standards, and the reengineering of IT including its procurement process. It is notable that Kost did not have any IT background at all (Kost). Kost maintains that it is more important that the CIO understand the business of government than have a strong technology background. Kost proceeded to do a major reengineering of IT in Michigan and claims that they successfully achieved many of the goals set by Engler. However, if the CIO does not possess strong technological skills as well as institutional knowledge about the IT system, he or she will need to have trusted and reliable staff who do have such skills in order to have the trust and respect of client agencies. One of the problems with the California DOIT was that it had little operational authority and was primarily an oversight agency. Thus, one of the recommendations of Anderson et al. (2003, p. 53) is to "transfer the majority of people with technical skills" from the finance and other departments to the new IT department so that it would be "properly staffed and positioned to provide technical approval."

Paul Strassmann served as director of defense information at the defense department from 1991 to 1993 where he was in charge of a $10 billion annual budget for IT and instituted major changes in the procurement process. Strassmann (1995) subsequently published a book entitled *The Politics of Information Management* in which he argues that managing IT is "primarily a matter of politics and only secondarily a matter of technology" (p. xxv). Strassmann goes on to hold that only the technical aspects of information can be safely delegated to computer specialists. He supports a "federalist approach to information management, delegating maximum authority to those who actually need to use the information" (p. xxix). Strassmann (1995) believes that it is the duty of the CEO to establish general principles: "Without a general consensus about the principles and policies of who does what, when, and how, you cannot create a foundation on which to construct information superiority" (p. 10). He says that the CEO should never delegate the responsibility for information management to a CIO because it is the CEO who must decide how to apply information systems.

In a majority of organizations of large size, there tends to be one or more advisory committees or groups set to assist the IT head in making decisions. In my experience, these advisory groups tend to fall into three categories: (a) end-user groups involving end users who are especially engaged in IT, (b) representatives of departments served by the IT department, who may or may not be heavy end users, and (c) external people who have substantial experience in IT. In Anderson et al.'s (2003) study of four states that are regarded as having successful IT departments (New York, Pennsylvania, Virginia, and Illinois), the state IT units generally had both internal groups made up of the line departments who represent the end users of IT and an external group consisting of private-sector IT heads who provided their expertise. I have known municipalities to use the same approach, and one municipal IT head told me of how the private-sector committee saved their community money with their advice about telecommunications strategy to obtain low-cost services from vendors. A politically adept CIO can make good use of these committees to build her or his political base.

Comparison of Politically Appointed vs. Career Administrators

As I have documented above, political skills are a necessary component of IT management. This has been documented at the federal level in studies of the jobs of federal CIOs carried out by the United States General Accounting Office (U.S. GAO, 2004). Given the importance of politics, would we expect politically appointed or career administrators to be more effective? The GAO study has some support for both positions. The GAO publication outlined what most CIOs considered to be major challenges, and they all involve the use of political skills: (a) implementing effective IT management, (b) obtaining sufficient and relevant resources, (c) communicating and collaborating internally and externally, and (d) managing change.

According to the GAO (2004) report, many thought that politically appointed CIOs would be more successful because they have more clout and access. However, others thought that skilled career administrators would be more successful because "they would be more likely to understand the agency and its culture" (p. 23). Another variable is that politically appointed CIOs (in federal agencies) had a shorter tenure of 19 months vs. 33 months for the career administrator, and the career administrators thought that this gave them a significant advantage because it can take a good deal of time to accomplish major tasks. Concerning their communication and collaboration skills, the GAO report concluded that it is critical for CIOs to employ these abilities to form alliances and build friendships with external organizations:

Our prior work has shown the importance of communication and collaboration, both within an agency and with its external partners. For example, one of the critical success factors we identified in our CIO guide focuses on the CIO's ability to establish his or her organization as a central player in the enterprise. Specifically, effective CIO's—and their supporting organizations—seek to bridge the gap between technology and business by networking informally, forming alliances, and building friendships that help ensure support for information and technology management. In addition, earlier this year we reported that to be a high-performing organization, a federal agency must effectively manage and influence relationships with organizations outside of its direct control. (p. 30)

Concerning the management of change, the GAO report (2004) found six CIOs (from the private sector) who said that dealing with government culture and bureaucracy was a major challenge and that they had to marshal resources to overcome resistance. A subsequent report by Lewis (2005) comparing political and career federal managers (this concerned all managers, not just CIOs) found that career managers were more successful based on the Bush administration's Program Assessment Rating Tools. A *Federal Computer Week* analysis found that only 16 out of 27 CIOs in office in 2004 remained in office in November of 2005 (Lunn, 2005). The most common reasons for leaving the job were to obtain better pay (19%) and a change in administration (16%).

The *Federal Computer Week* magazine (Hasson, 2004) conducted a survey that obtained responses from 129 CIOs concerning the career vs. political issue and found similar results. Two thirds of the CIOs agreed that the political appointee would have better access and one third agreed that they would be able to raise the profile of the IT department. One former CIO argued that the political appointee could be more aggressive while the career administrator "had to find a champion" to push projects through the legislature. However, two thirds also thought that the career CIO would have a bigger impact because of the longer tenure. In 2004, there was a number of prominent CIOs who left high-profile public positions, including the top IT officers of the states of Virginia and Florida (Towns, 2004). It is safe to say that in order to be effective, regardless of whether they are political appointees or career administrators, people in these positions will need to exercise effective political skills. Peled (2000) presents two case studies involving information technology leaders in Israel. In one case, a prestigious scientist with outstanding technical skills was called in to solve transportation problems by employing computing technology. This technologist viewed his job in technical terms and attempted to develop a project without communicating with and building support from other key actors. He refused to share information with other departments working on related projects and consequently encountered resistance leading to his resignation and the end of the project. His failure was largely due to deficiencies in communication and lack of knowledge of organizational politics such as the need to obtain support from others in order to develop a project. Peled provides a second case study in which a leader without any major technical skills was able to solve serious problems of building a land-registry database. He used a training system to develop a core of users and people committed to the new database and negotiated with unions about wage demands for using the new system. This leader viewed his project in political terms from the very beginning and this approach helped him to achieve success. The importance of political skills in managing e-government has been confirmed by a study by Corbitt (2000) of the eight factors associated with the failure of e-government projects—only one of these concerned technical problems. The others included the absence of a champion of change, lack of managerial support and attention, poor attitudes toward the IT department, lack of education and training, and a discrepancy between IT staff and the end users of the system. I am not saying that IT skills are not important. Indeed, in the numerous small governments that have only a tiny number of staff, they cannot afford to hire a nontechnical IT manager. Nevertheless, even in these small organizations, political skills are also an essential component to IT leadership.

Politics and the Purchasing Process

As we saw in the California DOIT case, the failure of a large IT system can often embroil an organization in politics, and it is crucial that generalist managers take measures to avoid such disasters. An underlying assumption of the current information resources management theory is that an organization's information system should be aligned with its business goals. A related assumption is that generalist managers must be involved in procurement and other important decisions. They need to specify what goals and functions should be achieved by purchasing new software and hardware. Sandlin (1996, p. 11) argues that managers need to be on guard for "technological infatuation." The author points out that generalist managers would never let the transportation departments buy a line of expensive cars, but they often allow equivalent purchases in the IT area.

The failure of expensive new computer systems is likely to expose governmental managers to political attacks. Even if knowledge of the failure remains internal, unsuccessful systems can undermine central management, IT, and other departments involved. The problem often begins with the failure of the internal management of the projects. For example, the Federal Retirement Thrift Investment Board fired and sued its contractor, American Management Systems (AMS) of Fairfax, Virginia (Friel, 2001), for breach of contract. The contractor defended itself by stating that the board had not determined system needs even 3 years after the beginning of the contract (Friel, 2000). The determination of system needs is primarily a managerial issue. The pervasiveness of contract failures has led to what some refer to as a "contract crisis" (Dizard, 2001) with political consequences: "Thanks to new project management techniques, improved oversight, employee training, and contract controls, several state CIO's reported that project failures are decreasing. But they agreed that the *political cost of bungled projects remains high* [italics added]."

The U.S. General Accounting Office (2002) studied the Department of Defense Standard Purchasing System and found that 60% of the user population surveyed were dissatisfied with its functionality and performance. If a failed project has high visibility, then often external political issues also develop, but even if not visible to outsiders, failed purchases weaken the credibility of IT staff and thus the purchasing process is one of the most critical areas for managers and IT staff to negotiate.

As end-user computing has grown, end users have enjoyed the freedom to innovate and strong centrifugal forces have resulted. Employees often have strong personal preferences and feelings concerning software and hardware purchases. Part of the ethos of end-user computing is the ability to make your own decisions about software and hardware. Allowing each end user (or end-user department) to make decisions about software is likely to lead to multiple hardware and software platforms.

A potentially major source of politics is developing due to the conflict between open-source software and proprietary software packages such as those of Microsoft. Recently, the CIO of the state of Massachusetts issued a policy that only open-source software would be used and that proprietary software would be phased out by 2007 (Towns, 2005). Other state agencies (e.g., secretary of state) questioned the authority of the CIO to mandate such policies for other agencies, and the state legislature considered a bill that would set up a state task force that would have to approve such mandates (Towns). Microsoft attacked the open-source policy labeling it as discriminatory. The Massachusetts CIO later resigned though the state maintained that it would continue its movement toward the open-source requirement (Butterfield, 2005). The movement to establish enterprise-wide standards as well as the move to open-source software promises to make the jobs of those involved in establishing IT standards politically sensitive in the future.

There are other trade-offs between allowing each department to use its preferred software and hardware vs. centralization. Multiple platforms complicate training, backup, and maintenance, too. The existence of "platform zealots" is not unusual and can lead to conflict (Hayes, 1996). In my experience, these problems with multiple platforms have led certain managers toward establishing a single platform and also centralized control over hardware and software acquisitions. Barrett and Greene (2001) make the point that leaders need to convince the end users of the strong advantages and rationale for the standardization of hardware and software. If they fail to take this step, they are likely to encounter directly or indirectly passive resistance to their policies. In some cases, formal control of IT purchases by the IT department is impossible if the funding source for hardware and software is from another level of government (e.g., state or federal funding). Regardless of what approach is taken, information management and generalist administrators need to provide the centripetal force needed to integrate information management in public organizations. If they do not do it, no one else will. However, this integrating role often runs into stiff resistance and it requires that the manager use powers of persuasion, negotiation, bargaining, and sometimes authority and threats.

Many generalist managers may want to establish standard policies that influence the purchasing choices of departments, such as the following: (a) Some governments take a position that data-processing functions should be privatized as much as possible, (b) many governments have instituted online purchasing and forms of purchasing pools that departments may be required to adhere to, and (c) some governments are establishing special arrangements with a small number of computer vendors with the idea of achieving advantageous pricing arrangements. Both the federal and many state governments have been revamping the purchasing process with more emphasis on speed and emphasizing value rather than lowest cost (Rocheleau, 2000). Kost (1996) believes that the CIO and CEO need to take charge of the purchasing process if they are to achieve goals such as privatization and "value purchasing": "For example, a policy advocating privatization is doomed unless the purchasing process allows privatization to occur... An intransigent purchasing director can often do more to thwart the direction of the administration than a policy-maker from the opposite political party" (p. 8).

At the federal level, Strassmann (1995) implemented a corporate information management (CIM) initiative that was aimed at streamlining the military's information system purchases such as the use of the same systems across the different services. Strassmann enunciated the following principle that the technicians were expected to follow: Enhance existing information systems rather than "opt for new systems development as the preferred choice" (p. 94). In one case, this CIM approach killed an $800 million Air Force system and replaced it with a similar one that was used by the Army (Caldwell, 1992). The Air Force had already spent $28 million on their system and resisted the move. Observers of the process noted that it was a "turf issue" and a GAO report concluded that CIM required centralization and a cultural change that were difficult for the defense department (Caldwell, pp. 12-13).

The acquisition and implementation of new systems often engender resistance. One of the basic principles of planning for new computer systems is to involve the people who will be using the system in its design, testing, and implementation phases. Indeed, there are entire books written concerning the principles of participatory design (Kello, 1996). An apparent example of user resistance occurred in Chicago when a new computer system was introduced to speed the building-permit process. After the system was implemented, lengthy delays drew widespread criticism (Washburn, 1998) and the delays caused a bottleneck during a time of booming construction. The new system tracked permit applications and allowed the scanning of plumbing, electrical, and other plans so that the plans could be viewed simultaneously on several screens. There were some technical problems acknowledged by city officials. For example, some staff had trouble seeing plan details on their screens and on-the-spot corrections were not possible due to the fact that applicants were not present when reviews were done. However, officials argue that many of the complaints were due to the fact that the system had changed the process of handling permits. Permits are now done on a first-in-first-out basis compared to the previous situation where expediters used to "butt into line," and, consequently, they feared loss of influence under the new system (Washburn). The contention was that the expediters deliberately spread false rumors about extensive delays in an attempt to "torpedo the new system."

The desire to standardize can create political resistance. For example, vendors who target their products to municipal governments begin with a basic general ledger and finance product and then expand to develop modules for other functions such as the permit and other processes. Managers see advantages to using the same vendor for all of their different modules such as ensuring interoperability among them as well as gaining favorable financial terms. However, in this author's experience, vendors who have strong financial modules often are weak in other areas and thus the desire to standardize on one vendor's software can lead to problems with end users who do not like these other modules of the vendor. In such cases, it is clear that generalist and information system managers will have to be sensitive to organizational politics and either accept the need for diversity in vendors or employ their personal and political resources to achieve change.

Computing, Sharing Information, and the Politics of Interdepartmental Relations

In addition to purchasing issues, there are many other interdepartmental issues that need to be dealt with by CIOs and generalist managers in order to establish an effective information system. For example, computing creates the possibility of free and easy exchange of information among governmental organizations. However, information is power and organizations tend to be sensitive about giving out information to outsiders, especially if it reflects on the quality of the organization's operations. Many agencies prefer to maintain autonomy over their data. For example, the author worked on a project with the job training agency of a state agency that was to employ databases drawn from several state agencies to evaluate the state's job training programs. However, despite obtaining verbal agreements from the top managers of the agency, the lower level programming staff delayed the sharing of data for months. It became clear that they saw our requests as an additional burden on them that would make their job more difficult if such requests were to become routinized.

Building e-commerce systems usually requires the cooperation and sharing of information among a number of different departments. Corbitt (2000, p. 128) conducted a case study of an organization that developed an e-commerce system and found that there were "important political and interest differences" among the departments as well as "differences in perceptions" that caused problems. In particular, Corbitt found competition between the data division and the e-commerce group about what needed to be done and who should exert leadership over it. Corbitt (p. 128) concludes that power is a "very substantial issue affecting implementation success."

Some new technologies such as geographic information systems (GISs) are forcing changes in computing structures and procedures among departments. Generalist managers may need to act to ensure that appropriate new structures are established. For example, although many geographic information systems are initiated by a single department, the systems are expensive and the software is relevant to many different departments. When Kansas City decided to build a GIS (Murphy, 1995), they found it necessary to form a GIS committee (made up of representatives of four participating departments—public works, water, city development, and finance) to conduct an interdepartmental needs assessment and resolve problems such as how to resolve conflicts in databases and how to minimize database development costs. Although such developments do force structural changes, there is still wide latitude in regard to the nature of the structure. Sharing data can lead to conflict. In a study of the exchange of information between municipal departments such as fire and police, this author (Rocheleau, 1995) found that a large percentage of departments fail to exchange information despite overlaps in their job responsibilities concerning problems like arson and emergencies. I studied one city where the fire department, clerk's office, and building department all shared information responsibility for entering information about buildings, but each department tended to point their finger at others when mistakes in the data were discovered. A major task of generalist and information managers is to deal with departmental concern with autonomy over the databases. If they defer to the status quo, information management will be less effective. Bringing about the change required to achieve integration may aggravate such conflicts. Overcoming these obstacles requires negotiating, political, and organizational skills.

Top managers may force the exchange of information via command. However, employees often find ways to resist change. For example, they may provide poor-quality information that renders the exchange useless. Markus and Keil (1994) provide a case study of a new and improved decision-support system designed to help salespersons that failed because it worked counter to underlying organizational incentives. The system was aimed at producing more accurate price quotes, but it hampered the sales staff's ability to sell systems, their most important goal, so the new system was used little.

The relationship between technology and individual career ambitions can lead to political aspects of information management. Knights and Murray (1994) have conducted one of the few detailed studies of the politics of information technology. In their case study of IT in an insurance company, they concluded that the success and failure of the systems were closely tied to the careers of mangers. Consequently, these managers often attempted to control the perception of the success of these systems because perception is reality (Knights & Murray, 1994):

The secret of success lies in the fact that if enough people believe something is a success then it really is a success ... it was vital for the company and for managerial careers that the pensions project was a success. (p. 172)

One of the key points made by Knights and Murray (1994) is that computing decisions become inextricably entangled with career ambitions and fears of individual employees and become embroiled in a very personal form of politics. Another detailed case study (Brown, 1998) of the implementation of a new computer system in a hospital found that different groups (the hematology ward, hematology laboratory, and information technology team) had very different perspectives on the reason for the failure of a new computer system. Moreover, each of the three groups used the common goal of patient care to legitimate their view of the system. Brown concludes that the study shows that participants were influenced by "attributional egotism" in which each person and group involved attributes favorable results to their own efforts and unfavorable results to external factors. Similar to Knights and Murray, Brown concludes that many of the actions are taken to protect individual autonomy and discretion.

Grover, Lederer, and Sabherwal (1988) borrowed from the work of Bardach (1977) and Keen (1981) to outline 12 different "games" that are played by those involved in developing new systems. They tested their framework by in-depth interviews with 18 IT professionals who confirmed that these games were played in their organizations. Most of their games involve interorganizational or interpersonal struggles similar to those I discussed above. For example, they discuss how in the "up for grabs" game, control over a new IT system involves struggles between the IT and other departments. They illustrate what they call the reputational game with a story about an IT manager who projected "a rough exterior" (Grover, Lederer, & Sabherwal, p. 153) in order to reduce demands on the IT department, but this approach led to a coalition against him and the IT department and resulted in the eventual dismissal of the IT manager.

The lesson of the above cases is that, prior to implementing new systems, information managers need to assess the organizational context and determine how proposed systems will be affected by incentives, informal norms, resistance to change and sharing, as well as other forms of organizational politics. A broad stakeholder analysis needs to be done. Many of these factors may be addressed by including end users in the planning process. Managers will often have to be involved in exerting political influence and engage systems outside their direct control in order to assure a successful outcome. For example, Kost (1996) describes how the Michigan Department of Transportation decided to change from a mainframe to a client-server environment, and this change endangered the jobs of a dozen mainframe technicians. The logical step was to retrain the mainframe technicians to do the new tasks, but the civil-service rules and regulations required that the mainframe workers be laid off and new employees be recruited to fill the client-server positions. Thus, in order to have an effective information system, generalist and information managers will often have to seek to change rules, procedures, and structures and, at the same time, alleviate as much as possible any perceived negative impacts of change. Still, change may bring information managers inevitably into conflict with other departments.

Computing and Communication Patterns

Information technology such as e-mail can affect organizational communication patterns. Changes in communication flows can be extremely political. For example, if a subordinate communicates sensitive information to others without clearing it with her or his immediate supervisor, strife is likely to result. While he was mayor of Indianapolis, Stephen Goldsmith encouraged every police officer and other public employees to contact him directly via e-mail (Miller, 1995a). He claims to have read 400 e-mail messages a day. Should mayors or managers encourage such use of direct contact from employees? Although such communication can and does occur via phone and face-to-face communication, e-mail communications are different from face-to-face communications—there is less rich information and many people portray themselves differently in e-mail than they do in person.

E-mail has become the dominant form of communication in many organizations and it has implications for organizational politics (Markus, 1994; Rocheleau, 2002b). Markus has shown that e-mail is routinely used as a device to protect employees in games in which they feel it is necessary to "cover your anatomy." E-mail now provides a digitized trail that can be used to support employees concerning their reasons for doing what they did. Employees often copy e-mail messages to their own or those of other superiors to let people know what they think is necessary, thus attempting to bring more pressure on the recipient of the e-mail (Phillips & Eisenberg, 1996). E-mail is now used for communicating bad news and even conducting negotiations. Many people (McKinnon & Bruns, 1992) are scornful of those who use e-mail for purposes such as reprimands and firings. However, some research now shows that e-mail may work better in communications that involve "dislike or intimidation" (Markus, p. 136).

The establishment of e-mail and other communication policies involves sensitive organizational issues. For example, if one employee sends a printed memo to an employee in another department concerning a matter of interest to his or her bosses, it is often expected that the sending employee will send a copy of the memo to the bosses. Should the same policy hold for e-mail exchanges? Is e-mail more like a formal memo where such a procedure is expected or more like an informal phone call where copying is not done? Such policies will likely lead to debate and perhaps conflict. Generalist and information managers need to be actively involved in making these decisions.

Technical leaders need to realize that keeping others informed in the organization is a crucial task, and devoting time to such communication often needs to take precedence over more technical issues as the following IT director for a local government described:

With the manager—I don't want any surprises and I don't want my manager to have any surprises. So if I see it [some problem] coming, I am up there communicating with him. This morning before you got here, I went to give a heads up to purchasing and to the manager's office to let them know "hey, this is going to be coming to you" [an unexpected expenditure]. I spend a lot of time doing things like that. ... The complaints that I hear a lot about are that people send things in and the manager's office doesn't know what is about. ... So communicating those things to grease the skids, and letting people know that what I would like to have happen—I spend a lot of time on it. And I think it pays big dividends in getting things done.

In short, good political skills concern the ability to communicate effectively with all of the key actors in the IT process from the end users to the top managers.

The External Politics of Information Management

There are several ways in which computing can become involved with external politics. Here are some examples.

1. Information technology is being used to determine the performance of governmental organizations as well as the presentation of these performance measures on Web sites. These online evaluative reports (e.g., report cards for school systems) can have much greater visibility and accessibility than previous evaluations and thus e-government can lead to greater citizen involvement. Access of the public to information about the performance of governments and other organizations (e.g., hospitals, health professionals) can often lead to controversy.
2. Legislatures, councils, and boards of public agencies may contest the purchasing decisions of public organizations. Likewise, the award of computer contracts may involve

political rewards. In the California case (Anderson et al., 2003), the legislature stepped in to weaken the amount of authority given to the new Department of Information Technology created in 1995, thus ensuring that the General Services Administration and Department of Finance would continue to dominate IT decision making.

3. It is possible that public computer records could be used for political purposes. The information could be used to schedule campaigns or find information that brings candidates into disfavor.
4. Information systems often involve sharing amongst different levels of government as well as private organizations. Often there is conflict among these organizations over basic issues such as how the information system should be structured and what data they should gather.
5. The rise of the World Wide Web and e-government has created the potential for politics. Political issues have erupted over the use of Web sites and other forms of CMC. A wide range of issues have developed such as the use of Web sites to attack governmental officials or their use for advertising purposes.
6. Major computer disasters or failures can bring negative attention to public organizations. These failures both hurt organizational performance and also threaten the jobs of staff.
7. A variety of computing technologies are viewed as a way of increasing citizen participation in government. These include interactive Web sites that allow citizens to easily post comments on proposed rules and other public issues. Public-participation GIS (PPGIS) and Web logs could also be used to achieve enhanced participation but at the same time can lead to controversy and increased conflict.

Online Accountability

The development of computing technology has had an important impact on measuring accountability and presenting this information to the public. Many governmental organizations are now posting information such as report cards of their performance online. The hope and expectation is that using Web sites to make such performance information available will make organizations accountable and also lead to more trust in government. Indeed, Mossberger and Tolbert (2005) have found some limited evidence that use of governmental Web sites has positive effects on perceptions of governmental responsiveness.

The extent to which online information can improve accountability depends in part on the accessibility of information. The Mercatus report (McTigue, Wray, & Ellig, 2004) analyzes accountability information on Web sites by separating them into desirable characteristics, such as breaking the report into downloadable sections (and multiple formats) if it is large, as well as the inclusion of contact information if anyone has questions concerning the report. It is instructive to look at the Mercatus report's comments on specific government agencies concerning transparency to observe in more depth what kinds of actions the authors regard as providing for good transparency. The U.S. Department of Labor scored highly and the positive comments include the fact that the accountability report is linked to the home page and is downloadable in "multiple PDF documents" (McTigue et al., p. 30), and the authors praise the report for being "clearly written" and providing trend data since 2000, though it notes that more trend data concerning key problems would have improved it. By way of contrast, the report critiques the Department of Health and Human Services (DHHS) for the obscurity of its report, stating that they could only find the report on the department's Web site via a "circuitous trip." It also criticizes the report for not providing information about the quality of the data of any of the 600 measures it uses. The Mercatus report acknowledges that an agency's actual performance may not be correlated directly with the Mercatus score. This brings up the issue of the quality of the data, and the Mercatus report states that the agency should indicate how confident it is in the quality of the data and, to ensure transparency, it should make the data available for independent verification.

The National Center on Educational Outcomes (NCEO) (Thurlow, Wiley, & Bielinski, 2003) assesses the reports done by state education agencies and has been assessing their Web sites for outcome information since 2001. Some of their evaluative criteria refer to the organization of the information on the Web site. Among the criteria they used to assess the state Web site were the following: (a) Are there clear words or links to get to the report on the agency Website, (b) how many clicks did it take to get from the agency's home page to the disaggregated results, and (c) what is the proximity of data on special education to the data for all students?

One of the major obstacles to the use of accountability efforts has been the argument that comparisons of performance are invalid—no two situations are the same. Consequently, organizations with poorer results have been able to point to factors that differentiate them from better performing organizations (Rocheleau, 2002a). The posting of information on the Internet makes this kind of information much more accessible than in the past when it would likely reside in obscure, hard-to-obtain governmental publications. These increased external forces often lead organizations to adopt strategies for resisting information or manipulating it so that negative information is not available to the public or oversight agencies (Rocheleau). For example, a U.S. General Accounting Office report (2002) found that abuse of nursing-home patients was rarely reported. Indeed, colleges, universities, health-maintenance organizations, and perhaps most organizations take steps to ensure that only positive information is reported via a number of strategies, especially if the information will affect high-stakes decisions (Bohte & Meier, 2001; Rocheleau). Information is power, and the demand for more external access to performance information makes the job of information management even more political. Now that this information is so accessible, organizations have to deal with demands for access to more information while other units may resist such demands and power struggles over information ensue. For example, the Tennessee teacher unions successfully resisted efforts to make public the scores of individual teachers based on a value-added system developed by the state (Gormely & Weiner, 1999). Other consequences include the likelihood that the data may be "cooked" in order to demonstrate high performance (Rocheleau) and the IT managers may face ethical issues concerning how to handle such situations.

The External Politics of Purchasing and Privatization

Many city managers attempt to defuse controversies over the purchasing process by involving council or board members in developing the proposals. Thus, decisions will not be brought up until strong council or board support exists. Achieving such a consensus may be more difficult these days because board members are more likely to be involved with computing in their own organizations (e.g., Pevtzow, 1989). When computing was restricted to mainframes and data-processing departments, council or board members were less likely to feel knowledgeable and able to challenge purchases. Rocheleau (1994) found that there could be conflicts over purchasing even if there is a consensus on what type of technology to use. Major contracts can be especially controversial during periods of budget strain, and expensive IT contracts may be viewed as taking away from services. For example, some state legislators argued that $52 million of a $90 million contract that California State University had awarded to Peoplesoft should be redirected to educational programs to compensate for cutbacks made by Governor Gray Davis (Foster, 2004). There also may be tensions about whether to purchase from local vendors vs. outside vendors. We saw earlier (Peterson, 2002) that major failures of a large IT project led to the creation of a state IT department, but controversies over another contract (with the Oracle Corporation) led to the end of the department after only 7 years of existence. Major failures in procurement can turn a project that begins as primarily an internal matter into a political football and such failures have led several IT directors to lose their jobs.

The move to privatize information systems can create external conflicts with legislative bodies as well as unions. For example, the state of Connecticut's administration decided to change the state's entire system from mainframes to the desktop and hire an external vendor to handle every aspect of the information function (Daniels, 1997). Later, it decided against the outsourcing because of a number of factors such as disputes with the state legislature and the union representing the IT employees, as well as reappraisals of the proposed contracts. Several other states have considered privatization including Indiana, Iowa, and Tennessee. In order to implement such plans, the managers will have to negotiate with legislatures and unions in order to reach agreements. For example, the Connecticut administration (Daniels) moved to assure the jobs of the state IT workers for a period of 2 years at the same salary and benefits in order to have the privatization move approved. In more recent times, the cities of Memphis and San Diego have moved to outsource their entire information services function. The absence of unions has facilitated the privatization of the Memphis operation (Feldman, 2000). However, as we have seen above, effective communication and political skills are required for an effective IT system, and turning over the entire operation to a private vendor could disrupt the communication patterns and power relationships necessary for the system to work smoothly. That is why in many cases the selected vendor often is an organization consisting of former employees or, as in the Memphis case, the winning vendor is expected to hire employees of the former municipal IT department (Feldman). One of the principles that Anderson et al. (2003, p. 31) found in their study of state IT structures was that "states with successful IT initiatives demonstrate commitment to employees during major changes." In some recent cases (Kahaner, 2004), state governments (e.g., New Jersey and Massachusetts) have passed laws that have outlawed outsourcing of call centers for services such as answering questions concerning electronic benefits like food stamps. The outsourcing would have saved money directly in lower contract costs, but some point to unemployment and other benefits that would have to be paid to the workers who would lose their jobs. In 2004, Florida technology officials became embroiled in a controversial outsourcing (that includes e-government) in which losing bidders complained about the bidding process (Towns, 2004), leading later to the resignation of the Florida CIO. The era of budgetary shortfalls leads to demands that CIO cut budgets and save money (Hoffman, 2004). This situation can lead to failure and resignations when CIOs are unable to meet these cutback expectations as occurred with the CIO of New Hampshire who resigned (Hoffman). Dealing with the implications or even expectations of personnel cutbacks due to IT decisions is one of the most sensitive and important tasks for IT leaders and the generalist administrators of governmental organizations.

Although privatization may be used to achieve positive goals, it can also be used for political rewards and result in problems. One such example occurred when a computer vendor, Management Services of Illinois Inc. (MSI), was found guilty of fraud and bribery connected to the state of Illinois awarding a very favorable contract to them (Pearson & Parsons, 1997). MSI had legally donated more than $270,000 in computer services and cash to Illinois Governor Edgar's campaign. The jury found that the revised contract had cheated taxpayers of more than $7 million. Campaign donations as well as the flow of governmental and political staff between government and private vendors can influence the awarding of contracts. More recently, there have been some suspicions that politics has been involved in the selection of no-bid contractors for "e-rate" contracts aimed at putting computers and other IT in Chicago's school system. Among the winners of no-bid contracts was SBC (then headed by Chicago Mayor Richard Daley's brother) and another company (JDL Technologies) headed by a friend of Reverend Jesse Jackson (Lighty & Rado, 2004).

At the same time that many states and municipalities are exploring the privatization of their information management function, there are several municipalities that are moving to become telecommunication owners and that leads to political controversy. For example, Tacoma's (Washington) municipal power company is aiming to build and provide cable services to homes and thus put it in competition with the local phone and cable companies (Healey, 1997). Many other cities including several small communities such as Fort Wright (Kentucky) are also planning to build telecommunication networks (Newcombe, 1997) in the United States and are providing telecommunication services for businesses and private homes in their communities. The rationale behind these moves is that the private cable and local phone companies have a poor record of providing up-to-date service (Healey). These moves have often been labeled as socialism and are opposed by the local cable and/or phone companies. However, in the Tacoma case, most local business leaders were backing the municipality because of the desire to have better technology (Healey). Some states (e.g., Texas, Arkansas, and Missouri) have prohibited municipal organizations from becoming telecommunication providers (Healey). The state of California is in the process of privatizing its state telecom system (Harris, 1998). In Iowa, many municipalities have been laying fiber to deliver cable in competition with cities, and the first suit brought by a phone company was found in favor of the municipality (Harris). More recently, the Supreme Court (Peterson, 2004, p. 27) "upheld the right of states to ban municipalities entering the broadband market." The Missouri Municipal League had argued that the Federal Telecommunications Act prevented states from passing laws to limit entry into providing services, but the court said that this provision of the law did not apply to governmental units. Thus, politics and the law are integrally related, but there are few fixed principles about law as it affects emerging computer technology.

When local governments decide to pursue broadband, they must be ready to be involved in a whole array of politics and will need to market their position to the public. One strategy of small local governments is to increase the viability of their position by collaboration with other municipalities. In Utah (Perlman, 2003), several municipalities joined together to form UTOPIA (Utah Telecommunications Open Infrastructure Agency) to provide high-speed fiber-optic networks to their communities. However, telephone and cable companies have allied with taxpayer groups to oppose the efforts. Eight of the 18 local governments that supported the initial feasibility study have withdrawn from UTOPIA, and the remainder has committed themselves to taxpayer-backed bonds to finance the required infrastructure. The feasibility issue depends heavily on assumptions about take-up rates: What percent of the targeted businesses and residents buy into the service, and will the project finance itself? According to Perlman, the take-up rate has averaged 40% in those local governments that have put "fiber to the home," and only a 30% take-up rate is required for the project to pay for itself.

A study by a conservative think-tank researcher (Lenard, 2004) found that none of the municipal entrants into the broadband market had been able to cover their costs and argues that they are not likely to do so in the future either. Defenders of municipal entry see it as a movement to bring services and competition to areas that telecommunication companies have poorly served or totally ignored and that broadband is now infrastructure needed to attract businesses to the community and is thus equivalent to building roads. Brookings researcher Charles Ferguson (2004) has labeled the broadband situation an example of "market failure" due to a lack of real competition. Defenders would argue that, by entering the market, municipalities will provide the competition to obtain the quality services that have been denied them. Lenard argues that the competitiveness of the telecom industry makes it likely that the private sector will meet these needs without governmental involvement. In short, the debate over local governmental provision of Internet services goes to the very heart of what government should be doing.

More recently, a similar controversy has erupted over the desire on the part of some local governments to provide "hotspots" or WI-FI (wireless fidelity) zones in downtown areas. These local governments view the provision of WI-FI capability as a service to their citizens as well as a way of assisting economic development. Opponents of municipal provision of hotspots view it as inappropriate public-sector competition with a service that is available from the private sector. For example, Philadelphia stated its intention to provide wireless service at very low prices (Peterson, 2005). A bill passed both houses of the legislature aimed at preventing other communities from taking similar steps (Peterson). In short, it is becoming clear that, although many information managers prefer to avoid controversy, telecommunications issues such as broadband and wireless may make it impossible for IT staff and governmental officials to avoid making politically sensitive decisions.

Computer Disasters and Information Management

It is likely that the majority of computer problems and disasters remain unknown to the public and even legislative bodies. However, certain disasters have so much impact on key operations that they do become public and create crises for generalist and information managers. For example, the delay in the opening of the new Denver Airport was due to software problems controlling the baggage system. Likewise, the state of Illinois' Medicaid program encountered many failures of the computer system with the system assigning patients to inappropriate health care providers (Krol, 1994).

Many disasters are beyond the control of managers and there is little they can do other than plan for emergencies. However, in many cases, disasters appear to result from overly high expectations for new computer systems and a lack of understanding on how difficult it is to implement a new system. This author has reviewed a large number of computer problems and failures (Rocheleau, 1997). Both the Federal Aviation Administration (FAA) and Internal Revenue Service (IRS) have experienced major failures that have led to threats from Congress to defund systems (Cone, 1998). Another example is the state of Florida's new human-services system that encountered much higher-than-projected costs and slower-than-expected implementation (Kidd, 1995). The perceived disaster led to the loss of the job of the state official in charge of the new system along with threatened legal action. However, over the long run, it appears that the system actually worked and has helped to reduce costs. Information management officials need to ensure that executives and the public have realistic expectations of system costs and performance. Computer problems and disasters are likely to occur more often as computing becomes central to governmental performance and communication with its constituents. In these situations, managers cannot avoid dealing with computing even if they have removed themselves from any decisions concerning it.

In contrast to the disaster cases, some politicians and managers make use of notable achievements in computing to boost their reputation for innovation and effectiveness. However, it can be dangerous for politicians or managers to claim success for large-scale new systems until the systems have been fully implemented and tested. For example, a former state of Illinois comptroller introduced a powerful new computer system that was aimed at speeding the issuance of checks as well as improving access to online information during June of 1997 (Manier, 1997). Soon afterward, there were complaints that checks were arriving behind schedule and that matters had not improved (Ziegler, 1997). The agency stated that it was just taking time for workers to get used to the new system. This is one case where there does appear to be a clear prescriptive lesson for managers: New computer systems that are large scale and introduce major changes usually experience significant start-up problems and claims of success should be muted until success can be proven.

Information Management and Interorganizational Struggles

Many of our largest governmental programs involve complex arrangements where administration and funding are shared by federal, state, and local governments. These governments are often at odds over how they view information systems. For example, the welfare-reform legislation passed in 1996 led to needed changes in how state and local governments gathered and analyzed data. Since welfare recipients move from state to state, the new welfare time limits require states to share and redesign their systems so they can calculate time periods on welfare and whether a recipient has exceeded state or federal limits (5 years total and 2 years consecutively for federal limits). Prior systems were oriented to yearly information and were concerned only with welfare activities in their own state. The federal government has established data-reporting requirements that many states view as burdensome and unnecessary (Newcombe, 1998). For example, they have to monitor the school attendance of teenage mothers and that requires sharing information with independent local school districts. The quality of the data submitted by the states is also an issue (Newcombe). The very purposes of the federal and state systems can be somewhat at odds. The federal government wants to use the system to determine the overall success of the program and to be able to compare the performance of states. The state governments often are opposed to the increase in the number of data elements required from 68 to 178 and the costs of gathering much new information (Newcombe). Similar disagreements can occur between state and local governments with the latter often feeling that states are too autocratic in how they implement information systems. The resolution of such disagreements will involve conflict and bargaining with creative solutions sought that meet the primary needs of all involved.

Schoech, Jensen, Fulks, and Smith (1998) provide a case study of a volunteer group in Arlington, Texas, who tried to create a data bank aimed at helping reduce alcohol and drug abuse. This volunteer group "discovered the politics of information." Hospitals opposed the identification of drug-affected births. When agency personnel changed, permission for access to the data had to be obtained over again. Changes in the structure of government and functions of office also led to the need to start again. They found that data gathering was not a high priority for other agencies involved and access to data was often delayed or not forthcoming.

The Center for Technology in Government (CTG) has conducted several case studies concerning the development of information systems that require the cooperation of several organizations. One case study involved the development of an information system for the homeless. First, several different actors needed to be involved in the development of the system including the New York State Office of Temporary and Disability Assistance, the Bureau of Shelter Assistance, the New York City Department of Homeless, the Office of Children and Family Services, the Department of Health, the Department of Labor, the Office of Alcohol and Substance Abuse Services, and the Division of Parole, and many independent nonprofit organizations contribute to and use information on the system. The huge number of actors meant that the development of the system had to be very deliberate and the first priority was to develop a sense of trust amongst them first before getting into technical design (CTG, 2000). The new information system was to be used to help set goals and thus affect evaluation. Consequently, there were issues that had to be settled about how ambitious to make goals. Consensus had to be forged on key definitions such as the "date of admittance" into the system because these definitions were important to funding of the end users of the system. Some agencies wanted a more detailed listing of ethnic options than others due to their federal funding requirements, so this detail needed to be negotiated, too. The basic points of this case study are that negotiation and trust are essential to the creation of interorganizational and intergovernmental systems. Organizations that believe that their financial interests and viability will be threatened by a new system will resist it regardless of how rational and sensible the policy looks on paper.

Another CTG (1999) study focused on attempts to build integrated criminal-justice systems. It found many conflicts over budgets, organizational relationships, and procedures. CTG concluded that these problems were not technical in nature but due to "conflicting visions" related to organizational and political interests. It also found that trust, participation, and understanding of the business were among those elements required for success. To achieve buy-in, they had to pay much attention to "interests and incentives" and use marketing and selling techniques. Political pressures played an important role in some cases. The study found that "turf is the biggest killer of integration": "Protecting turf can be particularly important when the potential loss of autonomy or control could benefit other agencies that are political or institutional adversaries" (p. 11).

There was also a need for a champion of the system who had major organizational or political influence that allowed this person to overcome the political barriers to integration. Bellamy (2000) found very similar results in a review of attempts to create criminal-justice systems in England. Although technical skills are always useful and sometimes essential, the development of successful interorganizational systems necessitates major use of political skills and resources. The Anderson et al. (2003) study of four states found that it was important to have general executive leaders who are champions of IT:

States with exemplary IT practices have executive leadership (governor and state CIO) who are champions of IT initiatives. All four of the states we visited exemplify this characteristic. These leaders emphasize the value of IT for the state in performing its missions. They view IT as an investment, rather than a cost ... (p. 33)

These studies and other literature on IT have two important implications: (1) Generalist administrators who want to have successful IT programs need to be engaged in IT, and (2) IT heads need to cultivate support from generalist administrators.

One of the current trends in information technology management is to encourage the easy sharing of information and to bring an end to information silos. Thus, states such as Pennsylvania (Chabrow, 2005) have hired deputy chief information officers with one of the major goals of their jobs being to enhance IT effectiveness by coordinating budgeting across agencies that share common functions (so-called communities of practice). For example, several agencies may have e-payment systems that could share a single application. Forcing agencies to share common applications is likely to lead to tension and political battles.

In some cases, the use of computing may help to reduce the amount of ad hominem politics and give more attention to the underlying facts of cases in development decisions (Dutton & Kraemer, 1985). They found that the computing models did not eliminate politics. Developers and antidevelopers employed competing models with different assumptions. However, the focus on the computer models helped to direct attention to facts of the case and away from personalities and unverifiable assumptions, thus facilitating compromise and agreement.

Information Management and the Politics of Databases

Most information managers prefer to avoid the release of information with political implications. However, often they cannot avoid releasing such information and need to have a defensible policy in this regard. The New York State Attorney General's Office (Yates, 2001) plan to track flows of donations to victims of the September 11th tragedy was resisted by organizations such as the Red Cross due to privacy and confidentiality issues. The Freedom of Information Act (FOIA) covers computer records in most states. Issues of privacy and public interest often collide and managers are often forced to make difficult choices. Although these problems existed prior to computers, the existence of computing has made it possible for outsiders to conduct very detailed critiques of the practices of public agencies with emphasis on pointing out their failures and questionable decisions. For example, the *Chicago Tribune* did a reanalysis of computerized information from the Illinois Department of Public Aid to do an exposé of fraud and waste in its Medicaid system (Brodt, Possley, &

Jones, 1993). The extent, magnitude, and speed of their analysis would have been impossible without access to computerized records. Consequently, it is not surprising that many governmental agencies resist FOIA requests. A series of articles (e.g., Mitchell, 1999) by the *News-Gazette* newspaper in Illinois revealed that many resist FOIA using a variety of reasons such as the fact that they would have to create a new document. The existence of a good computerized system can serve to weaken the argument that the obtaining of records would pose too great a burden on the governmental agency.

Generalist and information managers picture technology as a way to better services, but they should be aware that the same technologies and databases can be employed for political purposes. For example, many municipal and state governments are now constructing powerful geographic information systems that are aimed at improving services to citizens through the mapping of integrated databases. However, GISs and their data are now being used for "cyber ward heeling" in the 1990s and facilitate such traditional functions like the mapping of volunteers, canvassing of voters, and location of rally sites (Novotny & Jacobs, 1997). Politicians are likely to seek data from these public information systems to conduct their political campaigns. For example, databases allow the targeting of campaigns so that candidates can use several different messages and conduct "stealth campaigns" without alerting their opponents. Thus, the more powerful and information rich local GISs become, the more attractive they will be as databases for political activities, which could lead to controversy.

Online Computer-Mediated Communication

The impact of the World Wide Web has especially important implications for politics. Researchers such as Robert Putnam (1995) point out a substantial decrease in some forms of civic participation on the part of the public. Many people are frightened to speak at public hearings. The Web offers a way of increasing public participation in community decisions (Alexander & Grubbs, 1998). Many people see us entering a new age of cyberdemocracy (Stahlman, 1995). Shy people and stutterers would be able to provide testimony electronically and their arguments would be judged based on content rather than their appearance or public speaking skills (Conte, 1995). Municipalities may help to develop useful networks such as senior-citizen discussions. Parents can use the system to update themselves on student homework assignments. However, as noted above, there are several drawbacks to teledemocracy and the development of interactive Internet applications.

- There is less inhibition in telecommunications than in in-person communications against intemperate statements. Consequently, electronic forums often degenerate into "flaming" wars. The originator of the Santa Monica (California) online discussion system argued that, if he were to do it over again, he would like to have a moderator for the system and charge user fees (Conte, 1995; Schuler, 1995).
- Some people lack the computing technology and/or skills to participate in these electronic discussions (Wilhelm, 1997).
- It is feared by some that easing access to public testimony and input to public officials may result in such a massive and discordant amount of input that democracy would be stymied and that gridlock would increase.

- The Internet raises fears about privacy. Efforts to improve access can often lead to resistance. For example, the Social Security Administration (SSA) made interactive benefits estimates available over the Internet but was forced to withdraw the service due to privacy issues (James, 1997; United States General Accounting Office, 1997). Social Security numbers are not very private and all someone needed was the number plus the recipient's state of birth and mother's maiden name to gain access to earnings and benefits information.
- Public online systems may become campaign vehicles for certain politicians.

A detailed account (Schmitz, 1997) of a discussion group concerning homelessness on Santa Monica's PEN system made the following points about the successes and failures of online groups: (a) The discussion group was successful in bringing people who would never have engaged in face-to-face meetings together for discussion purposes on an equal basis, but (b) electronic media demand keyboarding and writing skills. Thus, there are many obstacles to the successful participation of the poor. One recent study by Gregson (1997) found that even politically active citizens were not able to transfer their activism to a community network without substantial training and experience. Hale (2004) did a study of the use of neighborhood Web pages, and his overall assessment was that the overall usage rates were low and that consequently these pages were not meeting the hope that they might help to "revitalize democracy."

Both generalists and information management staff need to give careful consideration to the possibilities and drawbacks of teledemocracy. If they decide to support electronic discussion groups, should they employ a moderator and, if so, who should act as a moderator? Would a moderator's censoring of input be a violation of the right to free speech? Fernback (1997) argues that most people accept moderation not as "prior restraint but as a concession" for the good of the collectivity. How can the argument that teledemocracy is elitist be handled? Is the provision of public places (e.g., in libraries) for electronic input sufficient to deal with this objection? In this author's experience, I have found strong resistance to the establishment of online discussion groups. Many local governmental officials believe that such communication is likely to result in contentious and strident behavior. Some also feel that increased participation would make governmental decision making slower and more difficult. Thus, West's (2001) finding that few governmental agencies allowed interactivity such as the posting of online messages, much less online discussion groups, may not be merely the result of their lack of technological sophistication, but may also be due to a calculated reluctance to sponsor CMC.

The Politics of Web Sites and Other Online Information

The creation of Web sites has produced a whole new set of opportunities for politics to occur. What kinds of information should be online? Who should decide what information should be online? These are issues that have been highlighted by the Web. Before, most public information resided in reports that few had access to or even knew existed. The ease of online access has changed that situation and can lead to controversies that would not have existed when information was restricted to paper reports. As a result of the events of

September 11th, some U.S. agencies have pulled from their Web sites information on hazardous waste sites. To many, these actions make good sense, but others have pointed out that the chemical industry has tried to keep such information private and that public access to this information can help to save lives ("Access to Government Information Post September 11th," 2002). The New York Attorney General's Office began to develop a database to track the distribution of money donated to victims of the attack, but the Red Cross raised privacy concerns (Yates, 2001). Placing some information on governmental Web sites can result in jeopardy for government officials. A U.S. Geological Survey contract employee was allegedly fired for posting a map on the U.S. government Web site that identified areas of the Arctic national wildlife as moose calving areas that the Bush administration would like to open to oil exploration (Wiggins, 2001). Recently, the American Education Association and American Library Association (Monastersky, 2004) have accused the Bush administration of politics in deleting information from its Web site and cited an internal memo that it was policy to remove information that was outdated or did not "reflect the priorities of the Administration." Thus, putting up certain information or omitting information from Web sites can become a controversial and political issue.

A related issue is to what extent should Web sites of governments act as, in effect, a campaign Web site for the top elected officials? There is great variability in state and local governments, but some governmental home pages appear to be campaign sites with photos of the top-level officials and their personal positions, and accomplishments of the elected official dominating much of the page. Such activities can stimulate opposition to political uses, and Franzel and Coursey (2004) report that the state of Florida banned the placement of almost any information other than basic personal and legal information on their Web sites. Coursey (in press) cites an interesting example in which Governor Jeb Bush ordered an e-mail link be placed at the top of all of the state's Web pages, but that led to thousands of responses that could not be answered in a timely fashion and the policy was eventually reversed.

Should advertising by private businesses be allowed on governmental sites? Many governments are resisting advertising. Some of the most successful such as Honolulu (Peterson, 2000) have used advertising to fund advanced electronic governmental systems but few others have followed. Honolulu put out a banner for a bank on its Web site. Peterson cites other governments as either being interested in advertising or not depending on whether they see it as necessary for funding. Decisions about opening government sites to advertising will involve ethical and practical issues.

Even decisions about what links to have on Web sites can become involved in political and legal controversy. A court case was brought by an online newsletter (*The Putnam Pit*) against the city of Cookville, Tennessee, due to its failure to provide a link to the newsletter despite the fact that several other for-profit and nonprofit organizations were linked to from the city's Web site (Anonymous, 2001). Many governments avoid making such links. The consensus is that if governments do make links, they need to have a carefully thought-out (nonarbitrary) policy for doing so.

E-Governance Issues

There is a large and rapidly growing literature about how IT will affect and change the nature of political decision making as well as partisan politics. I will note some significant aspects of how e-governance can affect those who manage IT. E-governance is defined by Carlitz and Gunn (2002, p. 389) as the "use of computer networks to permit expanded public involvement in policy deliberations." One particular form of e-governance is to create an e-docket that is aimed at increasing participation in the creation of administrative rules so that they will be less likely to be challenged in courts. Coglianese (2004, p. 2) has defined e-rule-making as "the use of digital technologies in the development and implementation of regulations." He argues that IT may help streamline processes and allow agency staff to "retrieve and analyze vast quantities of information from diverse sources." He cites early examples of rule making such as the Bureau of Land Management's scanning of 30,000 comments concerning a proposed rangelands rule. Coglianese describes how the Department of Transportation and Environmental Protection Administration (EPA) created entire e-docket systems that provide "access to all comments, studies, and documents that are available to the public." Indeed, the EPA has been designated the managing partner in an interagency project to establish a common Internet site for all federal regulatory issues to help the public find and comment on proposed regulations. There is a governmental Web site that describes this initiative at http://www.regulations.gov/images/eRuleFactSheet.pdf and has links to all of the federal-agency e-rule-making sites. Carlitz and Gunn (2002) have described how the e-rule-making process ideally would work:

An online dialogue takes place over a several week period. The dialogue is asynchronous, so participants can take part at their convenience, with ample time to reflect on background materials and the postings of other participants. Although in our experience a properly structured event is typically very civil, the dialogue is moderated to deal with the rare cases in which the discussion becomes too heated and to help keep the conversation focused. (p. 396)

However, the authors go on to acknowledge that there are many legal concerns about abridgement of First Amendment rights through attempts to moderate discussions and make them more civil. Carlitz and Gunn (2002, p. 398) said they were advised by the EPA's Office of General Counsel not to use their "usual prerogatives" as moderators because of these legal concerns, though they note that other departments have taken different positions.

A more recent study by Shulman (2005) raises serious questions about the ability of e-rule-making's ability to contribute useful input to the rule-making process. He studied randomly selected electronic contributions to e-rule-making and found that e-mail contributions give little deliberative input. The vast majority (more than 98%) of electronic contributions were exact or almost exact duplicates of form letters with little in the way of additions. Moreover, the process used by the private outfits to analyze the electronic communications is likely to miss the few meaningful contributions according to Shulman. He concludes that the few thoughtful communications are likely to be drowned out by the huge electronic output.

Concerning the local-government level, Chen (2004) studied the involvement of local officials with e-mail and the Internet in the Silicon Valley area. She surveyed city and county officials. Her findings were somewhat surprising to me in that they found that these officials rated e-mail ahead of traditional "snail mail" in importance to their office, though the absolute differences were small. It is surprising to me because I would think that the act of writing and mailing a letter takes considerably more effort than sending an e-mail message so that letters would be assigned a higher priority by officials. However, Chen also found that only 9% of the officials checked their own e-mail, leaving this to their assistants. The burden of incoming e-mail was relatively small compared to the huge amount that goes to members of Congress—just over 50% said they got more than 50 e-mails per day (Chen). Although they used e-mail to communicate with the public, they were careful about using broadcast features of e-mail that would allow them to send mass mailings to the public because of the resentment that spam can cause in citizens. Many governmental sites now do provide the voluntary opportunity for residents to sign up for various forms of electronic communication. In Chen's study, some local officials had suffered from "spoofing" in which opponents sent offensive messages to the public that appeared to embarrass them. The overall assessment of e-mail contact by most local officials was that e-mail is useful once a good relationship had been established via in-person or phone contact.

A study by Ferber, Foltz, and Pugliese (2003) of state Web sites found few interactive political input opportunities such as public forums, chat rooms, or Listservs. Similar to my expectations, Ferber et al. put less weight on e-mail communications than traditional phone or snail-mail communications. One of the issues politicians have with e-mail is that it is difficult if not impossible to determine the location of the e-mailers and thus politicians do not know if the e-mailers live in their political districts. One approach to this problem is to require that the person fill out a Web form giving his or her address prior to accepting the electronic communication. However, as Ferber et al. note, the vision of computer-generated e-mails receiving computer-generated responses does not meet what constitute "increased democratic participation in the political process."

There are a number of other developments in which IT is being used to encourage greater participation on the part of the public. For example, public-participation GISs are now being designed to enhance citizen participation by allowing members of the public to be able to visualize consequences of development decisions by studying maps and accompanying images (Krygier, 1998). Likewise, blogs (logs of individual opinions on issues recorded on the Internet) are being employed by some managers and could be used by governments to provide an additional method for citizens to express their viewpoints about public issues. The use of blogs in government has been rare until now. The State of Utah CIO (Harris, 2005) offered blogs to employees in order to encourage more open communication. However, much of the material in such blogs appears to have high public-relations content, and if the blog becomes the subject of political controversy, the blog may be ended. For example, a blog established by a metropolitan transportation-planning agency in Orlando, Florida, ended its blog when the blog was used by opponents to attack the policies of the agency as well as the agency's officials (Crotty, Dyer, & Jacobs, 2005). Use of the Internet is now viewed by some as a way of reorganizing government according to the views of certain ideologies. For instance, some conservatives view the Internet and Web sites as a way of increasing "choice-based policies" (Eggers, 2005) in areas such as education because, they argue, information would be presented about the efficacy of providers of education (both public and private schools) so that citizens could make an informed and voluntary choice.

Conclusion

Failure to become engaged and knowledgeable about internal politics can undermine the efficacy of information managers. I know of cases in which managers with good technical skills lost their jobs due to their failure to master organizational politics. Information managers need to negotiate, bargain, dicker, and haggle with other departments. They may need to form coalitions and engage in logrolling in order to achieve their goals. A good information manager needs good political skills to be effective. I have drawn from a number of resources to illustrate the politics of information management, but there exists little systematic research concerning the topic as Strassmann (1995) has pointed out. We need more research concerning the crucial issue—both in-depth qualitative case studies and surveys concerning how managers employ politics in their dealings with information technology.

The lessons of our review are clear. As Fountain (2001) has pointed out, generalist managers can no longer afford to ignore IT. Fountain sees the urgency for generalist-administrator involvement:

Public executives and managers in networked environments can no longer afford the luxury of relegating technology matters to technical staff. Many issues that appear to be exclusively technical are also deeply political and strategic in nature. In some cases, new use of technology furthers an existing agency or program mission. But in others, using the Internet can play a transformative role and lead to expansion or rethinking of mission and change in internal and external boundaries, accountability, and jurisdiction. (p. 249)

Likewise, IT managers can no longer afford to ignore politics. Internal political issues such as those I have discussed above (structures, purchasing, sharing information, and electronic communication) have become so central that managers will find that questions about these issues demand attention and decision. External political issues will continue to grow in number and importance as the Web and cyberpolitics become more prominent. For better or worse, information managers are going to have to possess effective political skills.

References

Access to government information post September 11th. (2002, February 1). *OMB Watch.* Retrieved April 2, 2005, from http://www.ombwatch.org/article/articleview/213/1/104/

Alexander, J. H., & Grubbs, J. W. (1998). Wired government: Information technology, external public organizations, and cyberdemocracy. *Public Administration and Management: An Interactive Journal, 3*(1). Retrieved from http://www.hbg.psu.edu/Faculty/jxr11/alex.html

Anderson, R., Bikson, T. K., Lewis, R., Moini, J., & Strauss, S. (2003). *Effective use of information technology: Lessons about state governance structures and processes.* Rand Corporation. Retrieved July 30, 2004, from http://www.rand.org/publications/MR/MR1704/index.html

Anonymous. (1994). The politics of information. *Logistic information management, 7*(2), 42-44.

Anonymous. (2001). Court rules that Web publisher may contest denial of link to city's Website. *3CMA: News of the City-County Communications & Marketing Association*, p. 1.

Bardach, E. (1977). *The implementation game: What happens after a bill becomes a law.* Cambridge, MA: MIT Press.

Barrett, K., & Greene, R. (2001). *Powering up: How public managers can take control of information technology.* Washington, DC: CQ Press.

Bellamy, C. (2000). The politics of public information systems. In G. D. Garson (Ed.), *Handbook of public information systems* (pp. 85-98). New York: Marcel Dekker, Inc.

Bohte, J., & Meier, K. J. (2001). Goal displacement: Assessing the motivation for organizational cheating. *Public Administration Review, 60*(2), 173-182.

Brodt, B., Possley, M., & Jones, T. (1993, November 4). One step ahead of the computer. *Chicago Tribune*, pp. 6-7.

Brown, A. (1998). Narrative politics and legitimacy in an IT implementation. *Journal of Management Studies, 35*(1), 1-22.

Butterfield, E. (2005, December 28). Mass. CIO Peter Quinn resigns. *Washington Technology.* Retrieved December 28, 2005, from http://www.washingtontechnology.com/news/1_1/daily_news/27656-1.html

Caldwell, B. (1992, November 30). Battleground: An attempt to streamline the Pentagon's operations has triggered a fight for control. *InformationWeek*, pp. 12-13.

Carlitz, R. D., & Gunn, R. W. (2002). Online rulemaking: A step toward e-governance. *Government Information Quarterly, 19*, 389-405.

Center for Technology in Government. (1999). *Reconnaissance study: Developing a business base for the integration of criminal justice systems.* Retrieved July 30, 2004, from http://www.ctg.albany.edu/resources/pdfrpwp/recon_studyrpt.pdf

Center for Technology in Government. (2000). *Building trust before building a system: The making of the homeless information management system.* Retrieved July 30, 2004, from http://www.ctg.albany.edu/guides/usinginfo/Cases/bss_case.htm

Chabrow, E. (2005, June 20). State CIO named governor's aide. *InformationWeek.* Retrieved June 23, 2005, from http://www.informationweek.com/story/showArticle.jhtml?articleID=164900717&tid=5979#

Chen, E. (2004, April). *You've got politics: E-mail and political communication in Silicon Valley.* Paper presented at the Annual Meeting of the Midwest Political Science Association, Chicago.

Coglianese, C. (2004). *E-rulemaking: Information technology and the regulatory process* (Faculty Research Working Paper Series No. RWP04-002). Cambridge, MA: Harvard University, John F. Kennedy School of Government. Retrieved January 18, 2006, from http://cbeji.com.br/br/downloads/secao/500122.pdf

Cone, E. (1996, August 12). Do you really want this job? *InformationWeek*, pp. 63-70.

Cone, E. (1998, January 12). Crash-landing ahead? *InformationWeek*, pp. 38-52.

Conte, C. R. (1995). Teledemocracy: For better or worse. *Governing*, 33-41.

Corbitt, B. (2000). Developing intraorganizational electronic commerce strategy: An ethnographic study. *Journal of Information Technology, 15*, 119-130.

Corbitt, B., & Thanasankit, T. (2002). Acceptance and leadership-hegemonies of e-commerce policy perspectives. *Journal of Information Technology, 1*, 39-57.

Coursey, D. (in press). Strategically managing information technology: Challenges in the e-gov era. In J. Rabin (Ed.), *Handbook of public administration* (3rd ed.). Marcel Dekker.

Crotty, R., Dyer, B., & Jacobs, T. (2005, September 25). MetroPlan drops blog after critics attack agency's plans. *Orlando Sentinel*, p. K2. Retrieved January 5, 2006, from http://pqasb.pqarchiver.com/orlandosentineal/

Daniels, A. (1997). The billion-dollar privatization gambit. *Governing*, 28-31.

Danziger, J. N. (1977, June). Computers and the frustrated chief executive. *MIS Quarterly, 1*, 43-53.

Danziger, J. N., Dutton, W. H., Kling, R., & Kraemer, K. L. (1982). *Computers and politics: High technology in American local governments.* New York: Columbia University Press.

Davenport, T., Eccles, R. G., & Prusak, L. (1992). Information politics. *Sloan Management Review*, 53-65.

Dickerson, C. (2002, November 4). The art of good politics. *Infoworld*, 52.

Dizard, W. P. (2001). CIOs labor over contracting crisis. *Government Computer News.* Retrieved November 21, 2004, from http://www.gcn.com

Dutton, W. H., & Kraemer, K. L. (1985). *Modeling as negotiating: The political dynamics of computer models in the policy process.* Norwood, NJ: Ablex Publishing Company.

Eggers, W. D. (2005). Made to order. *Government Technology.* Retrieved February 11, 2005, from http://www.govtech.net/magazine/story.php?id=92875

Eiring, H. L. (1999). Dynamic office politics: Powering up for program success. *The Information Management Journal*, pp. 17-25.

Fabris, P. (1998, November 15). Odd ducks no more. *CIO Magazine.* Retrieved April 13, 1999, from http://www.cio.com/archieve/111598_government_content.html

Feldman, J. (2002). Politics as usual. *Networking Computing, 13*(5).

Ferber, P., Foltz, F., & Pugliese, P. (2003). The politics of state legislature Web sites: Making e-government more participatory. *Bulletin of Science, Technology, & Society, 23*(3), 157-167. Retrieved January 7, 2006, from http://java.cs.vt.edu/public/projects/digitalgov/papers/Ferber.Statewebsite.pdf

Ferguson, C. (2004). *The broadband problem: Anatomy of a market failure and a policy dilemma.* Washington, DC: The Brookings Institution.

Fernback, J. (1997). The individual within the collective: Virtual ideology and the realization of collective principles. In S. G. Jones (Ed.), *Virtual culture: Identity and communication in cybersociety* (pp. 36-54). London: Sage Publications.

Foster, A. L. (2004, May 18). Faculty petition criticizes Cal Poly campus's plan to install PeopleSoft software. *Chronicle of Higher Education.* Retrieved May 18, 2004, from http://chronicle.com/daily/2004/05/2004051801n.htm

Fountain, J. E. (2001). *Building the virtual state.* Washington, DC: Brookings Institution Press.

Franzel, J. M., & Coursey, D. H. (2004). Government Web portals: Management issues and the approaches of five states. In A. Pavlichev & G. D. Garson (Eds.), *Digital government: Principles and best practices* (pp. 63-77). Hershey, PA: Idea Group Publishing.

Freedman, D. H. (1994, March 1). A difference of opinion. *CIO*, pp. 53-58.

Friel, B. (2001, July 18). TSP board fires, sues computer modernization firm. *GovExec.com.*

Gormely, W. T., Jr., & Weimer, D. L. (1999). *Organizational report cards.* Cambridge, MA: Harvard University Press.

Gregson, K. (1997). Community networks and political participation: Developing goals for system developers. In *Proceedings of the ASIS Annual Meeting* (Vol. 34, pp. 263-270).

Grover, V., Lederer, A. L., & Sabherwal, R. (1988). Recognizing the politics of MIS. *Information & Management, 14*, 145-156.

Gurwitt, R. (1996, December). CIOs: The new data czars. *Governing Magazine.* Retrieved February 3, 2003, from http:governing.com

Hale, M. (2004, April). *Neighborhoods on-line: A content analysis of neighborhood Web pages.* Paper presented at the Annual Meeting of the Midwest Political Science Association, Chicago.

Harris, B. (1998). Telcom wars. *Government Technology, 11*, 1, 38-40.

Harris, B. (2005). The coming of blog.gov. *Government Technology.* Retrieved February 25, 2005, from http://www.govtech.net/

Hasson, J. (2004, May 17). No easy answer to career vs. political standing. *Federal Computer Week.*

Hayes, M. (1996, August 19). Platform zealots. *InformationWeek*, pp. 44-52.

Healey, J. (1997). The people's wires. *Governing*, 34-38.

Hoffman, T. (2004, February 16). Turnover increase hits the ranks of state CIOs. *Computer World.* Retrieved January 10, 2006, from http://www.computerworld.com/governmenttopics/government/story/0,10801,90236,00.html

James, F. (1997, April 10). Social Security ends Web access to records. *Chicago Tribune*, pp. 1, 12.

Kahaner, L. (2004). A costly debate. *Government Enterprise*, 8-14.

Keen, P. G. W. (1981). Information systems and organizational change. *Communications of the ACM, 24*(1), 24-33.

Kello, C. T. (1996). Participatory design: A not so democratic treatment. *American Journal of Psychology, 109*(4), 630-635.

Kidd, R. (1995). How vendors influence the quality of human services systems. *Government Technology*, 42-43.

King, J. L. (1983). Centralized versus decentralized computing: Organizational considerations and management options. *Computing Survey, 15*(4), 319-349.

Knights, D., & Murray, F. (1994). *Managers divided: Organisation politics and information technology management.* Chichester, UK: John Wiley & Sons.

Koskinen, J. A. (1996, July 15). Koskinen: What CIO act means to you. *Government Computer News*, p. 22.

Kost, J. M. (1996). *New approaches to public management: The case of Michigan* (CPM Report No. 96-1). Washington, DC: The Brookings Institution.

Kraemer, K. L., & King, J. L. (1976). *Computers, power, and urban management: What every local executive should know.* Beverly Hills, CA: Sage Publications.

Krol, E. (1994, June 24). State's health plan for poor comes up short. *Chicago Tribune*, pp. 1, 15.

Krygier, J. B. (1998). *The praxis of public participation GIS and visualization.* Retrieved April 2, 2005, from http://www.ncgia.ucsb.edu/varenius/ppgis/papers/krygier.html

Laudon, K. C. (1974). *Computers and bureaucratic reform: The political functions of urban information systems.* New York: Wiley.

Lenard, T. M. (2004). *Government entry into the telecom business: Are the benefits commensurate with the costs* (Progress on Point 11.4)*?* The Progress Freedom Foundation. Retrieved July 24, 2004, from http://www.pff.org/publications/

Lewis, D. E. (2005). *Political appointments, bureau chiefs, and federal management performance.* Retrieved January 7, 2006, from http://www.wws.princeton.edu/research/papers/09_05_dl.pdf

Lighty, T., & Rado, D. (2004, September 12). Schools' Internet bungle. *Chicago Tribune.*

Lucas, H. C. (1984). Organizational power and the information services department. *Communications of the ACM, 127*, 58-65.

Lunn, F. E. (2005). *Survey by FCW Media Group confirms that frustration among public CIOs prompts short tenures despite long-term challenges.* Retrieved January 7, 2006, from http://www.fcw.com/article91325-11-07-05-Print

Manier, J. (1997, June 27). State endorses powerful machine. *Chicago Tribune*, pp. 1, 5.

Markus, M. L. (1983). Power, politics, and MIS implementation. *Communications of the ACM, 26*, 430-444.

Markus, M. L. (1994). Electronic mail as the medium of managerial choice. *Organizational Science, 5*(4), 502-527.

Markus, M. L., & Keil, M. (1994). If we build it, they will come: Designing information systems that people want to use. *Sloan Management Review, 35*, 11-25.

McKinnon, S. M., & Bruns, W. J., Jr. (1992). *The information mosaic.* Boston: Harvard Business School Press.

McTigue, M., Wray, H., & Ellig, J. (2004). *5th annual report scorecard: Which federal agencies best inform the public?* Mercatus Center, George Mason University. Retrieved July 14, 2004, from http://www.mercatus.org/governmentaccountability/category.php/45.html

Miller, B. (1995a). Interview with Indianapolis mayor, Stephen Goldsmith. *Government Technology*, 24-25.

Miller, B. (1995b). Should agencies archive e-mail? *Government Technology*, 22.

Mitchell, T. (1999, July 26). Stiff-armed, but not by law. *The News Gazette.* Retrieved October 2, 2004, from http://news-gazette.com/OpenRecords/monmains.htm

Molta, D. (1999). The power of knowledge and information. *Network Computing, 33*(1), 23-24.

Monastersky, R. (2004, November 25). Research groups accuse education department of using ideology in decisions about data. *The Chronicle of Higher Education.*

Mossberger, K., & Tolbert, C. (2005). *The effects of e-government on trust and confidence in government.* Retrieved January 7, 2006, from http://www.digitalgovernment.org/dgrc/dgo2003/cdrom/PAPERS/citsgovt/tolbert.pdf

Murphy, S. (1995). Kansas city builds GIS to defray costs of clean water act. *Geo Info Systems*, 39-42.

Newcombe, T. (1995). The CIO: Lightning rod for IT troubles? *Government Technology*, 58.

Newcombe, T. (1997). Cities become telecomm owners. *Government Technology.* Retrieved from http://www.govtech.net/

Newcombe, T. (1998). Welfare's new burden: Feds tie down states with data reporting requirements. *Government Technology, 11*(4), 1, 14-15.

Northrop, A., Kraemer, K. L., Dunkle, D. E., & King, J. L. (1990). Payoffs from computerization: Lessons over time. *Public Administration Review, 50*(5), 505-514.

Novotny, P., & Jacobs, R. H. (1997). Geographical information systems and the new landscape of political technologies. *Social Science Computer Review, 15*(3), 264-285.

Overton, K., Frolick, M. N., & Wilkes, R. B. (1996). Politics of implementing EISs. *Information Systems Management, 13*(3), 50-57.

Pastore, R. (1995, December 1). CIO search and rescue. *CIO*, pp. 54-64.

Pearson, R., & Parsons, C. (1997, August 17). MSI verdicts jolt Springfield. *Chicago Tribune*, pp. 1, 12.

Peled, A. (2000). Politicking for success: The missing skill. *Leadership & Organization Development Journal, 21*, 20-29.

Perlman, E. (2000). Moving IT out. *Governing*, 58.

Perlman, E. (2003). Plug me in. *Governing Magazine*. Retrieved July 24, 2004, from http://www.governing.com/

Peterson, S. (2000). This space for rent. *Government Technology, 14*, 140-141.

Peterson, S. (2002). End of the line. *Government Technology*. Retrieved July 23, 2004, from http://www.govtech.net/magazine/story.php?id=25335&issue=10:2002

Peterson, S. (2004). Broadband battle. *Government Technology*, pp. 27-30.

Peterson, S. (2005, December 8). The golden egg. *Government Technology*. Retrieved December 10, 2005, from http://www.govtech.net/magazine/story.php?id=97502&issue=12:2005

Pevtzow, L. (1989, July 8). Bitterly divided Naperville leaders decide computer strategy. *Chicago Tribune*, p. 5.

Phillips, S. R., & Eisenberg, E. M. (1996). Strategic uses of electronic mail in organizations. *Javnost, 3*(4), 67-81.

Pitkin, G. M. (1993). Leadership and the changing role of the chief information officer in higher education. In *Proceedings of the 1993 CAUSE Annual Conference* (pp. 55-66).

Poulos, C. (1998). Mayor Stephen Goldsmith: Reinventing Indianapolis' local government. *Government Technology*, 31-33.

Putnam, R. D. (1995). Bowling alone revisited. *Responsive Community, 5*(2), 18-37.

Quindlen, T. H. (1993, June 7). When is e-mail an official record? Answers continue to elude feds. *Government Computer News*, pp. 1, 8.

Rickert, C. (2002, April 7). Web site look-alike causes concerns. *Daily Chronicle*, p. 1.

Rocheleau, B. (1993). Evaluating public sector information systems: Satisfaction versus impact. *Evaluation and Program Planning, 16*, 119-129.

Rocheleau, B. (1994). The software selection process in local governments. *The American Review of Public Administration, 24*(3), 317-330.

Rocheleau, B. (1995). Computers and horizontal information sharing in the public sector. In H. J. Onsrud & G. Rushton (Eds.), *Sharing geographic information* (pp. 207-229). New Brunswick, NJ: Rutgers University Press.

Rocheleau, B. (1997). Governmental information system problems and failures: A preliminary review. *Public Administration and Management: An Interactive Journal, 2*(3). Retrieved from http://www.pamij.com/roche.html

Rocheleau, B. (2000). Guidelines for public sector system acquisition. In G. D. Garson (Ed.), *Handbook of public information systems* (pp. 377-390). New York: Marcel Dekker, Inc.

Rocheleau, B. (2002a, March). *Accountability mechanisms, information systems, and responsiveness to external values.* Paper presented at the 2002 Meeting of the American Society for Public Administration, Phoenix, AZ.

Rocheleau, B. (2002b). E-mail: Does it need to be managed? Can it be managed? *Public Administration and Management: An Interactive Journal, 7*(2). Retrieved from http://pamij.com/7_2/v7n2_rocheleau.html

Ruppe, D. (2001). *Some US agencies pull data from Web.* Retrieved from http://dailynews.yahoo.colm/h/abc/20011005/pl/wtc_internetsecurity_011004_1.html

Sandlin, R. (1996). *Manager's guide to purchasing an information system.* Washington, DC: International City/County Management Association.

Schattschneider, E. E. (1983). *The semisovereign people: A realist's view of democracy in America.* Fort Worth, Texas: Holt, Rhinehart, and Winston.

Schmitz, J. (1997). Structural relations, electronic media, and social change: The public electronic network and the homeless. In S. G. Jones (Ed.), *Virtual culture: Identity and communication in cybersociety* (pp. 80-101). London: Sage Publications.

Schoech, D., Jensen, C., Fulks, J., & Smith, K. K. (1998). Developing and using a community databank. *Computers in Human Services, 15*(1), 35-53.

Schuler, D. (1995). Public space in cyberspace. *Internet World*, 89-95.

Severin, C. S. (1993, October 11). The CEO should be the CIO. *InformationWeek*, p. 76.

Shulman, S. (2005, September). *Stakeholder views on the future of electronic rulemaking.* Paper presented at the 2005 Annual Meeting of the American Political Science Association, Pittsburgh, PA. Retrieved October 16, 2005, from http://erulemaking.ucsur.pitt.edu

Stahlman, M. (1995, December 25). Internet democracy hoax. *InformationWeek*, 90.

Standing, C., & Standing, S. (1998). The politics and ethics of career progression in IS: A systems perspective. *Logistics Information Management, 11*(5), 309-316.

Strassmann, P. A. (1995). *The politics of information management.* New Canaan: The Information Economics Press.

Thurlow, M., Wiley, H. I., & Bielinski, J. (2003). *Going public: What 2000-2001 reports tell us about the performance of students with disabilities* (Tech. Rep. No. 35). Minneapolis, MN: University of Minnesota, National Center on Educational Outcomes. Retrieved July 15, 2004, from http://education.umn.edu/NCEO/OnlinePubs/Technical35.htm

Towns, S. (2004). Year in review: People. *Government Technology.* Retrieved December 15, 2005, from http://www.govtech.net/magazine.story.php?id=97500&issue=12:2005

Towns, S. (2005, December 7). The year in review: 2005. *Government Technology.* Retrieved December 10, 2005, from http://www.govtech.net/magazine/story.php?id=97500&issue=12:2005

United States General Accounting Office. (1997). *Social Security Administration: Internet access to personal earnings benefits information* (GAO/T-AIMD/HEHS-97-123).

United States General Accounting Office. (2002). *DOD's standard procurement system* (GAO-02-392T). Washington, DC.

United States General Accounting Office. (2004). *Federal chief information officers' responsibilities, reporting relationships, tenures, and challenges* (GAO-04-823).

Vittachi, I. (2005, March 25). Western Springs set to welcome wi-fi zone. *Chicago Tribune.*

Washburn, G. (1998, May 18). Building-permit delays spur city shakeup. *Chicago Tribune*, pp. 1, 10.

West, D. (2001, August-September). *E-government and the transformation of public sector service delivery.* Paper presented at the Meeting of the American Political Science Association, San Francisco.

Wiggins, L. (2001). Caribou and the census. *URISA News*, 5.

Wilhelm, A. G. (1997). A resource model of computer-mediated political life. *Policy Studies Journal, 25*(4), 519-534.

Yates, J. (2001, September 29). N.Y. plans to track flow of donations: Red Cross raises privacy concerns. *Chicago Tribune.*

Zeigler, N. (1997). Agencies say kinds worked out of new state computer system. *Daily Chronicle* (DeKalb, Illinois), p. 1.

Chapter IV

Reconciling Information Privacy and Information Access in a Globalized Technology Society

George T. Duncan, Carnegie Mellon University, USA

Stephen F. Roehrig, Carnegie Mellon University, USA

Abstract

Government agencies collect and disseminate data that bear on the most important issues of public interest. Advances in information technology, particularly the Internet, have created a globalized technology society and multiplied the tension between demands for ever more comprehensive databases and demands for the shelter of privacy. In reconciling information privacy and information access, agencies must address a host of difficult problems. These include providing access to information while protecting confidentiality, coping with health information databases, and ensuring consistency with international standards. The policies of agencies are determined by what is right for them to do, what works for them, and what they are required to do by law. They must interpret and respect the ethical imperatives of democratic accountability, constitutional empowerment, individual autonomy, and information justice. In managing confidentiality and data-access functions, agencies have two basic tools: techniques for disclosure limitation through restricted data and administrative procedures through restricted access.

Introduction

We continue to see advances in the technology of computing and communications allowing the capture of enormous amounts of data, storage in very large databases, complex analyses, and the dissemination of information products to individuals, governments, businesses, and other organizations. This technology has increased the tension between information privacy and information access, adding significant stress to those, like government statistical agencies, that broker between data providers and data users. With technology as its driver, the context of this dynamic is an ever-globalizing society exemplified by e-commerce across national boundaries, international outsourcing, and worldwide terrorism. This article will examine the implications of a globalized technology society for confidentiality and privacy, exploring technical, ethical, and policy issues.

A globalized technology society could raise the spectre of a world devoid of humanity, its sparse landscape dominated by robotic automatons. Or it could facilitate meaningful and productive human interaction across the globe. Perhaps less to the extremes, a globalized technology society might promote economic efficiency while testing privacy through new tools for data capture, storage, integration, and dissemination.

As far as privacy and confidentiality are concerned, a globalized technology society is one that processes information in ways radically different than the world has ever experienced—not that globalization itself is a new phenomenon. In fact, history suggests a remarkable series of stages of globalization, each made possible by quite different technological advances. Specifically, building on Thomas Friedman's (2005) formulation in *The World is Flat: A Brief History of the Twenty-First Century*, we put forward three stages:

1. **Globalization 1.0** (1400 A.D. to WWI) with changes in transportation technology allowing the great explorers like Vasco de Gama (1460-1524) and Christopher Columbus (1451-1506) and culminated in steamships and airplanes
2. **Globalization 2.0** (WWI to 2000) with changes in communication key, giving us telephones, fax, radio and TV, and e-mail
3. **Globalization 3.0** (2000 to now) with computing power key, allowing PCs (personal computers) to be linked by fiber optics and the initiation of GRID computing (an emerging global, distributed parallel processing infrastructure)

Each stage has had more impact on the way people can work with information. Globalization 1.0 allowed mail packets—now dubbed "snail mail" packets—to be sent around the world in months and then days. Globalization 2.0 made communication electronic and cut the global circuit to seconds. Globalization 3.0 not only makes communications links quicker, but also makes them more complex, increasing the density of the web of connections.

The CSID Data Process

Taking a broader view of what happens with information, we can examine the CSID data process (Duncan, 2004).

- Capture (the process of obtaining data from individuals, households, organizations, and enterprises)
- Storage
- Integration
- Dissemination

The purpose of the CSID data process is to get information in the hands of analysts who can turn data products into information that serves the legitimate needs of a democratic, free-market society. However, the information organization (IO) must also keep information from those who use the data to violate the rules and harm individuals and society. Today, technology enables ever more sophisticated data snoopers to engage in the following:

- Identity theft
- Plagiarism, intellectual property violation
- "Phishing" attacks
- Breaching confidentiality
- "Peeping Tom" surveillance

Much of the data that are used to develop statistical information products by government are obtained directly from respondents in surveys and censuses or through building systems of administrative records based on a variety of citizen interactions with government. Surveys include the following:

- **Face-to-face interviews**, as with the National Longitudinal Surveys of Young Women conducted by the Bureau of Labor Statistics (BLS)
- **Telephone surveys**, as with the Behavioral Risk Factor Surveillance System conducted by the Center for Disease Control, which estimates current cigarette smoking and use of smokeless tobacco
- **Mail-back responses** (including electronic mail), as with the reactions to their Web site obtained by Inland Revenue of the Government of New Zealand (see http://www.ird.govt.nz/survey.htm)

Administrative records include the following:

- **Employer-furnished data**, as with Social Security Administration earnings records
- **Licensing data**, as with state departments of motor vehicles and local building permits
- **Individual- and firm-submitted data**, as with Internal Revenue Service (IRS) tax returns

The Internet has accelerated the demand for access to government information services, primarily by broadening the range of potential data users. Access demand is in commensurate tension to concerns about privacy and confidentiality. The National Science Foundation (Duncan, 1998, p. 1), in its digital-government program announcement, affirms, "Given the inexorable progress toward faster computer microprocessors, greater network bandwidth, and expanded storage and computing power at the desktop, citizens will expect a government that responds quickly and accurately while ensuring privacy." This chapter focuses on ways the public sector can resolve the growing tension between the demand for government data and concerns for privacy protection.

Government databases are rich in information and have evident practical utility for planning, marketing, and research. Still, many would-be users complain they cannot obtain the data they need, often thwarted by confidentiality concerns (Smith, 1991). On the other hand, privacy advocates warn of the dangers of unfettered data capture and dissemination. Their arguments are ethically based:

> *Individuals in the Western world are increasingly subject to surveillance through the use of data bases in the public and private sectors, and these developments have negative implications for the quality of life in our societies and for the protection of human rights.* (Flaherty, 1989, p. 1)

From a more utilitarian standpoint, public policy would be affected by declining survey participation rates. In 1992, for example, 31% of Americans refused to answer at least one survey, compared with 15% in 1982 (Leftwich, 1993; also see Dalenius, 1993). Acrimonious public debates rage about the proper use of mailing lists, credit records, medical histories, and Social Security numbers (Flaherty). The media highlights privacy concerns in the use of ever larger computer databases—those of terabyte size, labeled data warehouses.

Broad access to data supports democratic decision making. Access to government statistical information supports public-policy formulation in areas ranging through demographics, crime, business regulation and development, education, national defense, energy, environment, health, natural resources, safety, and transportation. Thrust against the evident value of data access is the counter value that private lives are requisite for a free society. This chapter deals with an important aspect of the tension between information privacy and data access: the proper handling of personal information that is collected by government. Other privacy issues such as video surveillance, telephone interception or bugging, Internet censorship, protecting children, encryption policy, and physical intrusion into private spaces are outside our purview.

A variety of governmental agencies have important roles in collecting, storing, analyzing, and disseminating information. Certainly this is the case with the federal statistical agencies, such as the U.S. Census Bureau, National Center for Health Statistics, and the Bureau of Labor Statistics. However, it is also true of state public-health organizations and functional agencies such as departments of motor vehicles. Each agency is tasked with the dual responsibilities of protecting privacy and confidentiality while disseminating information to client users, often including the general public (Duncan, Jabine, & de Wolf, 1993; Duncan & Pearson, 1991).

The Internet is becoming the richest source for data collected by government agencies and academic researchers. The following is a very short list of government Web sites, compiled to illustrate the ever-widening variety.

- **U.S. Census Bureau:** The *American FactFinder* Web site allows the construction of detailed tables of age, sex, race, and so forth for geographic areas as small as a block group. Hundreds of thousands of ready-made tables, indexed by data source, geography, and other characteristics are available (http://factfinder.census.gov).
- **National Center for Educational Statistics (NCES):** Data files, such as the Local Education Agency (School District) Universe Dropout Data, are available for download in a variety of machine-readable formats. NCES also offers data licenses giving authorized researchers access to more than a dozen restricted-use data products such as the National Educational Longitudinal Survey (http://www.nced.ed.gov/pubsearch).
- **California Dept of Corrections:** A list of all prisoners on death row (http://www.cdc.state.ca.us/issues/capital/capital9.htm)
- **Arizona Secretary of State:** Searches and listings of liens, associated filings, debtors, and secured parties (http://www.sos.state.az.us/Business_Services/UCC_full_Index.htm)
- **Inter-University Consortium for Political and Social Research:** Hundreds of data files from government-funded surveys on topics such as natality, drug-abuse treatment, and the American Housing Survey (http://www.icpsr.umich.edu/)

What are some contentious issues in the clash between demands for information privacy and data access? To address this question we can examine some of the bills submitted in the U.S. Congress. For all bills, the basic source is the Library of Congress site *Thomas* at http://thomas.loc.gov. Targeted to issues of information privacy and information access, including pending legislation, nongovernment Internet sites are maintained by the following:

- Electronic Privacy Information Center (EPIC) at http://epic.org/privacy
- Center for Democracy and Technology (CDT) at http://www.cdt.org
- Direct Marketing Association (DMA) at http://www.the-dma.org/privacy/

Some key issues addressed in the 519 bills introduced to the 109th Congress that mention privacy include the following:

- Require federal agencies to prepare privacy-impact statements for proposed rule making
- Calling for the Privacy and Civil Liberties Oversight Board as an independent body within the executive branch
- Prohibiting a business enterprise from disclosing personally identifiable information without a privacy notice
- Providing for the Universal Health Privacy Declaration, which once executed by an individual would prohibit a health care provider from using or disclosing the individual's protected health information for treatment without obtaining the prior consent of the individual
- Enhancing penalties for identity theft or misuse of information by data brokers

In the next section we examine legislation that is already on the books that both attempts to protect government data and constrains its use by agencies.

Relevant Legislation

Some agencies are guided by specific legislation. The U.S. Census Bureau, for example, is governed by Title 13 of the U.S. Code, which provides for tight controls on individually identifiable data. In particular, Section 9 provides the following:

(a) Neither the Secretary, nor any other officer or employee of the Department of Commerce or bureau or agency thereof, or local government census liaison, may, except as provided in section 8 or 16 or chapter 10 of this title—(1) use the information furnished under the provisions of this title for any purpose other than the statistical purposes for which it is supplied; or (2) make any publication whereby the data furnished by any particular establishment or individual under this title can be identified; or (3) permit anyone other than the sworn officers and employees of the Department or bureau or agency thereof to examine the individual reports. No department, bureau, agency, officer, or employee of the Government, except the Secretary in carrying out the purposes of this title, shall require, for any reason, copies of census reports which have been retained by any such establishment or individual. Copies of census reports which have been so retained shall be immune from legal process, and shall not, without the consent of the individual or establishment concerned, be admitted as evidence or used for any purpose in any action, suit, or other judicial or administrative proceeding.

Similarly for the Internal Revenue Service, Section 6108(c) of the U.S. Internal Revenue Code of 1986 stipulates that:

[n]o publication or other disclosure of statistics or other information required or authorized by subsection (a) or special statistical study authorized by subsection (b) shall in any manner permit the statistics, study, or any information so published, furnished, or otherwise disclosed to be associated with, or otherwise identify, directly or indirectly, a particular taxpayer.

The U.S. Social Security Administration enjoys comparable legislative protection of its data through Section 1106 of the Social Security Act.

On the other hand, the Bureau of Labor Statistics has no specific statutory protection to preserve the confidentiality of identifiable information. Instead, the Bureau's confidentiality policy is established by Commissioner's Order 3-93, *Confidential Nature of BLS Records*, which in Section 7(a) states that data "... collected or maintained by, or under the auspices of, BLS under a pledge of confidentiality shall be treated in a manner that will ensure that individually identifiable data will be used only for statistical purposes and will be accessible only to authorized persons." (See de Wolf, 1995, for a discussion of BLS confidentiality policy.)

Chapter 5 of Duncan, Jabine, et al. (1993) presents a comprehensive view of the legislative environment of federal statistical agencies. Under current legislation, the degree of data protection depends on the agency that holds it without regard to the sensitivity of the information. Also, data sharing among agencies is difficult because agencies with a high degree of legislative protection of their data are reluctant to share with those with a low degree of protection. For example, the National Agricultural Statistics Service (NASS) has had a complicated relationship involving the sharing of lists of farms for the census of agriculture, which is conducted every 5 years. The census of agriculture was officially moved from the Census Bureau to NASS on December 31, 1997. In conducting the 1997 agriculture census, the Census Bureau provided NASS employees (many of whom were Census Bureau employees) with farm-list information. To do this, they were made special sworn employees. Restricted access is provided through the Census Bureau's Bowie computer center to NASS headquarters and the regional centers. In 1992, the Census Bureau swore in a limited number of NASS employees to see data collected under the auspices of Title 13, but they were required to come to the Census Bureau's headquarters in Suitland, Maryland, to see the data. The official transfer of the program was delayed until the end the 1997 so that the Census Bureau would continue to have the authority to get IRS tax-return data to construct the list (NASS had no such authority). Beginning in 2002, NASS will have to work with the IRS to get access by amending the tax code or through proposed data-sharing legislation if it becomes law.

In addition to this legislation, there is a variety of other relevant laws and regulations at the federal level. They include, but are not limited to, the Computer Security Act of 1987 (http://www.epic.org/crypto/csa/), the Copyright Act of 1976 (http://www.law.cornell.edu/copyright/copyright.table.html), the National Archives and Records Administration Regulations, the Freedom of Information Act (http://www.aclu.org/library/foia.html), the Information Technology Management Reform Act of 1996 (http://www.itpolicy.gsa.gov/mke/capplan/cohen.htm), the Paperwork Reduction Act of 1995 (http://www.law.vill.edu/chron/articles/ombdon.htm), and the Privacy Act (http://www.eff.org/pub/Legislation/privacy_act_74_5usc_s552a.law). A discussion of each of these can be found on the Web at the indicated uniform resource locators (URLs).

What Principles Should Guide Data Stewardship?

The principles set forth here for data stewardship expand on those given in Duncan, Jabine, et al. (1993). The United States, and a growing list of other countries, embraces a freedom that recognizes pluralism, public decision making based on representative democracy, and a market-oriented economy. Consistent with this ethos, an ethics of information can be built on four principles: democratic accountability, constitutional empowerment, individual autonomy, and information justice. These principles can provide a useful guide for assessing the societal impacts of information policies of any organization, whether in the public sector or not. They have particular interpretations for government agencies that are explored as follows.

Democratic accountability is the assurance through institutional mechanisms, culture, and practice that the public obtains comprehensive information on the effectiveness of government policies. Prewitt (1985) explored this concept. The technology of the Web is the most exciting development for fostering democratic accountability. A quick click to the Social Security Administration's Web site at http://www.ssa.gov yields SSA's Accountability Report for FY (fiscal year) 1997. It provides full disclosure of SSA's financial and programmatic operations. This Web presentation gave the agency the right to assert,

With its publication on November 21, 1997—less than 2 months after the close of the fiscal year—SSA became the first Federal agency to publish its FY 1997 Accountability Report. FY 1997 marks the eleventh year that SSA has published audited financial statements, the fourth year that SSA has received an unqualified opinion on its financial statements and the third year that SSA has been authorized by the Office of Management and Budget [OMB] to streamline and consolidate statutorily required financial reporting into a single Accountability Report.

Constitutional empowerment is the capability of citizens to make informed decisions about political, economic, and social questions. Constitutional practice emphasizes restraints on executive excess and broad access to the political process through the direct election of representatives, as well as through separation and balance of power. Many government agencies have seized upon the Web as a vehicle for providing information broadly to the citizenry. A prominent development in this regard is the plan announced on June 25, 1998, by Bruce A. Lehman, Commissioner of Patents and Trademarks, to create the largest government database on the Internet. More than 2 million patents will be searchable by key word. Including trademark information, the database will comprise more than 21 million documents and require 1.3 terabytes of storage.

Individual autonomy is the capacity of the individual to function in society as an individual, uncoerced and cloaked by privacy. Individual autonomy is compromised by the excessive surveillance sometimes used to build databases (Flaherty, 1989), the unwitting dispersion of data, and the willingness by those who collect data for administrative purposes to make them available in personally identifiable form. Government agencies have both ethical and pragmatic reasons to be concerned about individual autonomy. Ethically, agencies ought to respect individual dignity and protect the personal information entrusted to them. Pragmatically, without attention to individual autonomy, agencies will find it difficult to enlist the voluntary cooperation that smoothes operations.

Information justice is the fairness with which information is provided to individuals and organizations. Information injustice can occur when data are captured from only some individuals and according to unacceptable selection criteria. It can occur when data are denied to some groups or individuals inappropriately. Ubiquity of access may well be an appropriate initial position affirming information justice. As the World Association for Christian Communication asserts (http://www.churchworldservice.org/news/archives/2004/08/230.html), "Information justice is built on the premise that information is a vital resource available for all rather than for the privileged few."

The Special Roles of Government Agencies: *Functional Separation*

In implementing the three fundamental principles of democratic accountability, constitutional empowerment, and individual autonomy, a government agency should affirm a policy of functional separation. This policy makes a distinction between administrative data and statistical data. The distinction is on the basis of use.

- Administrative data are used so that data on an individual has a direct impact on that individual.
- Statistical data are used to create aggregate measures that have an impact on individuals only through substantial group membership.

Thus, Joe Brown's liquor license application is initially part of administrative data since it is used to determine whether to issue Joe a license. When a database of such applications is used to determine whether females are issued liquor licenses as frequently as males, this constitutes a statistical use. Conceivably, such a study might affect administrative practice about license issuance in which case it might affect the chances of Joe Brown's subsequent application. This impact is due solely to Joe being male and is not determined by his particular data.

Agencies accept responsibility for protecting privacy and confidentiality for several ethical and pragmatic reasons. First, it is the right thing to do. Generally accepted ethical standards in the profession of data collection require attention to privacy and confidentiality (American Statistical Association, Committee on Professional Ethics, 1989; International Statistical Institute, 1986). Second, they get better data this way. Professionals believe that confidentiality pledges lower nonresponse rates and improve the quality of responses (Singer, Mathiowetz, & Couper, 1993). Third, the law requires it. Privacy and confidentiality protection is often mandated by legislation or regulation (Duncan, Jabine, et al., 1993).

The Privacy Protection Study Commission (1977, p. 574) proposed,

[t]hat the Congress provide by statute that no record or information contained therein collected for a research or statistical purpose under Federal authority or with Federal funds may be used in individually identifiable form to make any decision or take any action directly affecting the individual to whom the record pertains, except within the context of the research plan or protocol, or with the specific authorization of the individual.

The National Research Council sponsored the Panel on Confidentiality and Data Access. In its report, *Private Lives and Public Policies*, the panel made the following recommendation (Duncan, Jabine, et al., 1993, p. 134):

Statistical records across all federal agencies should be governed by a consistent set of statutes and regulations meeting standards for the maintenance of such records, including the following features of fair statistical information practices:

(a) *a definition of statistical data that incorporates the principle of functional separation as defined by the Privacy Protection Study Commission,*

(b) *a guarantee of confidentiality for the data,*

(c) *a requirement of informed consent or informed choice when participation in a survey is voluntary,*

(d) *a requirement of strict control on data dissemination,*

(e) *a requirement to follow careful rules on disclosure limitation,*

(f) *a provision that permits data sharing for statistical purposes under controlled conditions, and*

(g) *legal sanctions for those who violate confidentiality requirements.*

This recommendation refers to federal agencies. The issues are similar for state agencies and for countries other than the United States. We would, therefore, maintain that these fair information practices are broadly applicable to government operations.

Privacy Legislation in the European Union

The discussion above illustrates the distinction in the United States between legislation and policies applicable to government agencies, and laws restricting the data activities of private entities such as insurers and marketers. A different model is provided by the European Union (EU). Information about Directive 95/46/EC (1995) of the European Parliament is available at http://www.cdt.org/privacy/eudirective.

This EU directive became effective in October 1998. It specifies policies for the protection of individuals that must be adhered to by all members of the EU. These policies apply uniformly to both the public and private sectors. Each member country is permitted to have stricter rules governing data related to their own residents; the EU directive provides a minimal set of protections across all members, and is designed to foster consumer confidence and encourage e-commerce. The directive also establishes rules to ensure that personal data are only transferred to countries outside the EU when their continued protection is guaranteed.

Looking first at the rules governing the movement of personal data, the directive delineates the kinds of data affected. In language similar to the U.S. Privacy Act, data processing is covered only if it is "automated or if the data processed are contained or are intended to be contained in a filing system structured according to specific criteria relating to individuals, so as to permit easy access to the personal data in question"; private correspondence and personal address books are excluded, for example. Further exemptions include data necessary for public safety, defense, state security, and "activities of the State in the area of criminal laws."

Data subjects must be informed of the data collected and the processing that will take place; furthermore, subjects are guaranteed access to their data, and have the right to request corrections. Again, this is similar to that of the U.S. Privacy Act. However, an additional consideration is that subjects may "object to processing in certain [unspecified] circumstances."

The issues of de-identification and data linkage (discussed later in this chapter) are also considered in the EU directive. The rules are essentially these: Data that have been de-identified ("rendered anonymous in such a way that the data subject is no longer identifiable") are explicitly not covered by the directive. However, in determining whether or not de-identification has in fact occurred, the owner of the data must take into account "all the means likely reasonably to be used" to identify the data subject.

Paragraph 34 of the EU directive also deals with one area of medical records privacy:

(34) Whereas Member States must also be authorized, when justified by grounds of important public interest, to derogate from the prohibition on processing sensitive categories of data where important reasons of public interest so justify in areas such as public health and social protection—especially in order to ensure the quality and cost-effectiveness of the procedures used for settling claims for benefits and services in the health insurance system—scientific research and government statistics; whereas it is incumbent on them, however, to provide specific and suitable safeguards so as to protect the fundamental rights and the privacy of individuals.

Here, and in other places, the directive allows member states to derogate from the general plan, provided explicit safeguards are in place.

Formally, the EU's Directive on Data Protection prohibits the transfer of personal data to non-EU nations unless protections in the receiving country are at least as strong as those provided by the directive. This poses a difficult barrier for U.S. firms conducting business with EU member states, since, as mentioned above, most U.S. privacy legislation is not directed toward business. The self-regulated rather than legislated status of privacy policy in the private sector resulted in the "safe harbor" framework devised by the U.S. Department of Commerce and the European Commission. Under the safe harbor rules, U.S. firms can be certified to conduct data transactions with EU nations by pledging to provide adequate privacy protection (see http://www.export.gov/safeharbor).

The safe harbor framework insists on seven privacy principles:

1. **Notice:** Individuals must be notified about the collection and use of their personal data.
2. **Choice:** Individuals have the opportunity to refuse to allow sensitive data about them to be transferred.
3. **Outward transfer:** A certified organization must ensure that data transfers to a third party must also adhere to safe-harbor principles.
4. **Access:** Individuals must have access to data about themselves, and must be given the opportunity to correct inaccurate information.
5. **Security:** Reasonable precautions must be taken to protect personal data from unauthorized access, loss, or misuse.
6. **Data integrity:** Personal data must be used for the purpose intended, and must be accurate, current, and complete.
7. **Enforcement:** There must be means to ensure compliance, including individual recourse, verification procedures, and clear sanctions and penalties.

The safe-harbor framework has not been widely embraced by U.S. firms. In June 2002, more than a year and a half after its adoption, only about 200 U.S. companies had been certified.

Problems and Opportunities in Ensuring Confidentiality and Data Access

The public sector faces a variety of predicaments as well as opportunities as it seeks to fulfill its responsibilities for both confidentiality and access to data. The problems and possibilities are accentuated by economic and cultural changes, and importantly by developments in information technology. This section will examine these factors as they affect the various stakeholders in the process: the data subjects, the data users, and the agencies that have stewardship of the data.

Data Subjects

Government agencies depend on individuals, firms, and organizations to provide data that accurately reflect some of the most personal and sensitive aspects of their lives and operations. Some of this data provision is mandated by legislation as public corporations are required to provide the Securities and Exchange Commission with filing information and individuals must file an income-tax return with the Internal Revenue Service. Other data provision is voluntary, as when the Substance Abuse and Mental Health Services Administration interviews people at their residence about use of licit and illicit drugs.

Anecdotal evidence suggests that response rates of federally funded demographic surveys have been declining. To address this issue, the Federal Committee on Statistical Methodology formed the Subcommittee on Nonresponse that collected information on 26 federally sponsored demographic surveys. Response and refusal rates remained relatively constant, but noncontacts fluctuated over the 10-year period of 1982 to 1991. The committee found a core proportion of the population that routinely refuses to participate in federally sponsored surveys.

The key ethical concepts related to data subjects are informed consent and notification. Informed consent is appropriate for truly voluntary surveys, while notification applies otherwise to data collection that is mandatory, such as the Decennial Census, or where benefits hinge on providing information, such as applications for welfare benefits. The Privacy Act requires that each person asked to supply information be informed of (a) the authority under which the information is requested, (b) the principal uses for the information, (c) the routine uses that may be made of the information, and (d) the implications, if any, of not providing the information. The National Research Council's Panel on Confidentiality and Data Access (Duncan, Jabine, et al., 1993) made recommendations for strengthening these stipulations. Noteworthy among these are requirements that data providers be notified of (a) nonstatistical uses of their data, (b) any anticipated record linkages for statistical purposes, and (c) the length of time the information will be retained in identifiable form.

Data Users

Data users span a diverse range of individuals and organizations. They include academic researchers at Cambridge University, policy analysts for the American Association for Retired Persons and the National Association of Home Builders, business economists for Wells Fargo bank, and statisticians for the Health Care Financing Administration. They include reporters for the *Toronto Star*, marketing analysts for L. L. Bean, advocates for the National Abortion Rights League, and medical insurance underwriters for Cigna. In general, data users employ the data they obtain for end uses such as policy analysis, commercial and academic research, advocacy, and education. The may also use data for various intermediate purposes such as the development of sampling frames for surveys and the evaluation of the quality of other data.

From a government agency's viewpoint, the primary concern of data users is obtaining access to data. Users desire data that are relevant, accurate, and complete. They also want a usable data format, easy accessibility (low price, little hassle, quick response), timeliness (automatic updates and corrections), and few limitations on use. All of these concerns are legitimate, and they are uncontroversial except for the last, which can raise serious confidentiality concerns.

Often unanticipated are data users who force access through legal action, often as part of a discovery process and involving a court-issued subpoena. As Duncan, Jabine, et al. (1993) notes, many statistical agencies lack adequate legal authority to protect identifiable statistical records from mandatory disclosures for nonstatistical uses. An example of this was the ruling that the Environmental Protection Agency could not protect company survey responses from the Department of Justice's Antitrust Division for use in compliance activities.

Many data users want to further disseminate the data to other users. This secondary data provision occurs in a variety of contexts. A government agency sponsoring a survey may share the information with another agency. A motor-vehicle service may pass on licensed-driver information to insurance companies for a fee. This data sharing requires careful managing of the advantages of more efficient data collection against the risks of confidentiality loss. Some laws governing the confidentiality of data prohibit or severely limit interagency sharing of data, even for solely statistical purposes.

Organizations Representing Stakeholder Interests

A number of organizations represent various configurations of the stakeholder interests described previously.

- The Association of Public Data Users (APDU; see http://www.apdu.org/) assists users in the identification and application of public data, establishes linkages between data producers and users, and brings the perspectives and concerns of public data users to issues of government information and statistical policy.
- The Council of Professional Associations on Federal Statistics (COPAFS; see http://members.aol.com/copafs/) represents academic and professional organizations interested in the production of federal statistical and research data. Member organizations

include professional associations, businesses, research institutes, and others interested in federal statistics. COPAFS seeks to accomplish the following:

  - Increase the level and scope of knowledge about developments affecting federal statistics
  - Encourage discussion within member organizations to respond to important issues in federal statistics
  - Bring the views of professional associations to bear on decisions affecting federal statistical programs

- The Council for Marketing and Opinion Research (CMOR; see http://www.cmor.org/) is a nonprofit trade association formed to protect the interests of the marketing and opinion research industry. It encourages respondent cooperation and lobbies lawmakers to protect research from restrictive legislation.
- The American Civil Liberties Union (ACLU; see http://www.aclu.org/) affirms both privacy rights and the public's right to know.
- Computer Professionals for Social Responsibility (CPSR; see http://www.cpsr.org) is a public-interest alliance of computer scientists and others concerned about the impact of computer technology on society.

Such organizations provide consequential input to government agencies in dealing with privacy and information issues.

Managing Confidentiality and Data-Access Functions

General Issues

Wide-ranging mechanisms exist to deal with conflicts about the capture and dissemination of data. They span federal legislation, interorganizational contractual arrangements, intraorganizational administrative policies, and ethical codes. They also include technological remedies such as the release of masked data that may satisfy data users' needs for statistical information while posing little risk of disclosure of personal information (Duncan & Pearson, 1991). In managing confidentiality and data-access functions, government agencies have two basic tools for responsible provision of information: restricted data and restricted access. As developed in Duncan, Jabine, et al. (1993), these concepts have the following interpretations:

- **Restricted data:** Data are transformed to have lower disclosure risk. This is accomplished through disclosure limitation techniques such as (a) releasing only a sample of the data, (b) including simulated data, (c) blurring the data by grouping or adding random error, (d) excluding certain attributes, and (e) swapping data by exchanging the values of just certain variables between data subjects.

- **Restricted access:** Administrative procedures impose conditions on access to data. These conditions may depend on the type of data user; conditions may be different for interagency data sharing from those for external data users. An example of an institutional arrangement for restricted access by external data users is the California Census Research Data Center maintained by UCLA (University of California, Los Angeles) and Berkeley (http://www.ccrdc.ucla.edu/).

Various restricted-access policies (Jabine, 1993a, 1993b) have been implemented in the last 20 years. Notable has been the fellowship programs run jointly by the American Statistical Association and the National Science Foundation together with four agencies: the Bureau of Labor Statistics, the Bureau of the Census, the U.S. Department of Agriculture, and the National Institute of Standards and Technology. The fellowship programs require that specific research projects and their data needs have to be evaluated. If approved, data users relocate to the agency to gain access to unrestricted data. In some cases of restricted access, for example, to the Panel Study of Income Dynamics and the National Longitudinal Survey of Youth (Jabine, 1993a), the researcher must post bond. The money will be forfeited if the researcher fails to honor the release agreement, say, by unauthorized sharing of the data or performing analyses not specified in the proposal.

The Bureau of the Census has long sought a mechanism by which it could make detailed census information more readily available to researchers, as well as connect census data to other important national data sets, such as those housed at the Environmental Protection Agency and the Department of Justice, while maintaining the integrity and confidentiality of that data. With this in mind, the bureau recently established a number of census centers housed at universities across the country. Through access to such valuable data, the center has attracted nationally renowned scholars to engage in interdisciplinary, collaborative research on important policy issues.

Data Linkage

As more detailed personal data become available, from an ever-widening collection of sources, the danger that innocuous individual data sets may be combined to reveal sensitive information increases. The merging of data sets over common attributes is called data linkage. Through data linkage, even a data set that has been de-identified by removing identifiers such as name, address, Social Security number, and other unique attributes may still pose a risk for disclosure. Consider the simple case of two files, the first containing such identifiers but no sensitive information, and a second file containing some sensitive data, but consisting only of de-identified records. If the two files have nonsensitive attributes in common, it may be possible to find records that have unique and equal values for the overlapping attributes. A pair of such records links the identifying information in one file to the sensitive information in the other.

Readily available software and a desktop computer make it quite easy for a computer-literate individual to attempt data linkage. Methods originally developed to purge mailing lists of duplicate entries, where true duplicates may be slightly different because of typographical error, have been improved and extended to provide probabilistic estimates of matching.

Precisely because these methods can be effective in the face of even substantial error in the data, they make the task of masking data to protect sensitive information more difficult.

A number of studies have shown that modern techniques for data linkage can be remarkably effective even in the presence of some types of data masking (Fuller, 1993; Kim & Winkler 1995; Winkler, 1994, 1995). Since re-identification is probabilistic for data that have been blurred before release, the best approach for protecting confidentiality is to ensure that this probability is low enough to prevent a data intruder from assuming a match is correct.

Institutional Mechanisms

Privacy or Information Advocate

A privacy or information advocate is a one-sided intervenor (Kaufman & Duncan, 1992) whose mandate is to counterbalance power and resource inequalities among parties of a data dispute. At the U.S. federal level, the Internal Revenue Service appointed Robert Veeder, a specialist in the Privacy Act and Freedom of Information Act at the Office of Management and Budget, to be the IRS's first privacy advocate. As an example of his activity, he presented a paper entitled *Making Information Accessible while Protecting Privacy* with Sara Hamer, associate commissioner of the Social Security Administration, to a seminar of the Federal Internet Institute in December 1997.

Privacy or information advocates act to right a power imbalance by championing the position of a weaker party. It is presumably difficult for an advocate to switch gears between privacy advocacy and data-access advocacy. Advocates are quite limited in their ability to address privacy and information disputes as advocacy is their only tool. Access to advocates may be hindered if they are located within a bureaucracy, hence obstructing their visibility.

Privacy and Information Clearinghouse

A privacy and information clearinghouse provides a forum for intervention in disputes between information organizations and data providers as well as between information organizations and data users. It provides education and advice to those having questions and concerns about privacy and data-access procedures.

An exemplar of such an institutional mechanism is the Privacy Rights Clearinghouse (PRC) in California. It offers information on how consumers can protect their personal privacy. It provides a Web site at http://www.privacyrights.org/ and also operates a telephone hotline for those who seek information about privacy issues. The Privacy Rights Clearinghouse was established with funding from the Telecommunications Education Trust, a program of the California Public Utilities Commission. Some of its materials were developed through the University of San Diego, Center for Public Interest Law, which administered the PRC from its inception in 1992 to October 1996. Given the need for such a clearinghouse to have a reliable source of funding and appropriate administrative support, its existence is contingent on highly specific circumstances. It may not be possible to replicate these conditions in other states.

Ombudsperson

Another intercessory mechanism with one-sided characteristics is an ombudsperson. An ombudsperson works within an agency to deal with complaints by data providers or data users.

The federal government has taken a step in the direction of using ombudsperson mechanisms for information and privacy disputes. The Office of Management and Budget Circular A-130 that was revised February 9, 1996 (see http://www.whitehouse.gov/WH/EOP/OMB/html/circulars/a130/a130.html), provides uniform government-wide information-resources management policies. In Section 9a (10) it says that an ombudsperson be designated by each agency. The ombudsperson is to be a senior agency official charged "to investigate alleged instances of agency failure to adhere to the policies set forth in the Circular and to recommend corrective action as appropriate."

An ombudsperson provides an alternative and generally easily identifiable complaint route for those in dispute with an information organization. This increases the power of data suppliers and data users. An ombudsperson can only be responsive to organization-specific disputes. This mechanism is limited in flexibility because the ombudsperson can only direct and articulate concerns. In particular, the ombudsperson typically does not have mediation or arbitration powers. Access may of course be limited if the ombudsperson is hidden within an IO's bureaucracy. It is essential that an ombudsperson be granted the authority to act with some neutrality.

Internal Privacy Review Board

The function of internal privacy review boards is akin to that of institutional review boards (IRBs) in universities. In fact, such IRBs could themselves provide "oversight mechanisms, with suitable definition of their scope to cover research uses of federal data sets, [to ensure that] adequate controls are in place to monitor compliance with data protection rules and regulations by users in the research community" (Duncan, Jabine, et al., 1993, p. 107).

The Bureau of the Census has the Microdata Review Panel that was formally chartered in 1981 (Cox, McDonald, & Nelson, 1986). It is charged with reviewing policies for the dissemination of microdata files, especially public-use data tapes and CD-ROM products. With a similar charge, the National Center for Education Statistics has the Disclosure Review Board, which was created in 1989. It is staffed by NCES employees and a Census Bureau representative. In the private sector, the Inter-University Consortium for Political and Social Research frequently convenes ad hoc privacy review boards consisting of research and privacy experts.

An internal privacy review board can vary in its influence on the balance of power in privacy and information disputes depending on how it sees its mandate. Some such boards may see their role as simply ensuring that extant administrative rules and legislative requirements are met. Others may be more actively involved in disputes between an information organization and their data providers or their data users. An internal privacy review board can be quite responsive to specific disputes. If it is granted adequate authority, it can employ a wide range of mechanisms to resolve disputes. Unless the board is specially constituted for this purpose, access to an internal privacy review board by data providers and data users may be quite limited.

Administrative Review Agency

An administrative review agency would derive its mandate from legislative or executive authority. It would function like the OMB Statistical Policy Office. While not having direct administrative responsibility for federal statistical agencies, the Statistical Policy Office provides long-range planning for statistical programs and coordinates statistical policy within the federal government. The office reviews all data-collection requests developed by the Census Bureau and the Bureau of Economic Analysis. Other federal agencies submit their data-collection requests, including those for statistical purposes, to OMB clearance officers who are not part of the Statistical Policy Office.

An administrative review agency typically would have as part of its charge the responsibility to ensure an appropriate balance of power among agencies, data providers, and data users. As constrained by its legislative mandate and resources, it could have wide-ranging responsiveness to privacy and information disputes, high flexibility in dealing with them, and potentially high access by concerned parties.

Data and Access Protection Commission

Perhaps the most elaborate institutional mechanism is an independent data and access protection commission. It would have legislative authority to regulate all stages of the information gathering and dissemination process by promoting accountability and fair information practices. In various ways such commissions have been implemented in Canada and several European countries (see Flaherty, 1989, for a detailed discussion of their operation). Canada has institutionalized a balance between data protection and data access. It has both a privacy commissioner and an information commissioner. The province of British Columbia combines these functions in a privacy and information commissioner. Both Australia and New Zealand have privacy commissioners, the United Kingdom has a data-protection registrar, Switzerland has a data-protection authority, Norway has a data inspectorate, and Spain has a privacy authority.

In the United States, most drafts of the Privacy Act of 1974 provided for a permanent privacy-protection commission, but this provision was deleted before final passage. Recently, interest in such a proposal has waxed and waned. In the House, Representative Wise introduced legislation in 1989 and 1991 to establish an independent data-protection board. In 1995, Representative Collins introduced H.R.184 to amend the privacy provisions of the Privacy Act to improve the protection of individual information and to reestablish the permanent Privacy Protection Commission as an independent entity in the federal government. In the Senate, Paul Simon introduced Senate Bill 1735 in November of 1993 to establish the Privacy Protection Commission. The bill provided for an advisory and independent commission of five members to be appointed by the president, with the consent of the Senate, to serve staggered 7-year terms. More recently, in February 2001, Representative Asa Hutchison introduced HR 583, mandating the establishment of a federal commission for the comprehensive study of privacy protection, to balance the protection of privacy while allowing appropriate uses of information. As of May 2002, no such bill had passed. Nonetheless, the repeated introduction of this provision and its support from various bodies, including the National Academy of Sciences Panel on Confidentiality and Data Access (Duncan, Jabine, et al., 1993), indicates the popularity of the concept.

Data-protection commissioners could also assist in reviewing data-provider-related disputes. The National Research Council's Panel on Confidentiality and Data Access recommended "an independent federal advisory board charged with fostering a climate of enhanced protection for all federal data about persons and responsible data dissemination for research and statistical purposes" (Duncan, Jabine, & de Wolf, 1993, p. 217). A data and access protection commission could have a legislative mandate giving it wide-ranging authority. Such authority might charge it with the responsibility to provide for a balance of power among stakeholders, to be broadly responsive to disputes about data, to be flexible in employing mechanisms to resolve disputes, and to provide easy access to all disputants. A commission could also serve as a liaison for the negotiation of international agreements regarding privacy and information issues.

Technical Procedures

Technical procedures for maintaining data confidentiality involve the release of restricted data; techniques developed over the past 20 years have been proposed in both the statistical and computer science literature (Duncan & Pearson, 1991; Fienberg, 1994; Keller-McNulty & Unger, 1993). Unlike restricting access, restricting data is a technical device. It involves such methods as removing explicit identifiers and masking the sensitive data, for example, grouping into categories or adding noise. By implementing data restrictions, agencies have operationalized statistical disclosure-limitation practices. Some guidelines used by the European statistical system (Eurostat, 1996) are summarized in its *Manual on Disclosure Control Methods*.

Statistical disclosure-limitation practices have allowed agencies to provide increasing amounts of data to the research community. Jabine (1993b) gives an excellent summary of statistical disclosure-limiting practices for selected U.S. agencies. The techniques proposed depend on the nature of the data, whether in tabular, microdata, or online form.

Tabular Data

From demographic surveys, frequency counts of variables such as age, sex, and race of responding individuals are tabled. Respondents can be identified, and so a disclosure occurs, with small counts in the cells of the table. If a table, for example, showed only one Asian female in a census tract and shows her as an orphan, then a disclosure has occurred. From establishment surveys, variables such as Standard Industrial Classification Codes and salary levels are used to create tables. For establishment data, the disclosure issue is to avoid releasing information that will identify characteristics of particular establishments. As noted by Cox and Zayatz (1995), there are four principal methods for disclosure limitation of tabular data: cell suppression (Willenborg & de Waal, 1996), rounding (Cox, 1987), perturbation (Duncan & Fienberg, 1998), and modification of the underlying microdata (Griffin, Navarro, & Flores-Baez, 1989). A recent survey of disclosure-limitation methods for tabular data is Duncan et al. (2001).

Microdata

Microdata are records directly on the unit of analysis, so they may involve data about specific individuals or establishments. Because of the demand for more information than can be obtained from tables, public-use microdata files have been developed by some agencies. A public-use microdata file provides unrestricted access to restricted data. Public-use microdata files have been limited to data concerning individuals. Data on organizations tend to be more highly skewed than for individuals. The skewed distributions coupled with the time series and longitudinal nature of organizational data make many data-restriction techniques, including top coding, difficult. Consequently, very few public-use microdata files for organizations have ever been released. An important exception to this is the U.S. Census of Agriculture, which beginning with the 1987 data, has released microdata that have been disclosure limited through high levels of geographic aggregation and data categorization as well as being a 5% sample of farms (Kirkendall et al., 1994).

Online Data

Technological advances in computers and communications offer both opportunities and threats: opportunities to capture, analyze, store, and disseminate large databases more efficiently, and threats of unauthorized access to individually identifiable data. Full use of the capability of today's information technology involves data access through online data query systems (McNulty & Unger, 1989). The data user directly requests all statistical analyses of interest. Steel and Zayatz (1998) lay out the technical procedures for disclosure limitation that will be used for the 2000 Census. In order that the released data products may have both higher quality and lower disclosure risk, they propose data-swapping procedures that will target the most risky records. To allow broader and easier access to data and to allow users to create their own data products, they are deploying the American FactFinder system. Users can submit requests electronically.

Conclusion

Agencies in the public sector play a key role in responding to the challenge of data dissemination and data privacy. They are central in collecting and disseminating data that bear on the most important issues in the public interest. Advances in information technology, particularly the Internet, have multiplied the tension between demands for ever more comprehensive databases and demands for the shelter of privacy. In mediating between these two conflicting demands, agencies must address a host of difficult problems. These include providing access to information while protecting confidentiality, coping with health information databases, and ensuring consistency with international standards. The policies of agencies are determined by what is right for them to do, what works for them, and what they are required to do by law. They must interpret and respect the ethical imperatives of democratic accountability, constitutional empowerment, and individual autonomy. They must keep pace with technological developments by developing effective measures for making information available to a broad range of users. They must both abide by the mandates of legislation and participate in the process of developing new legislation that is responsive to changes that affect their domain. In managing confidentiality and data-access functions, agencies have two basic tools: techniques for disclosure limitation through restricted data and administrative procedures through restricted access. The administrative procedures can be implemented through a variety of institutional mechanisms, spanning privacy advocates, internal privacy review boards, and a data and access protection commission. The technical procedures for disclosure limitation involve a range of mathematical and statistical tools. The challenge of developing and implementing these administrative and technical tools is great, and the value to society of the information that agencies can provide is hard to overestimate.

References

American Statistical Association, Committee on Professional Ethics. (1989). *Ethical guidelines for statistical practice.* Alexandria, VA: Author.

Cox, L., McDonald, S.-K., & Nelson, D. (1986). Confidentiality issues at the United States Bureau of the Census. *Journal of Official Statistics, 2*, 135-160.

Cox, L. H. (1987). A constructive procedure for unbiased controlled rounding. *Journal of the American Statistical Association, 82*(398), 520-524.

Cox, L. H., & Zayatz, L. (1995). An agenda for research in statistical disclosure limitation. *Journal of Official Statistics, 11*(2), 205-220.

Dalenius, T. (1993). Discussion: Informed consent and notification. *Journal of Official Statistics, 9*, 377-381.

De Wolf, V. A. (1995). *Procedures for researcher access to confidential microdata at the Bureau of Labor Statistics.* Washington, DC: Office of Research and Evaluation, Bureau of Labor Statistics.

Drucker, P. F. (1992). The new society of organizations. *Harvard Business Review*, 95-104.

Duncan, G. T. (1990). Disclosure limitation research and practices: A commentary on two agencies' perspectives. *Proceedings of the Seminar on Quality of Federal Data Council of Professional Associations on Federal Statistics.*

Duncan, G. T. (1997). Data for health: Privacy and access standards for a health care information infrastructure. In A. R. Chapman (Ed.), *Health care and information ethics: Protecting fundamental human rights* (pp. 299-339). Sheed & Ward.

Duncan, G. T. (2004). Exploring the tension between privacy and the social benefits of governmental databases. In P. M. Shane, J. Podesta, & R. C. Leone (Eds.), *A little knowledge: Privacy, security, and public information after September 11* (pp. 71-88). New York: The Century Foundation.

Duncan, G. T., & de Wolf, V. A. (1990). Mediating confidentiality and data access. *Chance, 3*, 45-48.

Duncan, G. T., de Wolf, V. A., Jabine, T. B., & Straf, M. L. (1993). Report of the panel on confidentiality and data access. *Journal of Official Statistics, 9*, 271-274.

Duncan, G. T., & Fienberg, S. E. (1998). Obtaining information while preserving privacy: A Markov perturbation method for tabular data. In *Eurostat: Statistical data protection 98*. Lisbon, Portugal.

Duncan, G. T., Fienberg, S. E., Krishnan, R., Padman, R., & Roehrig, S. F. (2001). Disclosure limitation methods and information loss for tabular data. In P. Doyle, J. I. Lane, J. J. M. Theeuwes, & L. V. Zayatz (Eds.), *Confidentiality, disclosure, and data access* (pp. 135-166). Amsterdam: Elsevier.

Duncan, G. T., Jabine, T., & de Wolf, V. (1993). *Private lives and public policies: Confidentiality and accessibility of government statistics*. Washington, DC: National Academy Press.

Duncan, G. T., & Lambert, D. (1986). Disclosure-limited data dissemination (with comments). *Journal of the American Statistical Association, 81*, 10-28.

Duncan, G. T., & Lambert, D. (1989). The risk of disclosure for microdata. *Journal of Business and Economic Statistics, 7*, 207-217.

Duncan, G. T., & Mukherjee, S. (1991). Microdata disclosure limitation in statistical databases: Query size and random sample query control. In *Proceedings of the 1991 IEEE Symposium on Research in Security and Privacy* (pp. 20-22).

Duncan, G. T., & Pearson, R. W. (1991). Enhancing access to data while protecting confidentiality: Prospects for the future. *Statistical Science, 6*, 219-239.

Engelage, C. (1992). *Statistical confidentiality in the context of community statistics: The legal framework* (Eurostat report). Luxembourg.

Eurostat. (1996). *Manual on disclosure control methods*. Luxembourg: Office for Official Publications of the European Communities.

Federal Committee on Statistical Methodology. (1994). *Statistical policy working paper 22: Report on statistical disclosure limitation methodology*. Washington, DC: U. S. Office of Management and Budget.

Fienberg, S. E. (1994). Conflicts between the needs for access to statistical information and demands for confidentiality. *Journal of Official Statistics, 10*, 115-132.

Flaherty, D. H. (1989). *Protecting privacy in surveillance societies*. Chapel Hill, NC: University of North Carolina Press.

Friedman, T. L. (2005). *The world is flat: A brief history of the twenty-first century*. New York: Farrar, Straus and Giroux.

Fuller, W. A. (1993). Masking procedures for microdata disclosure limitations. *Journal of Official Statistics, 9*, 383-406.

Griffin, R., Navarro, A., & Flores-Baez, L. (1989). Disclosure avoidance for the 1990 census. In *Proceedings of the Section on Survey Research Methods, American Statistical Association* (pp. 516-521).

International Statistical Institute. (1986). Declaration of professional ethics. *International Statistical Review, 54*, 227-242.

Jabine, T. B. (1993a). Procedures for restricted data access. *Journal of Official Statistics, 9*, 537-590.

Jabine, T. B. (1993b). Statistical disclosure limitation practices of United States statistical agencies. *Journal of Official Statistics, 9*, 427-454.

Kaufman, S., & Duncan, G. T. (1992). A formal framework for mediator mechanisms and motivations. *The Journal of Conflict Resolution, 36*(4), 688-708.

Keller-McNulty, S., & Unger, E. A. (1993). Database systems: Inferential security. *Journal of Official Statistics, 9*, 475-500.

Kim, J. J., & Winkler, W. E. (1995). Masking microdata files. In *Proceedings of the Section on Survey Research Methods, American Statistical Association* (pp. 114-119).

Kirkendall, N. J., Arends, W. L., Cox, L. H., de Wolf, V. A., Gilbert, A., Jabine, T. B., et al. (1994). *Report of the Subcommittee on Statistical Disclosure Limitation Methodology, Federal Committee on Statistical Methodology.* Washington, DC.

Lambert, D. (1993). Measures of disclosure risk and harm. *Journal of Official Statistics, 9*, 313-331.

Leftwich, W. (1993). How researchers can win friends and influence politicians. *American Demographics*, 9.

McNulty, S., & Unger, E. (1989). The protection of confidential data. In *Computer Science and Statistics: Proceedings of the 21st Symposium on the Interface* (pp. 215-219). American Statistical Association.

National Science Foundation. (1998). *Digital government program announcement: Directorate for Computer and Information Science and Engineering.* Washington, DC.

Norwood, J. (1990). Statistics and public policy: Reflections of a changing world (Presidential address). *Journal of the American Statistical Association, 85*, 1-5.

Prewitt, K. (1985). Public statistics and democratic politics. In J. J. Smelser & D. R. Gerstein (Eds.), *Behavioral and social science: Fifty years of discovery.* Washington, DC: National Academy Press.

Privacy Protection Study Commission. (1997, July). *Personal privacy in an information society* (Report). Retrieved February 23, 2007, from http://www.epic.org/privacy/ppsc1977report/

Regan, P. M. (1984). Personal information policies in the United States and Britain: The dilemma of implementation considerations. *Journal of Public Policy, 4*, 19-38.

Singer, E., Mathiowetz, N. A., & Couper, M. P. (1993). The impact of privacy and confidentiality concerns on survey participation: The case of the 1990 U.S. Census. *The Public Opinion Quarterly, 57*(4), 465-482.

Smith, J. P. (1991). *Data confidentiality: A researcher's perspective. Panel on Privacy and Confidentiality.* Paper presented at the Annual Meeting of the American Statistical Association, Anaheim, CA.

Steel, P., & Zayatz, L. (1998). *Disclosure limitation for the 2000 Census of Population and Housing.* Paper presented at the Annual Meeting of the American Statistical Association, Dallas, TX.

Willenborg, L., & de Waal, T. (1996). Statistical disclosure control in practice. In *Lecture notes in statistics* (Vol. 111). New York: Springer Verlag.

Winkler, W. E. (1994). Advanced methods for record linkage. In *Proceedings of the Section on Survey Research Methods, American Statistical Association* (pp. 467-472).

Winkler, W. E. (1995). Matching and record linkage. In B. G. Cox (Ed.), *Business survey methods* (pp. 355-384). John Wiley.

Chapter V

Privacy-Sensitive Tracking of Behavior with Public Information Systems: Moving Beyond Names in a Globalizing Mass Society

Chris C. Demchak, University of Arizona, USA

Kurt D. Fenstermacher, University of Arizona, USA

Abstract

This chapter explores the roles of names and name equivalents in social tracking and control, reviews the amount of privacy-sensitive databases accumulating today in U.S. legacy federal systems, and proposes an alternative that reduces the likelihood of new security policies violating privacy. We focus on the continuing public-authority reliance on unique identifiers, for example, names or national identity numbers, for services and security instead of dissecting a better indicator of security threats found in behavior data. We conclude with a proposed conceptual change to focusing the social-order mission on the behavior of individuals rather than their identities (behavior-identity knowledge model, BIK). It is particularly urgent to consider a different path now as increased interest in biometrics offers an insidious expansion of unique identifiers of highly personal data. E-government can be wonderful for central government's effectiveness and efficiency in delivering services while also being a disaster for both privacy and security if not regulated legally, institutionally, and technically (with validation and appeal processes) from the outset.

Introduction

For a nation, ensuring security and acceptable social order are key missions of central governments; identifying and then restraining those who act contrary to societal rules are essential to these missions. Thus, reaching back to China's Han Dynasty or William's Domesday Book in 1086, governments have developed census and other list-making activities intended to monitor compliance. As populations grew and people moved, clan, geographic, marriage, or professional affiliations slowly became names to uniquely identify individuals. Being known to a legal system by a name, however, does not mean one's behavior is known. To know one citizen or many, nations like organizations have long used names as the key tracking device, but have needed to use local surveillance by the individual's community to monitor an individual's behavior. As long as communities were relatively immobile, the name-plus-neighbors system worked sufficiently well to allow small, local police units to supplant centuries of military-based internal social control in Europe.[1]

Today, however, the emerging global mobility of the information and terrorism age has changed the parameters of governments' roles in social control, the security needs of Westernized societies, and the privacy of citizens of democratic states. To address the first two factors, some governments have lessened the third: Privacy has routinely been sacrificed for security in a perceived privacy-security trade-off. Today, the firewall of "practical obscurity unless a criminal suspect" has been massively eroded by public authorities handing out private information in the name of public openness, inadvertent community data sharing, technological oversights in releasing data, and even commercial interests and seeking data for private market purposes. The legacy systems of most governments—including state and federal agencies in the United States—have not adapted to the new circumstances, especially the increasing shortcomings of overreliance on unique, persistent identifiers such as names and Social Security numbers (SSNs) when the goal is behavior monitoring. Public information systems need a new regime that will support the changing goals of governments while preserving the privacy that citizens expect in democratic societies.

This chapter explores the roles of names and name equivalents[2] in social control, reviews what kinds of data are being accumulated today in legacy systems, and proposes an alternative focused on the security goal that reduces the likelihood of violating the privacy of ordinary citizens. We focus on the Achilles heel in the government's efforts to both ensure social order and yet minimize the loss of privacy: the continuing reliance on unique identifiers, for example, names or national identity numbers. We conclude with a discussion about a conceptual change to focusing the social-order mission on the behavior of individuals rather than their identities. Such a system variation would be an intermediate step to a world of citizens' control of their private identities while allowing traceable public behavior. It is particularly urgent to consider a different path now as increased interest in biometrics (Alterman, 2003; Barton, Byciuk, Harris, Schumack, & Webster, 2005; Boukhonine, Krotov, & Rupert, 2005; Harris, 2003; Milone, 2001; Zorkadis & Donos, 2004) offers an insidious expansion of unique identifiers. While names and other identifiers can change, identifying an individual with an unchangeable physical attribute can transform transient mistakes into permanent ones. Moreover, if biometrics become casually linked to privately held data, it will be exceptionally difficult to return the genie to the bottle to ensure privacy at even minimal levels within or beyond our borders. The increasing use of radio technologies, such as radio-frequency identification (RFID), further exacerbates privacy and identification issues by sharing data wirelessly (Anonymous, 2006; Juels, 2006; Nisbet, 2004; Pottie, 2004; Roberts, 2006; Soppera & Burbridge, 2005).

Names Work with Less Mobile Populations

Tracking an individual's behavior begins with an identifier uniquely associated with that individual, which then serves as an index that points to a record of violations of required or proscribed behavior; traditionally, the identifier is a label permanently attached to the individual. Given the typical characteristics of social tracking schemes (listed in Table 1),

Table 1. Attributes of social tracking processes

Attribute	Attribute Definition
Direct or Indirect	Tracking is focused on the individual (direct) or on surrogate indicators (indirect), such as a criminal record or inferred crimes.
Obvious or Obscured	Tracking is apparent to those tracked through normal means, such as through sight or hearing, or by a cop on beat (obvious), or tracking is hidden from view, such as through camouflaged, remote cameras (obscured).
Ubiquitous or Selected	Tracking functions are everywhere, such as speed-limit enforcement cameras being at all intersections (ubiquitous), or tracking takes place in some places, such as cameras being only at major street intersections (selected).
Task Related or Membership Related	Tracking focuses on task-specific behaviors, such as monitoring professional activities through licensing requirements (task related), or tracking focuses on monitoring, for example, as practiced in specific religious groups (membership related).
Continuous or Intermittent	Tracking is continuous or is limited to specific periods (intermittent).
Intrusive or Nonintrusive	Tracking requires some accommodation by the observed, such as an electronic monitoring bracelet for suspects freed on bail (intrusive), or tracking requires no accommodation, such as a software-driven, embedded tag turned off simultaneously with purchase (nonintrusive).
Acknowledged or Unacknowledged	Tracking requires the individual's acceptance, such as picking up a name tag at individual events (acknowledged), or tracking is embedded routinely in all organizational activities, for example, by routinely wearing a uniform (unacknowledged).
Scalable or Fixed Data	Tracking data are easily expanded, for example, additional information may be added at will through, say, colors and patterns in clothing choices within a minority group (scalable), or tracking data are restricted to fixed inputs, such as clothing restricted to certain kinds of cloth and color choices (fixed).

name equivalents work well so long as the tracked individuals do not move frequently. In fact, when German government officials sought to prevent a further rise in the terrorism of the 1970s, they mandated extensive name gathering (and in some cases, forced name modifications) as part of the 1993 census for just this purpose—a move that upset and confused many German citizens (Benjamin, 1993). Table 1 lists many of the variations found in social tracking regimes that names or their equivalents embody.

Historically, names—which often encoded clan and familial relationships—served the purpose of labeling individuals relatively well across these characteristics. In small, relatively fixed communities, a single name sufficed for centuries; however, as populations grew and mobility increased, the need for more specificity in individual identity produced second, third, and even fourth names. With further population growth, names were augmented with clans, birth dates, birthplace names, and even occupations. Because individuals rarely moved around, they could not easily forsake their given names and so names served as early biometric equivalents. One could not simply dump the birth name of "John the miller's son" without going a long, dangerous distance from home.

In limited-social-mobility environments with few opportunities to adopt a new personal label, names are effective tracking mechanisms. These labels are usually direct, ubiquitous, and durable. All people have a name given directly by their parents. If the person is socialized into sharing this label on social encounters, the name becomes widely known in small societies. Once a name is taken for granted, it is also usually nonintrusive and generally unacknowledged. Even if initially rooted in an occupation, names eventually become membership-tracking mechanisms; John the Miller has a great-grandson named Joseph Miller, who may well work as an optometrist. In some groups, people carry multiple middle names to allow more distinct identification and familial affiliation, as with the patronymics common in many societies. Moreover, names are durable in the sense that even a person in a coma continues to hold her name; names are thus well suited for use in continuous tracking processes. Because all modern societies are rooted in histories in which names (or their equivalents) worked well for social tracking, they have continued as a popular choice for government tracking to this day—even when they may no longer be appropriate.

Static Name Equivalents Dominate Governmental Systems

Managing the descriptive data defining an individual, society, or subordinate group in modern governance means enormous influence over the social options available to the individuals or groups. For example, being able to categorize who is and is not unemployed has been a longstanding political struggle in the United States. During the political debate surrounding the formation of the Department of Labor at the turn of the 20th century, conservative business elements succeeded in making the new cabinet-level department weak and ineffectual by deliberately keeping the Bureau of Labor Statistics out of the new department. Without the ability to control the lists of who is and is not employed, the department was routinely unable to win legislative battles critical to its institutional and political influence. The labor department is known to be one of the least influential departments of the cabinet (Kimeldorf & Stepan-Norris, 1992). In another example, the small religious party Shas of Israel demanded the Ministry of Education post in any alliance government. In this role, the Shas minister could define the terms under which hundreds of religious Israelis could be classified as students. By consistently labeling Shas members as such, the central government supported them modestly for most of their lives through several simultaneous scholarships. (The leader of the Shas eventually went to jail for fraud for issues related to the practice of counting every three credit hours as full-time religious study, enabling each student to get three scholarships at once [Amado, 2001]. The scholarships, however, did not cease following the leader's arrest—the labels stuck.)

The Emergence of Name Equivalents

As the size of societies has grown, however, name equivalents have emerged to augment and sometimes replace birth names. Surnames alone are too variable in spelling and too often duplicated to support effective tracking in modern mass societies with redistributive functions like welfare or national income taxes. Furthermore, names in immigrant communities are often simply what the new immigrant says it is. With easily changed or lost papers, names as identifiers become less and less accurate. Names easily change at borders—often shortly before border officers ask for one. Linguistic differences mean names can be misunderstood, profoundly affecting the lives and social acceptance of those whose names are altered. The relative immutability of names in the Westernized world contrasts sharply with the fluidity of names from Middle Eastern cultures. While family and clan associations are not usually flexible, what a person calls himself varies according to his internal version of his life's progress (Peteet, 2005). This tendency is not a problem in relatively immobile societies because other members of the society can identify the individual and recall his various names if asked.

In the globalizing, more mobile world, however, name fluidity causes routine identification difficulties for passport officers, a difficulty exploited by those operating covertly across borders. American families have many stories of a grandmother's familial name changing upon entry into the United States as an immigrant. In an informal interview, a colleague explained that an Ellis Island immigration official changed his Catholic Hungarian family name into a historically Jewish name in the 1930s. The family story is that the officer could not pronounce the immigrants' birth name and generously wanted to help the family out by giving them a good "American" name. In so doing, the officer changed the way the whole family was accepted within New York City's largely Jewish Bronx borough for the next two generations.

Altering a name is often enough more than a mistake or a convenience. In modernizing nations, changing clan or religious affiliations can mean the difference between survival and starvation. In several small eastern Indian states bordering the largely Muslim country of Bangladesh, a remarkable preponderance of Christians exists in the data files of public authorities. Being Muslim in Hindu India and an illegal immigrant is a double burden for those filtering into India from Bangladesh for economic survival. A Bangladeshi Muslim could not easily pass as Hindu but could pass as a Christian, the now dominant political elite in these smaller states. Hence, illegal immigrants declare themselves to be Christians to be accepted. They must also change their Muslim given names, not only for acceptance but also for physical safety since the Koran authorizes all Muslims to kill at will any Muslim who has converted. The story becomes more complex if they attempt to go home for a visit. They need to seamlessly shed the new names and trappings for a while, and then resume them without calling attention to themselves once back in India. The frequent name changes can cause a good deal of confusion for both Indian and Bangladeshi border authorities (Indian Press Trust, 2005).

National Identification Numbers

In the middle of the past century, for most of the Westernized world, identity cards with names and often with numbers (the Social Security number in the United States) began to be employed to identify a country's nationals, not always for the best of purposes. The first nationwide national identity number in France was issued by the Vichy government in 1940 to help identify Jewish French citizens intended for deportation to Germany (Wikipedia, 2006). In smaller Western nations with a relatively homogeneous and immobile population, such as Sweden, the national identification number (*personnummer*) encodes an individual's date of birth and sex along with a digit sequence and a checksum (Riksskatteverket, 1999). In the much larger, immigrant-fed, decentralized United States, SSNs are allocated by region and so the number reveals (indirectly) the postal code of the application's origin, which might or not be the birthplace of the applicant. In both instances, the number is meant to provide singularly identifying tracers, which, when linked to other distinctive data, provide the authorities with the information needed to make allocation decisions across the society. For example, Britain issued, in both world wars, national identity cards to ease rationing that were then discarded after the wars. National identification numbers are not universal but, along with the biometric passport, are gaining popularity due to the terrorism threat from visiting or immigrant populations. Even Britain will have a mandatory national identity card for all its citizens by 2013; today, 21 of 24 members of the European Union have some form of identity card, though not all centrally managed (Wikipedia).

Collection of Privacy-Sensitive Data by the United States Government

The use of a name and an associated number as tracking mechanisms has grown across agencies and nations. In the United States, for example, it is difficult to estimate the total number of databases across state and federal agencies that collect and use privacy-sensitive data on citizens, almost always associated with a name and/or the ubiquitous Social Security number. The 1974 Privacy Act (5 U.S.C. 552a, Public Law 93-579) requires agencies of the federal government to report annually on what files they maintain that may have privacy implications and the rules by which the data was accumulated. Searching online through http://www.fedgov.gov, however, it is difficult to find more than a smattering of these agency filings. Announcements of these file holdings must be published in the Federal Register as Privacy Act notices; since 1995, the notices have also been published online at the U.S. Government Printing Office's Web site (*GPOAccess*, http://www.gpoaccess.gov/privacyact/index.html; U.S. Government Printing Office, 2004). However, visiting *GPOAccess* does not easily provide compilations across agencies; rather, the entries tend to be annual, dated, and inconsistent—several of the links to agency files were broken with no alternative link provided.

Despite the difficulty in accessing these lists, government agencies do host a wide variety of privacy-sensitive databases, some of which are listed online. The Department of Defense listed in 2003 that it had 166 such databases across 19 subagencies, including each of the major services. The latest list available for the Department of the Interior was published in 2001 and listed 95 privacy-sensitive databases. Other agencies with compilations noted are large and small. For the year 2002, the Tennessee Valley Authority said it had 29 such databases while the Railroad Retirement Board said it had 33 privacy-sensitive files in 2003. The oldest list came from the Fish and Wildlife Service, dated 1992, which noted roughly 40 distinct databases, while a list for the Commodities Futures Trading Commission dated 1999 also listed about 40 privacy-sensitive collections. Many lists have no discernible year and are quite large. For example, the Department of Interior claims 191 privacy-sensitive collections, but there is no year given. Similarly, the Office of Personnel Management notes it has 39 privacy-related databases, but no year is provided (U.S. Government Printing Office, 2004).

Complicating an effort to estimate the breadth of the holdings in the federal government are the waivers and exemptions taken by the agencies. Very often, a posted Privacy Act notice concerns the intent of the agency to not discuss the contents of a database. For example, the Transportation Security Administration (TSA) notes that it had roughly 15 privacy-sensitive databases (U.S. Government Printing Office, 2004), including several it had just added. However, the bulk of the discussion explained why the TSA did not list what private data were actually collected in that database for reasons of national security. It is instructive to note that not only are these filings difficult to locate and document conveniently online, but the most precisely dated and cumulated lists are older, predating the current post-9/11, security-conscious policy environment.[3]

Another way to gain an appreciation of the magnitude of the growing personal data banks held by public authorities for a variety of reasons is to look at databases in terms of communities. For example, in the days of paper, applicants for jobs, grants, or consulting contracts would provide personal data that eventually were discarded to save space. Those not hired, awarded, or selected would certainly have their data cleansed from the system periodically. That an application is rejected today is no guarantee that all that data did not go into several general digital files easily stored and shared inside an agency or among agencies. In Congressional hearings in late 2004, for example, the Congress was told that 80% of all applicants for positions in the federal government were submitting applications online, directly into databases.

Being a mere contractor is no protection. In the same hearing, the representative of the National Science Foundation (NSF) said 100% of grant proposals were now digitally submitted and processed, including not only personal data on the principal investigating professors but also on associated scholars, consultants, and often graduate students. For job applicants and individuals associated with an NSF grant application, the physical firewall created by paper applications and dispersed filing cabinets accumulating personal data has vanished; moreover, there is no particular information on where else that data might migrate to or for what other purposes that exceptionally personal data might be used (U.S. House of Representatives, 108th Congress, 2004).

Regional Government Data Collection

State or provincial governments make the threats more complex as well. In the United States, even if state governments have only a fraction of the inventory of privacy-sensitive databases held by the central government, there are still another 50 complex, multilevel entities collecting data on individuals who reside, work, or request any kind of service in the state. Beyond the state level are still more counties, cities, and even public-utility districts. Today in the city of Tucson, Arizona, a local city policy to combat the methamphetamine epidemic now requires ordinary purchasers of any over–the-counter medicine containing pseudoephedrine to fill out a form giving personal identifying data. Who receives, transcribes, stores, and then uses that data is not clear, but no purchase is authorized unless the data are surrendered. This sign-in-before-you-buy policy is spreading across the United States (Colker, 2005) in an otherwise admirable effort to stop the methamphetamine epidemic sweeping the country's teenagers, but it squarely and unrestrictedly places otherwise private data in the unprotected files of hundreds of pharmacies across the same locations.

In a Mobile, Digitized World, "Practical Obscurity" No Longer Saves Privacy

For most of the 20th century, the effort to amass privacy-sensitive data was expensive and time consuming. Citizens and public authorities endured fragmented policies and paper-based or proprietary databases dispersed across states, counties, jurisdictions, and departments within agencies. Policies were often inconsistent across states and agencies, even when data collection was driven by national initiatives mandating consistency such as aid-to-dependent-children programs or the sharing of gang-related data across police agencies. Furthermore, increasing mobility exacerbated the accuracy of social tracking by names or name equivalents; population movements tended to reinforce the practical difficulties of accumulating data on many people at once. Privacy was thus assured simply because people could move easily but personally revealing data about them could not.[4]

Two global trends converged in the late 1990s to make legacy systems less useful while simultaneously trading off privacy without a commensurate, but expected, improvement in security. First, the widespread and enthusiastic embrace of easily exchanged digital data leapfrogged the normal processes of social and legal adjustment. Over just a few years, public authorities across the Westernized world moved rapidly from cumbersome paper or platform-bound legacy files to the easily distributed and accessible data hoards, seeking unprecedented benefits and unleashing unforeseen threats. Second, a now clearly globalizing social movement has emerged to pursue an ancient "war of a thousand strikes" from within the normal operations of increasingly heterogeneous Westernized societies (writ large) and the United States and allies in particular. In response, the United States and other Westernized nations are moving to restrict privacy to find terrorists. The first government trend inadvertently sacrifices privacy; the second deliberately reduces personal privacy in hopes of preempting catastrophic violence in Western communities.

Abuse of Personal Information

Beyond governments and terrorists are commercial entities in corporations and criminals, all seeking to gain some monetary advantage from an individual's personal information. Today the commercial world is collecting data at a previously unimaginable rate and making the results available to anyone willing to pay for access. Credit reports are regularly sold to businesses but little is done to control what happens to those reports once delivered. In 2004, investigators for the Social Security Administration (SSA) arrested a former manager of a local furniture store using stolen credit reports from the store's credit clients to conduct elaborate frauds involving consumer loans worth hundreds of thousands of dollars, according to the testimony of Patrick P. O'Carroll of the SSA (U.S. House of Representatives, 2004).

However, it is the unprecedented presence of American public records on the Web, open not only to citizens but also to the wider world, that offers criminal access into the private lives, finances, and future prospects of individuals in the United States. The easy complacency borne of an era of fragmented, hard-to-find databases made acceptable this rush to accelerate connections across previously unconnected databases with private citizen information. It tends to come as a shock to most citizens when they find out just how easily they are completely identifiable by databases now offered openly. For example, the abused wife who bought her own house 5 years ago might find today that her purchase information, including her new home address, is now online courtesy of her local county tax assessor. In the intervening 5 years, county records were neatly archived digitally and published with nearly no effort to protect privacy since they were, often by law, "public" records. For most of these newly published county records, there is no log of who reads them, who copies the data, or even who spends hours searching and collating hundreds of those records. Instead, these data are simply tossed out in cyberspace for human and software collectors to gather, sell, or use for any purpose.

Identity Theft

The result for the United States is an extraordinary rate of identity theft, phishing attacks (whereby victims are lured into entering personal information on online forms that masquerade as legitimate enterprises), and other crimes enabled by this wealth of private information. In 2004, investigators for the Social Security Administration were startled to see an advertisement on an online auction Web site offering about 250,000 Social Security numbers for sale. The individual had pilfered them from the files of a large state university that used Social Security numbers to track students, staff, and contractors (U.S. House of Representatives, 2004). In such thefts, the costs are rarely calculated, but across all categories, it is millions of dollars and senseless losses in time and lives disrupted.

Balancing Privacy and Social Control

The difficulty for a government in addressing the situation is defining the correct balance in ensuring social control. How far should public authorities be able to intrude into private identity in order to enforce rules in a mass society? What compensating policies and sanctions are required to ensure a remaining firewall of individual protections of privacy-sensitive data? Governments do need to accumulate data on citizens to analyze mass trends to set the nation on better paths for future prosperity. Security forces do need to track down violators of the social order quickly and firmly, within the rule of law. Businesses do need a way to anticipate consumer wants and prepare in advance to meet those evolutions in needs and desires. However, the digital frontier transferred the convenience of access from the small pool of visitors to local public-records archives to the world beyond our borders without pause to consider what to protect and what was not necessary to share with, say, sociopaths. A good example of this lesson not learned is the MySpace Internet community (http://www.myspace.com) that encouraged youngsters to meet online, only to discover pedophiles took the opportunity as well. The system is not designed to keep these predators from lying about their age and luring credulous youngsters into private chats (Rawe, 2006). As a result, the openness and convenience of MySpace is now strongly constrained by the Web site owners seeking protection from lawsuits and from parents now concerned about a previously unheard of vulnerability for their children.

The abrupt rise in Web-available data has parallels with the historical shift in the United States from the frontier communities of the early Tennessee Valley, where surviving depended on skill and luck, to the frontier communities of gold-rush California, where schools, prostitutes, opium dens, murderous fights, libraries, and duplicitous storekeepers sat beside one another. The difference is ease of access to one's private life. Now both governments and criminals across layers and layers of unaccounted-for name-based intermediary and invisible databases can reach out to touch or disrupt citizens. E-government (known as i-government in Europe) can be a wonder drug for central government's effectiveness and efficiency in delivering services while also a disaster for both privacy and security if not regulated legally and physically (with validation and appeal processes) from the outset.

Track Actions, Not Labels

"[A]nonymity and pseudonymity are not all-or-nothing qualities but can be achieved in degrees and through layers of cloaking" (Nissenbaum, 1999, p. 144).

In a mobile, digitized, globalizing world, neither security nor privacy is ensured by simply linking legacy systems more efficiently. Names and labels say only who someone is, not what they are doing; in other words, they convey status but not action. In a world that demands preemption of bad behaviors for security but also asks for the protection of privacy, names are no longer good even as a primary point of reference. Security threats come from behaviors, and names are poor behavior-tracking mechanisms. They are only useful if public security forces already using other behavioral monitors separately monitor individual actions. However, names, national identity numbers, birth dates, and addresses freely handed out in the wild function well as attractors of criminal behaviors. Thus, using identity as a first choice in tracking terrorist behavior fails to pinpoint the bad actors for security purposes but handily exposes citizens' private lives to other kinds of criminals. If security requires a focus on behavior and privacy requires a firewall around critical personal identities, then a different conceptual approach is necessary, one that emphasizes the tracking of undesirable behavior over tracking a name.

To balance security and privacy but retain the tradition and advantages of names, one needs not an absolute name but a conditional one—one that becomes a true name only upon triggering behavioral signals of public interest. In short, the alternative behavior-based policy would produce identifying information much like names but would mask the real name, birthplace and birth date, national identity number, home address, and maiden name for the sake of privacy. This new identifier (a pseudonym, rather than a true name) would, like names in the less mobile era, be direct, ubiquitous, and durably connected to an individual's true identity. With the issue of a new identifier, we could enable a new tracking mechanism that would be (usually) nonintrusive, unacknowledged, membership specific, and scalable. The relative obviousness of the mechanism would vary according to the circumstances. A practical alternative to the current regime would replace a name for the purposes of public tracking until the name becomes necessary for public threat reduction.

A system of traceable pseudonymity seems the most promising alternative if it can be seamlessly inserted into the existing structure of societal interactions that currently provides all this freely floating data. A looming current difficulty is the growing use of uniquely identifying biometrics. Shortly after the September 11th attacks, a car-loan agency was reported to have required fingerprints, a critical biometric, from customers before they were allowed to rent cars (Scheeres, 2001). Disney theme parks—purportedly the happiest places on earth—now require finger scanning for park admission (Local 6 News, 2005). An alternative behavior-based system of monitoring is urgently needed before all these loosely secured databases become linked equally freely to immutable individual biometric data carried around on, say, easily read credit-card magnetic strips.

The behavior-identity knowledge (BIK) system (Demchak & Fenstermacher, 2004a, 2004b, 2005) begins conceptually with the notion that private data are kept by the owning individual and masked for public consumption by a trusted third party (TTP). The TTP is envisioned as a nonprofit, networked group of institutions that use technical and social engineering to secure critical data. The TTP would maintain the linkage between an individual's masked identity (associated with one or more pseudonyms) and the individual's true identity. The system would depend on a public-key infrastructure (Adams & Lloyd, 1999) or PKI. A PKI builds on the notion of public-key cryptography (Salomaa, 1996), which uses two complementary, numeric keys to encrypt and decrypt information. One key is known to all participants and so is labeled the public key; the second key is known only to its individual owner (and, in the BIK world, to the TTP) and so is called the private key. A system can encrypt data with either the public or private key and the encrypted data can then only be decrypted with the complementary key. Chaum (1992), Chaum and Evertse (1987), and Chaum (1981, 1987) describe in detail how a PKI can be used to support data privacy in a BIK-like world.

In a BIK world, a person would access a secure portal, perhaps at his or her local motor-vehicle office, to use a private key—known only to the person and the TTP—to create a masked identity, which is then used for all public interactions such as making purchases or presenting credentials. The needed information is encoded on a smart card, a more secure version of the widely used magnetic-stripe technology currently used for credit and debit cards. Smart cards have been studied extensively for their advantages in protecting personal information while supporting more advanced transactional infrastructures (Bailey & Caidi, 2005; Banaszak & Rodziewicz, 2004; Bolchini & Schreiber, 2002; Chan, Chanson, Ieong, & Pang, 2000; Fan, Chan, & Zhang, 2005; Yang, Han, Bao, & Deng, 2004). Biometrics are also encrypted, used only to verify that the individual offering the smart card is the individual

Figure 1. BIK framework

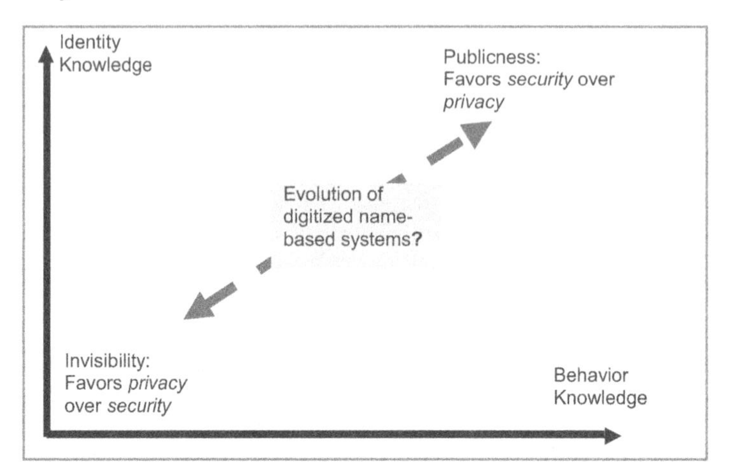

behind the masked identity; biometric data are never stored away from the TTP or on a particular card. The masked data are then collected just as data are today with an additional data storage site at the TTP. This is further made anonymous for sale to commercial entities who receive only aggregate data, saving them the need to pursue databases at the individual level and preserving the individual from the tricks of the commercial world to winnow out personal data; Figure 1 portrays this alternative.

In this system, only patterns of behavior, not names, elevate a masked identity out of the crowd for further inspection by authorities. Under legally prescribed conditions, authorities then must validate the accuracy of the linkage between the behavior and the masked identity and then request of the TTP information on the true identity of the citizen. In this process, much stricter legislation on the collection, storing, and use of true personal data will be necessary.

Over time, the current static identifying data of name-equivalent legacy systems become useless as the world changes but legacy data does not. For example, because Americans move on average every 2 years; the John Doe who was living on 1st Street may or may not be the 264735BBB living in the neighborhood now. The identity data now in the wild become less and less useful as the masked identity slowly becomes disconnected from previous uncontrolled databases commercially and publicly. Behavior data, however, are more easily available to both government and legitimate commercial entities, with true identities hidden behind technical, institutional, legal, validation, and appeal requirements. Figure 2 demonstrates how BIK blends into the current credit-card transaction infrastructure.

While the BIK model does not solve all the issues of privacy and security trade-offs, it has in principle the advantages of fitting relatively seamlessly into existing systems. It could well offer an intermediate step to a world of citizen control of their private identities while allowing traceable public behavior to ensure security against terrorism and other globalizing threats.

Figure 2. Living in a BIK world

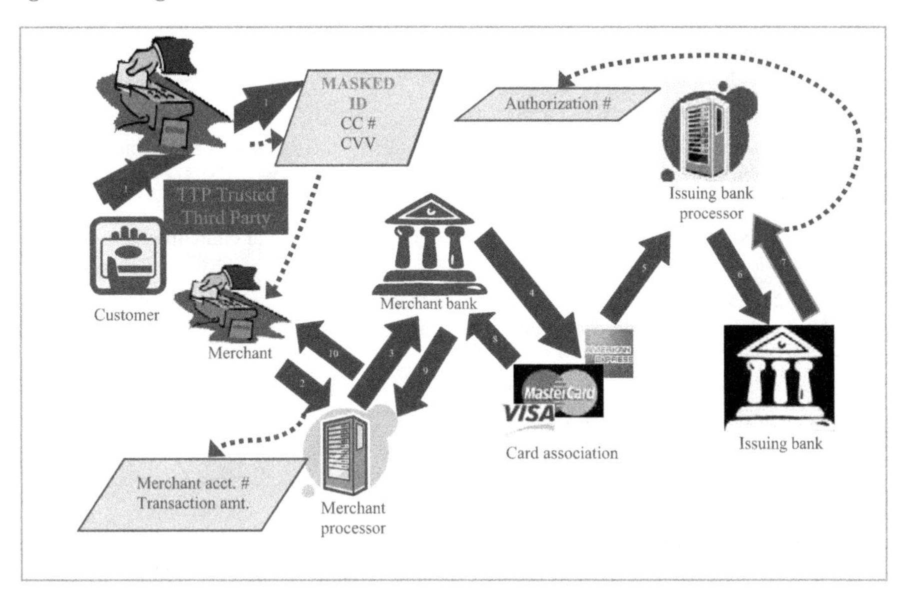

References

Adams, C., & Lloyd, S. (1999). *Understanding public-key infrastructure: Concepts, standards, and deployment considerations*. Macmillan Technical Publishing.

Alterman, A. (2003). "A piece of yourself": Ethical issues in biometric identification. *Ethics and Information Technology, 5*(3), 139.

Amado, R. (2001). Checks, balances, and appointments in the public service: Israeli experience in comparative perspective. *Public Administration Review, 61*(5), 569-584.

Anonymous. (2006). MIT and HID to address RFID privacy concerns. *Sdm, 36*(5), 40.

Bailey, S. G. M., & Caidi, N. (2005). How much is too little? Privacy and smart cards in Hong Kong and Ontario. *Journal of Information Science, 31*(5), 354-364.

Banaszak, B., & Rodziewicz, K. (2004). Trust and security: Digital citizen cards in Poland. *Electronic Government, 3183*, 342-346.

Barton, B., Byciuk, S., Harris, C., Schumack, D., & Webster, K. (2005). The emerging cyber risks of biometrics. *Risk Management, 52*(10), 26.

Bayley, D. H. (1975). The police and political development in Europe. In C. Tilly (Ed.), *The formation of national states in Western Europe* (pp. 328-379). Princeton, NJ: Princeton University Press.

Benjamin, D. (1993, July 15). What's in a name greatly interests German officialdom. *Wall Street Journal*, p. 1.

Bolchini, C., & Schreiber, F. A. (2002). Smart card embedded information systems: A methodology for privacy oriented architectural design. *Data & Knowledge Engineering, 41*(2-3), 159-182.

Boukhonine, S., Krotov, V., & Rupert, B. (2005). Future security approaches and biometrics. *Communications of the Association for Information Systems, 16*, 1.

Chan, P. N., Chanson, S. T., Ieong, R., & Pang, J. (2000). Smart card payment over Internet with privacy protection. *Smart Card Research and Applications, 1820*, 98-104.

Chaum, D. (1992). Achieving electronic privacy. *Scientific American, 267*(2), 96-101.

Chaum, D., & Evertse, J. H. (1987). A secure and privacy-protecting protocol for transmitting personal information between organizations. In *Advances in Cryptology—CRYPTO '86: Proceedings* (LNCS 263, pp. 118-167). Springer-Verlag.

Chaum, D. L. (1981). Untraceable electronic mail, return addresses, and digital pseudonyms. *Communications of the ACM, 24*(2), 84-90.

Chaum, D. L. (1987). *Security without identification: Card computers to make big brother obsolete.* Retrieved May 10, 2006, from http://www.chaum.com/articles/Security_Wthout_Identification.htm

Colker, A. C. (2005). Restricting the sale of pseudoephedrine to prevent methamphetamine production. *NCSL Legisbrief, 13*(7), 1-2.

Demchak, C. C., & Fenstermacher, K. D. (2004a). *Balancing security and privacy in the 21st century.* Paper presented at the Second Conference on Intelligence and Security Informatics, Tucson, AZ.

Demchak, C. C., & Fenstermacher, K. D. (2004b). Balancing security and privacy in the information and terrorism age: Distinguishing behavior from identity institutionally and technologically. *The Forum, 2*(2), Article 6.

Demchak, C. C., & Fenstermacher, K. D. (2005, September). *Ensuring security, enhancing privacy.* Paper presented at the Eighth Public Management Research Conference, Los Angeles.

Fan, C. I., Chan, Y. C., & Zhang, Z. K. (2005). Robust remote authentication scheme with smart cards. *Computers & Security, 24*(8), 619-628.

Harris, S. (2003). Biometrics need a measure of security. *Government Executive, 35*(9), 70.

Indian Press Trust. (2005). *"Abnormally high" growth of Christians in NE.* Retrieved August 1, 2006, from http://www.expressindia.com/fullstory.php?newsid=41957

Juels, A. (2006). RFID security and privacy: A research survey. *IEEE Journal on Selected Areas in Communications, 24*(2), 381-394.

Kimeldorf, H., & Stepan-Norris, J. (1992). Historical studies of labor movements in the United States. *Annual Review of Sociology, 18*, 495-517.

Local 6 News. (2005). *Finger scanning at Disney parks causes concern.* Retrieved August 1, 2006, from http://www.local6.com/news/4724689/detail.html

Milone, M. G. (2001). Biometric surveillance: Searching for identity. *The Business Lawyer, 57*(1), 497.

Nisbet, N. (2004). Resisting surveillance: Identity and implantable microchips. *Leonardo, 37*(3), 210-214.

Nissenbaum, H. (1999). The meaning of anonymity in an information age. *The Information Society, 15*(2), 141-144.

Peteet, J. (2005). Words as interventions: Naming in Palestine-Israel conflict. *Third World Quarterly, 26*(1), 153-172.

Pottie, G. J. (2004). Privacy in the global e-village. *Communications of the ACM, 47*(2), 21-23.

Rawe, J. (2006). *How safe is MySpace?* Retrieved from http://www.time.com/time/archive/preview/0,10987,1207808,00.html

Riksskatteverket. (1999). *Personnumer (SKV 704)*. Retrieved July 31, 2006, from http://skatteverket.se/download/18.b7f2d0103e5e9ecb0800074/70407.pdf

Roberts, C. M. (2006). Radio frequency identification (RFID). *Computers & Security, 25*(1), 18-26.

Salomaa, A. (1996). *Public-key cryptography*. Springer.

Scheeres, J. (2001). *No thumbprint, no rental car*. Retrieved July 30, 2006, from http://www.wired.com/news/privacy/0,1848,48552,00.html

Soppera, A., & Burbridge, T. (2005). Wireless identification privacy and security. *BT Technology Journal, 23*(4), 54.

U.S. Government Printing Office. (2004). *Privacy Act issuances: About*. Retrieved July 31, 2006, from http://www.gpoaccess.gov/privacyact/about.html

U.S. House of Representatives, 108th Congress. (2004). *Electronic government*.

Wikipedia. (2006). *Identity document*. Retrieved from http://en.wikipedia.org/wiki/Identity_document

Yang, Y. J., Han, X. X., Bao, F., & Deng, R. H. (2004). A smart-card-enabled privacy preserving e-prescription system. *IEEE Transactions on Information Technology in Biomedicine, 8*(1), 47-58.

Zorkadis, V., & Donos, P. (2004). On biometrics-based authentication and identification from a privacy-protection perspective: Deriving privacy-enhancing requirements. *Information Management & Computer Security, 12*(1), 125.

Endnotes

[1] It is rarely recognized how recently most Westernized states were internally stable enough to create police forces rather than use military units to control the internal peace. The 18th century saw the emergence of the community police force while military units focused more on outside threats. (Bayley, 1975)

[2] Name equivalents are also known as “isonames” or by the Latin, “isonym.”

[3] The U.S. Internal Revenue Service is excluded from this discussion because it has long been known that taxpayer data is routinely transcribed from paper to digital files and the advent of online filing merely transfers the locus of data entry. Furthermore, regulations regarding the use of that data are much better established than the growing number of other databases scattered across other agencies.

[4] The scope of this piece does not allow a comparison of the relative mobility of European citizens versus Americans and its impact on privacy, but a piece linking privacy, mobility, and convergence or divergence in U.S.-Europe data and privacy policies is planned for the future.

Chapter VI

E-Government: An Overview

Shannon Howle Schelin, University of North Carolina at Chapel Hill, USA

Abstract

Information technology has fundamentally altered many aspects of daily life, including interactions with the government. The role of the Internet continues to increase as more citizens use it to find pertinent information, purchase goods and services, and participate in virtual communities. By capitalizing on the Internet revolution, governments can create new channels of communication and new methods for participation via e-government. The changing environment, coupled with citizen and business demands, encourages government involvement in e-government initiatives and related uses of information technologies. Clearly, the role of information technology in the public sector has changed rapidly over the past decade. The computer systems that were once a luxury investment for wealthy states and local governments are now supporting almost every function of local government. In virtually all local governments across the nation, information-technology investments are becoming an increasingly important area of attention for elected officials and administrative leadership alike.

Technology-Enabled Government

The public sector has made incredible strides in technology over the past decade. The investments in IT have brought the many states, counties, and municipalities into standing with other leading private-sector companies across the nation. In virtually every governmental jurisdiction, information technology is playing a vital role in each department and function of the organization. In fact, while information technology only comprised between 1 and 3% of the general fund budget, on average, the technology investments support approximately 98% of the work conducted by governments. In order to assist the public sector in moving beyond the status quo and leveraging technology as a means of delivering more efficient and effective services, as well as to maintain and gain a competitive economic-development advantage, it is important to recognize the positive technological advancements that have made significant impact on service delivery over the past 6 years. By establishing this strong technological foundation, many governments are poised to reap the rewards associated with greater investment in technology.

The scope of technology in governments across the United States can be grouped into three basic categories: infrastructure, hardware, and applications. The first two categories have seen a plethora of growth as states, counties, and municipalities have installed various communication media to assist with connecting disparate locations. One common infrastructure and hardware solution across many states has been the installation of fiber between government facilities or around jurisdictions. Other governments have chosen to use wireless technologies to connect remote locations without physically laying fiber. As service becomes less location dependent, it is essential for government employees and off-site departments to have high-quality, secured access to records and databases. Fiber-ring and secured wireless initiatives have provided a critical first step to this effort.

Another major technological impact has been generated by the powerful communication medium of e-mail. Using e-mail technologies as collaborative tools, employees are able to generate significant cost savings by the reduction of travel time for communication purposes. E-mail and other electronic communications media, such as instant messaging, allow almost instantaneous communication across departments, jurisdictions, and states and have been credited with quicker problem resolution, increased data sharing, and easier information and personnel access. Furthermore, the use of e-mail has significantly improved citizen relationships by allowing another channel of communication.

The third important technology investment is a World Wide Web presence. By creating and maintaining an outstanding Web site, cities and counties have been on the cutting edge of citizen, visitor, and business engagement. In fact, many economic-development experts indicate that a high-quality Web presence is the most important tool in a local government's development efforts (Horrigan, 2003). The National Governor's Association and the National Conference of State Legislatures offer publications concerning the value of technology investments and impact on economic development. Furthermore, Web sites provide a unique and timely vehicle for communication with citizens and visitors alike. It is an essential feature in any local government.

A final critical technological impact has been garnered through the investment in geographical information systems (GISs). The use of GIS greatly enhanced the work of local governments. For example, GIS has improved internal government efforts by aiding in meter and pipe location. It has improved the efforts of planning departments by enabling staff members to access a comprehensive database that spatially represents areas and items of interest. In addition, GIS is one of the most productive and useful tools in technology revolution. The power and breadth of the software is being used in crime fighting, fire fighting, and countless other ways.

Most public-sector entities have seen an incredible amount of change over the past decade. Many have progressed from nontechnical to successful, technology-friendly governments with many of the investments previously mentioned. However, some governments have only recently begun to invest or consider investing in technology. These smaller or more financially challenged governments have faced a variety of issues but have created unique opportunities to leverage existing infrastructure and assets.

The term *digital divide* was originally coined to describe the division between economic groups regarding access to information technology. However, there is another type of digital divide facing governments. The majority of rural counties and municipalities in many states are experiencing significant economic hardships. These local governments are strained for the basic resources and investments in information technology are not deemed mission critical. Therefore, the gap between the technological haves and have-nots in local governments is widening.

One interesting solution to this governmental digital divide is being pioneered in small communities throughout the nation. The use of cross-boundary collaboration has become a best practice among economically strained local governments. In the collaborative model, local governments partner together in an attempt to leverage demand aggregation, economies of scale, and staff capacity enhancement in the technological arena. For example, demand aggregation can occur when a county offers a centralized procurement process that includes its municipalities in order to increase the overall value of the bid or order while enhancing the negotiating power of the governments. In order to capitalize on economies of scale, counties may seek to share IT staff on a rotational model, which allows for increased technical capacity due to greater monetary incentives.

The current economic stress facing the states mandates that new models be employed in order to address the gaps in service delivery, and cross-boundary collaboration is one such model.

There are numerous ways to invest in public-sector information technology. Some of the most common investments are focused on electronic communications and department-specific applications. However, new cross-departmental, and even cross-jurisdictional, efforts are beginning to emerge. These efforts follow the trends in the private, federal, and state sectors and create greater value for the strategic investments.

Trends

A variety of management trends have occurred in the last few years related to information-technology adoption and implementation in the public sector. This section highlights some of the most critical trends, including the advent of e-government and the move to enterprise-wide technology efforts. Both of these trends offer significant advantages to citizens, businesses, employees, and visitors. In addition, the trends have proven to generate cost savings, increased efficiencies, and greater effectiveness in a variety of cases. The final trend explored is the advent of the chief information officer (CIO) and the increasing professionalism of the role of information-technology staff.

E-Government

E-government (electronic government) has become a buzzword in local, state, and federal government. The era of e-commerce and e-business began with the widespread adoption of the Internet in the mid-1990s and today many citizens expect the same responsiveness and access to government services as found in the private sector. Not only are citizens looking for improved ways to interact with the government, elected officials demand improved services to enhance their legacies. Competitive governments do not want to be seen as slow adopters of e-government (Sprecher, 2000).

E-government has become a mainstay in local, state, and federal government. According to the 2004 International City and County Managers Association (ICMA) e-government survey, over 91% of municipalities with populations larger than 2,500 have Web sites, compared to 73% in 2002. In North Carolina local governments, the Center for Public Technology has found that over 80% of all local governments have Web sites and more governments are developing a Web presence each month. Additionally, the 2003 Pew Internet and American Life Project (Horrigan, 2003) indicates that 77% (97 million people) of American Internet users have accessed at least one governmental Web site. Not only are citizens looking for improved ways to interact with the government, but elected officials demand improved services to enhance their legacies.

Although there is widespread interest in the topic, e-government lacks a consistent, widely accepted definition. It is often related to revolutionizing the business of government through the use of information technology, particularly Web-based technologies, that improve internal and external processes, efficiencies, and service deliveries. The American Society for Public Administration (ASPA) and the United Nations Division for Public Economics and Public Administration (UNDPEPA) have defined e-government as "utilizing the Internet and the world wide web for delivering government information and services to citizens" (UN & ASPA, 2001). Based on this working definition of e-government, this chapter seeks to examine the theoretical constructs, historical premises, and associated typologies of e-government.

Historical Premises

E-government has evolved from the information-technology revolution. Information technology enables new methods of production, increases the flow and accuracy of information, and may even replace traditional standard operating procedures (Landsbergen & Wolken, 2001). Information technology in government has long been acknowledged as a method for improving efficiency and communication (Kraemer & King, 1977; Norris & Kraemer, 1996). However, until the advent of the Internet, the use of technology in government primarily dealt with the batch processing of mass transactions using mainframe computers. Now, IT developments such as electronic mail (e-mail) have changed interpersonal communications to eliminate the constraints of geography, space, and time with profound organizational consequences (Rahm, 1999). The ability to buy and sell goods and services via the Internet has led to new private-sector industries, constituting a new business model that the public sector now seeks to emulate. In addition, IT has promoted globalization, which also changes the environment within which public agencies function (Kettl, 2001).

The main concerns of e-government now focus not only on the electronic dissemination of public information arising from traditional agency functions, but even more on reinventing agency processes to fully exploit the potential of information technology. As Fountain (2001) has noted, the reinvention process requires overcoming the rigidities and limits of traditional bureaucratic forms. Specific objectives may include the centralization of public data and the improvement of internal processes and communications (Alexander & Grubbs, 1998).

In detailing the history of e-government, it is important to note the distinctions between public-sector usage of information technology and e-government initiatives. Although there is a distinction between public-sector information technology and e-government, they are often interdependent and difficult to quantifiably separate, but e-government is that subset of public information technology that involves the delivery of government services and information to citizens, as defined by UN and ASPA (2001). This delivery of services and information also involves the integration of government networks and databases to allow for cross-agency communication and interaction, which is an internal technology application (Moon, 2002).

One of the first comprehensive visions of e-government is found in the 1993 National Performance Review report, *Creating a Government that Works Better and Costs Less: Reengineering Through Information Technology* (Gore, 1993; Kim & Wolff, 1994). This report laid the groundwork for new customer- and client-oriented ways for agencies to engage citizens via technology, involving both improved agency processes and improved methods of delivery. Most reinventing-government literature has cited the need to rely on information technology to improve citizen-centric government services (Kettl, 1998; Osborne & Gaebler, 1992). It is common for current works to refer to the symbiotic relationship between the earlier reinventing-government movement and the current e-government movement (Ho, 2002; Scavo & Shi, 2001). Although the two movements are conjoined, the prospects are that the focus of public administration on e-government will endure for the foreseeable future, outlasting the reinventing-government movement.

The 1995 amendment of the 1980 Paperwork Reduction Act (PRA) was another important milestone in the history of e-government. This amendment offered specific policies on issues associated with managing electronic information, including the establishment of standards, mandates for cross-agency information-technology initiatives, and technology investment guidelines (Relyea, 2001). By outlining guidelines for information technology, the amended PRA solidified the government's commitment to improving citizen services via new channels based on technology. Several other pieces of federal legislation focusing on information technology and e-government followed the PRA, including the Electronic Freedom of Information Act (EFOIA) Amendment of 1996.

The 1996 Electronic Freedom of Information Act amendment added a new level of clarity to the issue of electronic records. This amendment extended the right of citizens to access executive agency records to include access to electronic formats and online information (Relyea, 2001). EFOIA also extended the time limit from 10 to 20 days to prohibit the use of agency backlogs as an excuse for noncompliance with information requests (Hammitt, 1999). The 1996 Clinger-Cohen Act further heightened the role of information technology in government. It established a CIO in every agency, making agencies responsible for developing an IT plan. Later, when e-government became a priority, the existence of the CIO strategic planning structure was an important element facilitating e-government implementation at the federal level.

Also in 1996, Congress passed the Personal Responsibility and Work Opportunity Reconciliation Act (PRWORA). This act, also known as the Welfare Reform Act, represented one of the first national pushes to incorporate the rhetoric of e-government with the routine services of agencies, specifically the administration of Temporary Aid to Needy Families (TANF). The act required interagency, interstate, and intergovernmental coordination of information-technology systems to ensure that no individual exceeded the allotted 5-year lifetime cap on assistance (Scavo & Shi, 2000). The failures associated with the PRWORA in terms of technology have been duly noted and offer a warning to e-government proponents about the harsh realities of creating interoperable, cross-platform systems.

In July 1996, President Clinton issued Executive Order 13011, which sought to improve the management of information technology at the federal level. It also provided broad support for coordinated approaches to technology application in the Executive Office (Relyea, 2001). Although this executive order mandated implementation of and adherence to the PRA and the Clinger-Cohen Act, it also focused on the alignment of technology goals with strategic organizational goals. The support for interagency coordination of technology is codified in Executive Order 13011. Mandated goal-alignment and technology-investment reviews are included in the directive as a method for reducing the failure rates and cost overruns associated with federal technology initiatives.

More recently, in 2001 the E-Government Act was offered for consideration in the U.S. Senate. This act, approved by the Senate in June 2002, mandated the establishment of an e-government administrator in the Office of Management and Budget, and also provided for considerable financial assistance to spur interagency e-government initiatives. Each of these legislative actions has strengthened the federal government's commitment to e-government.

One of the most significant information-technology developments at the federal level occurred after the tragedies of September 11th. The terrorist acts committed against the United States included the utilization of massive technology, notably the commercial aviation infrastructure, to perpetrate the crimes of mass destruction (Bonvillian & Sharp, 2001). The attack against America forced government officials to reexamine their information-technology policies, infrastructure, and systems. The newly established Office of Homeland Security and its associated directives comprise the largest centralization and consolidation effort involving governmental databases in the history of the United States. A recent General Accounting Office report (2002) highlights this effort and its challenges by examining the types of necessary data, the amount of data, and the transmission format of the data across vertical and horizontal governmental lines. The report also notes the challenges associated with traditional agency "stovepipes" and the need to move beyond this approach toward a centralized enterprise initiative (p. 8). The lack of connectivity and interoperability between databases and agency technologies is another crucial challenge that must be overcome, it is argued, in order to create a comprehensive infrastructure to deal with issues of terrorism.

Another example of the critical need to centralize and consolidate government information in order to mitigate future terrorist attacks is found in the Chambliss-Harman Homeland Security Information Sharing Act (HR 4598). This act, passed by the U.S. House of Representatives in June 2002, mandates the dissemination of critical intelligence information from federal agencies, such as the Central Intelligence Agency (CIA) and Federal Bureau of Investigation (FBI), to state and local governments (Dizard, 2002). The goal of this act is to further reduce the vertical stovepipes that exist between federal, state, and local governments with respect to information access and to encourage data sharing across all branches and levels of government in order to foster coordination and collaboration.

Another result of the September 11th attacks is the renewed concern about security issues related to information technology. These increased concerns have caused governments to reexamine the information they provide online and, in some cases, to restrict access to such information. Section 204 of the Homeland Security Act of 2002 allows the federal government to deny FOIA requests regarding information voluntarily provided by nonfederal parties to the Department of Homeland Security. Additionally, many government Web sites have removed potentially sensitive information. One example of the removal of sensitive governmental information involves the Department of Energy (DOE). According to the watchdog nonprofit organization OMB Watch, the DOE has removed detailed maps and descriptions of nuclear facilities with weapons-grade plutonium and uranium from its Web site (OMB Watch, 2002).

Although the effects of September 11th have impacted the use of information technology in the public sector in a variety of ways, there is little doubt that citizen demand for electronic information and services is likely to continue the trend of e-government adoption and expansion. According to the 2002 Pew Internet and American Life Project, Americans continue to use the Internet to access government information, research policy issues, contact elected officials, and participate in e-democracy in increasing numbers (Larson & Raine, 2002).

Theoretical Constructs

E-government has been viewed in a variety of ways. One context for examining e-government centers on recognition that e-government is more than just a shift in communication patterns or mediums. At least potentially, it involves a transformation of the organizational culture of the government. Recent authors argue that governments are mandated by citizen and business demands to operate within new structures and parameters precipitated by information technology (Bovens & Zouridis, 2002; Heeks, 1999; Ho, 2002; Osborne & Gaebler, 1992). Current demands require crosscutting services, which in turn require government to improve communication and interaction across traditional bureaucratic lines (Alexander & Grubbs, 1998). These new requirements, which fundamentally alter the nature of government, are made possible through the strategic use of information technology.

Garson (1999) has divided the theoretical frameworks of e-government into four main areas: decentralization and democratization, normative and dystopian, sociotechnical systems, and global integration theories. The first two suffice to explain basic variations in e-government theory. The decentralization-democratization theory of e-government revolves around the progressive potential of technology and focuses on the positive governmental advances associated with e-government. Normative-dystopian theory emphasizes the high rate of conflict and failure associated with information-technology applications and counters the positivist progressivism of the decentralization-democratization theory with a realist view of inherent technological limits and contradictions. Each school of thought has its proponents as well as its critics. While neither framework can be considered fully descriptive, taken together they provide a useful delineation of the theoretical literature on e-government.

Decentralization-democratization theory is the most commonly held orientation associated with e-government. In fact, articles beginning with Bozeman and Bretschneider's seminal article in 1986 refer to the transformational, progressive nature of technology adoption in the government sector. Others have concluded that e-government and associated information-technology adoption lead to shifting paradigms, extending beyond the notion of simple progress (Ho, 2002; Reschenthaler & Thompson, 1996). Reschenthaler and Thompson are among theorists who note the revolutionary potential of information technology in government. They contend that the power of e-government technology lies in its ability to even the playing field for all sizes and types of governments. Additionally, they see in e-government the basis for reengineering the business of government, refocusing its work on the needs of the citizens and returning government to its core functions (Reschenthaler & Thompson). Other authors have also used the decentralization theory to highlight the transformational components of information technology and e-government.

Some authors examine the traditional bureaucratic model of public-service delivery (the Weberian model), which highlights specialization, departmentalization, and standardization (Ho, 2002; Weber, 1947). This traditional model has created the stovepipes associated with government—those departmental silos that resist functioning across agency boundaries. The main promise of the Weberian model is to ensure that all citizens are treated equitably with the utmost efficiency. However, in the 1990s, the reinventing-government movement sought to alter the core focus of government, moving from departmentalization and centralization to citizen-centric decentralization (Osborne & Gaebler, 1992), much in contrast to the Weberian model.

According to Ho (2002), the e-government paradigm, which emphasizes coordinated network building, external collaboration, and customer services, is slowly replacing the traditional bureaucratic paradigm and its focus on standardization, hierarchy, departmentalization, and operational cost efficiency. The new paradigm mirrors many of the tenets of the reinventing-government movement, including user control and customization, flexibility in service delivery, horizontal and vertical integration culminating in one-stop shopping, and innovative leadership focused on the end user (Ho; Osborne & Gaebler, 1992). This paradigm shift is precipitated by the advent of the Internet, which provides government the ability to use technology to impact customers directly instead of simply reengineering internal processes (Scavo & Shi, 1999).

A core concept of e-government, which falls under the decentralization orientation, is the idea of using technology as a linkage between citizens and government. According to the work of Milward and Snyder (1996), the governmental use of information technologies to engage and interact with the public further enhances the value of technology adoption, which in turn leads to greater e-government penetration. IT can be substituted for other governmental institutions to link citizens to government services. This linkage causes both the citizens and the government to become reliant upon IT functions, increasing the penetration and adoption rates of e-government.

As governments—federal, state, and local—become more involved in information technology and gain recognition for their e-government efforts, other jurisdictions not currently engaged in e-government will increasingly become interested in and adept at this form of information-technology usage (Norris, 1999). That is, the decentralization theory predicts that e-government's diffusion will snowball as its benefits to citizens and to the agencies themselves is demonstrated.

Technology and e-government adoption rates are a key interest of decentralization proponents. In fact, the argument is virtually tautological: In order to define the factors that increase e-government penetration and adoption, one must believe in the positive potential of the concept. In terms of information-technology adoption and implementation, the greater the number of governments planning for and using information technologies and e-government approaches, the more legitimacy the technology gains (Fletcher, 1999). It is evident from the proliferation of e-government activities at the federal and state level that the widespread benefits of using information technology to provide more timely, seamless government services is a legitimate, even preferred, method of action. Citizen and business demands for e-government applications have extended to the local-government level and as more local governments begin using e-technologies, the legitimacy of the applications will further increases (Norris, Fletcher, & Holden, 2001).

By adopting new technologies, governments may be able to respond to the changing environment with improved service delivery, increased efficiency, and reduced costs (West, 2000). The use of e-government applications allows governments to more readily engage citizens and businesses in a virtual world that is thought to be more responsive and accountable to the needs of the customer. However, there is a marked lag in government adoption of new technologies and methodologies compared to implementation in the private sector.

The structural inertia that afflicts most governments, as well as risk aversion, means that governments are slow to adopt and implement new technologies. According to Herbert N. Casson, the government was one of the final sectors to adopt the telephone, approximately 10 to 15 years after its distribution to the public at large (Relyea, 2001). The lag associated with governmental adoption of new technologies continues today. In fact, research decades ago by Kraemer, Danziger, and King (1978) indicated that local government, in general, has had about a 10-year lag time between the introduction of a new technology and its adoption and use in localities. Furthermore, smaller local governments have a lag adoption time of 15 years or more. For government information-technology adoption and implementation, the inertia that exists within the public sector means that organizations are often less willing and able to engage new technologies (Bretschneider, 1990). The inherent tension between the need for reliability and accountability contrasted with reliance on maintaining organizational status quo leads to the increased adoption lag time in governmental organizations.

Support for e-government is allied with the decentralization theory, which offers optimistic prospects for the future of virtual governance. However, the stark reality of public technology and e-government failures has been repeatedly noted in news media. The concerns associated with privacy, security, and the digital divide are all common threads under normative-dystopian theory, which offers a critical approach to evaluating e-government. Proponents of this view of e-government are traditionally concerned with the dehumanizing and isolating aspects of information technology. Recent concerns about the digital divide, the technology gap that exists between certain subpopulations in the United States, highlight the problems of fairly implementing technology as a mode of communication and service delivery. Fisher (1998) notes the disparity in power and access to technology that occurs in distinct subpopulations when analyzing grassroots social movements. His findings are consistently supported by other studies that note racial, regional, educational, gender, and age disparities among Internet users and technology owners (Norris, 2001; Novotny, 1998).

Bovens and Zouridis (2002) highlight the dystopian framework of e-government by examining the inherent problems associated with the application of information technology and the shift toward an e-government paradigm in real-world settings. They contend that the emerging emphasis on information technologies as the medium for citizen interaction with government fundamentally alters the role of the bureaucrat. The traditional Weberian model uses street-level bureaucrats to interact with citizens and to determine the proper services and service levels to assist these citizens (Lipsky, 1980). However, as technology becomes more integrated into government agencies, computer programs are often used to interface with clients, determine eligibilities, and decide upon proper levels of service (Boven & Zouridis). As a result of this new computer-based assessment trend, the street-level bureaucrats are losing their discretionary power. Bovens and Zouridis do not presuppose to judge the new, nondiscretionary technology models but rather use the paradigm shift to focus on the risks of arbitrariness and associated threats to the legitimacy of governmental actions at the street level.

Types of E-Government

Although several typologies have been developed to explain the progression of e-government (Layne & Lee, 2001; Moon, 2002), the UN and ASPA (2001) definition of the stages of e-government maintains continuity with the working definition set forth at the outset of this chapter. It is also important to note that the stages to be discussed do not represent a true linear progression, nor are they specific block steps. Rather, the stages are a continuum in which governments can be within the same stage with very different functionalities and service offerings.

According to the UN and ASPA (2001), there are five main stages of e-government. The lack of an organizational Web site is not defined by a stage, but may be considered Stage 0. Stage 1 is the emerging Web presence, which involves static information presented as a type of online brochure. The main goal of the emerging Web stage is to provide an online mechanism for communicating key general information about the government to interested citizens and entities. The Web site lacks information about services and is not organized in a citizen-focused manner. Typically, the government has used a "go it alone" approach, which visually represents the stovepipes or silos that exist within agencies—there is little coordination across agencies and levels of government in Stage 1 Web sites.

In Stage 2, enhanced Web presence, the role of the Web site becomes associated with information on services, although it is still organized by departments rather than by user groups. Enhanced Web presence sites typically have e-mail as a means of two-way communication. However, rarely are forms available for download. Stage 2 offers limited communication and greater information about the services of the government, but it does not meet the citizen-centric approach that has been advocated for e-government.

Stage 3, interactive Web presence, begins to move into the citizen-centric realm of e-government. Typically, the information is portrayed by intuitive groupings that cross agency lines. For example, the Web site might use a portal as the single point of entry into various departments and service areas. The portal would offer major groupings like businesses, new residents, seniors, children, or other standard groups. Then, the end user would select the grouping that applies and be launched into a new section of the portal where the most common services requested for the group are located. The services would not be listed by departmental areas, but rather by functional areas. Stage 3 sites have downloadable forms with online submissions, e-mail contact for various governmental employees, and links to other governmental Web sites.

Stage 4, transactional Web presence, offers the ability to conduct secure online transactions. This stage is also organized by user needs and contains dynamic information. The Web site may offer a variety of transactions, including paying for services, paying bills, and paying taxes. Transactional Web presence includes the online submission of forms, many downloads, e-mail contact, and several links to other governments. The use of digital signatures also falls under Stage 4.

The final stage, Stage 5, involves seamless government. Although this stage represents an ideal, there is no real example of its application. Stage 5 involves a cross-agency, intergovernmental approach that only displays one front, regardless of service area. For example, a seamless Web site would offer local, state, and federal government services via the state

Table 1. E-government typology (Sources: Adapted from UN and ASPA, 2001; Ho, 2002)

Stages	Orientation	Services	Technology	Citizens
Stage 1: Emerging Web Presence	Administrative	Few, if any	Only Web	Going it alone
Stage 2: Enhanced Web Presence	Administrative, Information	Few forms, no transactions	Web, e-mail	Links to local agencies
Stage 3: Interactive Web Presence	Information, Users, Administrative	Number of forms, online submissions	Web, e-mail, portal	Some links to state and federal sites
Stage 4: Transactional Web Presence	Information, Users	Many forms and transactions	Web, e-mail, digital signatures, PKI (public-key infrastructure), portals, SSL	Some links to state and federal sites
Stage 5: Seamless Web Presence	Users	Mirror all services provided in person, by mail, and by telephone	Web, e-mail, PKI, digital signatures, portal, SSL, other available technologies	Crosses departments and layers of government

portal without the end user recognizing what level of government provides the service. A Stage 5 site would offer vertical and horizontal integration and would require true organizational transformation with respect to administrative boundaries.

With a working knowledge of the typology associated with e-government, it is easy to assess the current status of the concept. Much of the literature indicates that Stage 2, enhanced Web presence, is the typical placement of an American local government on the e-government continuum. Alexander and Grubbs (1998) note, "Few sites capitalized on the interactive nature of the Internet to conduct public discussions, maintain bulletin boards, or provide data and information available for download." However, local governments are making strides in this arena, spurred by the recognition of the Web site as a vital economic development tool.

A review of the 2004 International City and County Managers Association's e-government survey finds that approximately 90% of cities and counties with populations over 2,500 are not offering transactional Web sites. Furthermore, based on the 2004 ICMA e-government survey, only 8.6% of cities and counties offer the online payment of taxes, 9.2% offer the online payment of utility bills, and 7.3% offer the online payment of fines and fees. The state and federal government offer more robust transactional services, but local governments are recognizing the need to offer electronic services to satisfy customers as well as to reduce the costs associated with traditional walk-in and mail service delivery. E-government will continue to be a critical area for local-government investment as citizens, businesses, employees, and visitors increasingly expect Internet-based options for governmental services.

E-government has been viewed in a variety of ways. One context for examining e-government centers on the recognition that e-government is more than just a shift in communication patterns or mediums. At least potentially, it involves a transformation of the organizational culture of the government. Recent authors argue that governments are mandated by citizen and business demands to operate within new structures and parameters precipitated by information technology. These new requirements, which fundamentally alter the nature of government, are made possible through the strategic use of information technology to accomplish enterprise goals.

Enterprise Approaches

The primary management goal for information technology is to support the business objectives of the local government and to facilitate departmental efforts to provide efficient and effective services to the citizens, businesses, and visitors. Information technology has become a strategic partner in the governmental efforts to provide high-quality, consistent, and equitable services. The driving vision for information technology within many governments includes the development of an enterprise-wide focus on IT, a focus on the customer, and the use of IT as an enabler in efficient and effective customer service. This vision marks a significant departure from the traditional government silo approach with its individualistic, department, or agency-centric efforts, as illustrated in the following chart.

Many future technology efforts will cross multiple local government departments with a single goal of providing services to the citizens, businesses, and visitors. In this new environment, technology is used as the basis for communication, interoperability, and data and

Figure 1. The changing landscape of technology-enabled government

Enterprise Approach	Silo Approach
Enterprise Focus	*Departmental Focus*
← Organizational strategic planning	← Planning done at department levels
← Comprehensive, cross-departmental projects	← Limited cross-departmental efforts
Hardware, Software, Architecture	*Hardware, Software, Architecture*
← Standardization	← No standardization
←Economies of scale and support	← Large support requirements
← Common applications	← Redundant or incompatible applications
Technology Skills	*Technology Skills*
← Sharing of technical skills	← Limited and isolated skill base
← Skill and knowledge transfer	← No sharing of resources
Enterprise Design	*Departmental Design*
← Shared data, relational databases	←Redundant data capture and storage
← Integrated applications	←Functional applications

resource sharing. Furthermore, technology is the vehicle through which cost reduction can occur by increasing efficiency and effectiveness of services through the use of an enterprise architecture and standards. State and local governments throughout the nation are using enterprise approaches to achieve high levels of return on investments, greater customer satisfaction, and increased cost savings.

Chief Information Officers and Professional Staffing

Information technology has fundamentally altered many aspects of daily life, including interactions with public and private sectors. The role of the Internet continues to increase as more citizens use it to find pertinent information, purchase goods and services, and participate in virtual communities. By capitalizing on the Internet revolution, governments can create new channels of communication and new methods for participation via e-government. The changing environment, coupled with citizen and business demands, encourages government involvement in e-government initiatives and related uses of information technologies.

CIOs emerged as a mechanism to connect the business units in an organization with the information-technology staff. In essence, CIOs are the linchpins between these two seemingly disparate, and often contentious, components of an organization. In the past few decades, CIOs have been revered as supreme organizational aligners and lamented as overtitled technocrats. Regardless of the hype and hyperbole surrounding the role of chief information officer, one thing is certain: The job of CIO is always demanding and often difficult. The CIO is responsible for disseminating the critical technology plans to senior executives in order to engender their support while maintaining one foot firmly entrenched in the realm of new and emerging technologies. The CIO must possess the vision for the future while maintaining an eye on the historical legacies of the organization. Too often, chief information officers are forced to take sides between the business units and the information-technology department when, in fact, their role is to build the bridges between these organizational silos. The role of the CIO is critical and the job requires skillful navigation of the various minefields and bear traps that can ensnare and destroy technology projects.

As established as the role of CIO is within the private sector, it is only just emerging in the public sector. The role of the CIO has been adopted from the private sector as one way to navigate the emerging reality of public-sector information technology and e-government. As early as 1981, the title of CIO emerged in the private-sector literature as the defined leadership role for information technology. Extensive research has been conducted on the attributes and characteristics of successful CIOs in the private sector. Some of the most commonly cited traits include being a generalist, having significant power and authority in the organization, and providing a common vision for the implementation of strategic information technology. Based on the success of the CIO in providing leadership and status to information-technology projects in the private sector, the federal public sector followed suit by institutionalizing the position with the passage of the 1996 Clinger-Cohen Act.

The 1996 Clinger-Cohen Act heightened the status of information technology in government. It established a chief information officer in every federal agency, making agencies responsible for developing an IT plan. Now as e-government becomes a priority at the federal, state, and local government levels, the existence of the CIO and a strategic planning structure becomes critical to facilitating e-government implementation. The importance of successful IT projects and their requisite investments is critical in both public and private sectors, as evidenced by the Clinger-Cohen Act and solidified by the rapid proliferation of CIOs in a variety of public and private organizations.

Conclusion

Information technology has fundamentally altered the way we interact in today's society. The role of the Internet continues to increase in society as connectivity becomes more readily available to disparate geographical and demographic sectors of the United States. E-government offers an opportunity to create new channels of communication and new methods for participation. The success of existing e-government efforts provides increased legitimacy for further information-technology adoption (Norris, 1999). The changing information environment and the movement toward the knowledge economy, juxtaposed against citizen and business demands, mandate that government become involved in e-government initiatives and related uses of information technologies (Ho, 2002).

It is clear that the advent of technology has fundamentally altered the way governments conduct business. States, cities, and counties are using technology to improve service delivery, enhance efficiency, and increase transparency and accountability. The citizen and business demand for electronic access to building permits, dog licenses, and birth certificates has heightened the need for investing in information technology. Even more importantly, many governments are moving away from traditional bureaucratic emphasis on departmental silos and information isolation to a new paradigm that emphasizes coordinated network building, external collaboration, and customer services due to information-technology investments. Furthermore, information technology is providing new opportunities for civic engagement and participation from a variety of citizens and groups. The public sector has been engaged in the information-technology revolution and the continued investment and support of these efforts is imperative.

References

Alexander, J. H., & Grubbs, J. W. (1998). Wired government: Information technology, external public organizations, and cyberdemocracy. *Public Administration and Management: An Interactive Journal, 3*(1). Retrieved from http://www.pamij.com/

Bonvillian, W. B., & Sharp, K. V. (2001). Homeland security technology. *Issues in Science and Technology*, 43-49.

Bourdourides, M. A. (2001, October). *New ICTs policies.* Paper presented at ICITA 2001 Conference, Berlin, Germany. Retrieved from http://www.math.upartas.gr/~mboudour/

Bretschneider, S. (1990). Management information systems in public and private organizations: An empirical test. *Public Administration Review, 50*(5), 536-545.

Dizard, W. (2002). House oks data-sharing bill. *Government Computer News, 21*(17). Retrieved from http://www.gcn.com/21_17/news/19182-1.html

Fisher, D. (1998). Rumoring theory and the Internet: A framework for analyzing the grass roots. *Social Science Computer Review, 16*(2), 158-168.

Fletcher, P. D. (1999). Strategic planning for information technology management in state governments. In G. D. Garson (Ed.), *Information technology and computer applications in public administration: Issues and trends* (pp. 81-97). Hershey, PA: Idea Group Publishing.

Fountain, J. (2001). *Building the virtual state: Information technology and institutional change.* Washington, DC: Brookings Institution.

Garson, G. D. (1999). Information systems, politics, and government: Leading theoretical perspectives. In G. D. Garson (Ed.), *Handbook of public information systems* (pp. 591-605). New York: Marcel Dekker, Inc.

General Accounting Office. (2002). *National preparedness: Integrating new and existing technology and information sharing into an effective homeland security strategy* (GAO-02-811IT). Washington, DC: Government Printing Office.

Gore, A. (1993). *Creating a government that works better and costs less: Reengineering through information technology* (Report of the National Performance Review). Washington, DC: Government Printing Office.

Hammitt, H. (1999). The Legislative Foundation of Information Access Policy. In G. D. Garson (Ed.), *Handbook of public information systems* (pp. 27-40). New York: Marcel Dekker, Inc.

Heeks, R. (1999). Reinventing government in the information age. In R. Heeks (Ed.), *Reinventing government in the information age* (pp. 9-21). New York: Routledge.

Ho, A. T.-K. (2001). Reinventing local government and the e-government initiative. *Public Administration Review, 62*(4), 434-444.

Horrigan, J. B. (2003). Consumption of information goods and services in the United States. *Pew Internet and American Life Project.*

Kettl, D. (2000, June). *The transformation of governance: Globalization, devolution, and the role of government.* Paper presented at the Spring Meeting of the National Academy of Public Administration. Retrieved from http://www.lafollette.wisc.edu/research/publications/transformation.html

Kim, P. S. L. W., & Wolff. (1994). Improving government performance: Public management reform and the National Performance Review. *Public Productivity & Management Review, 18*(1), 73-87.

Kraemer, K. L., Danziger, & King. (1978). Local government and information technology in the United States. In *Local government and information technology.* Paris: OECD Informatics Studies #12.

Landsbergen, D., Jr., & Wolken, G., Jr. (2001). Realizing the promise: Government information systems and the fourth generation of information technology. *Public Administration Review, 61*(2), 206-220.

Larsen, E., & Raine, L. (2002). The rise of the e-citizen: How people use government agencies' Web sites. *Pew Internet and American Life Project.* Retrieved from http://www.pewinternet.org/reports/pdfs/PIP_Govt_Website_Rpt.pdf

Layne, K., & Lee, J. (2001). Developing fully functional e-government: A four stage model. *Government Information Quarterly, 18*(2), 122-136.

Milward, H. B., & Synder, L. O. (1996). Electronic government: Linking citizens to public organizations through technology. *Journal of Public Administration Research and Theory, 6*(2), 261-276.

Moon, M. J. (2002). The evolution of e-government among municipalities: Rhetoric or reality? *Public Administration Review, 62*(4), 424-433.

Moulder, E. (2001). E-government … If you build it, will they come? *Public Management, 83*(8), 10-14.

Norris, D., & Kraemer, K. (1996). Mainframe and PC computing in American cities: Myths and realities. *Public Administration Review, 56*(6), 568-576.

Norris, D. F. (1999). Leading edge information technologies and their adoption: Lessons from US cities. In G. D. Garson (Ed.), *Information technology and computer applications in public administration: Issues and trends* (pp. 137-156). Hershey, PA: Idea Group Publishing.

Norris, D. F., Fletcher, P. D., & Holden, S. H. (2001). *Is your local government plugged in? Highlights of the 2000 Electronic Government Survey.* Paper prepared for the International City and County Managers Association and Public Technologies, Incorporated. Retrieved from http://icma.org/download/catIS/grp120/cgp224/E-Gov2000.pdf

Norris, P. (2001). *Digital divide: Civic engagement, information poverty, and the Internet worldwide.* Cambridge: Cambridge University Press.

Novotny, P. (1998). The World Wide Web and multimedia in the 1996 presidential election. *Social Science Computer Review, 16*(2), 169-184.

OMB Watch. (2002). *Access to government information post-September 11th.* Retrieved August 24, 2002, from http://www.ombwatch.org/article/articleview/213/1/104/

Osborne, D., & Gaebler, T. (1992). *Reinventing government: How entrepreneurial spirit is transforming the public sector.* Reading, MA: Addison-Wesley.

Rahm, D. (1999). The role of information technology in building public administration theory. *Knowledge, Technology, and Policy, 12*(1), 74-83.

Relyea, H. C. (2001). E-gov: The federal overview. *The Journal of Academic Librarianship, 27*(2), 131-148.

Reschenthaler, G. B., & Thompson, F. (1996). The information revolution and the new public management. *Public Administration Research Theory, 6*(1), 125-143.

Rocheleau, B. (1999). The political dimensions of information systems in public administration. In G. D. Garson (Ed.), *Information technology and computer applications in public administration: Issues and trends* (pp. 23-40). Hershey, PA: Idea Group Publishing.

Scavo, C., & Shi, Y. (1999). World Wide Web site design and use in public management. In G. D. Garson (Ed.), *Information technology and computer applications in public administration: Issues and trends* (pp. 246-266). Hershey, PA: Idea Group Publishing.

Scavo, C., & Shi, Y. (2000). The role of information technology in the reinventing government paradigm: Normative predicates and practical challenges. *Social Science Computer Review, 18*(2), 166-178.

Seneviratne, S. J. (1999). Information technology and organizational change in the public sector. In G. D. Garson (Ed.), *Information technology and computer applications in public administration: Issues and trends* (pp. 41-60). Hershey, PA: Idea Group Publishing.

Sprecher, M. H. (2000). Racing to e-government: Using the Internet for citizen service delivery. *Government Finance Review, 16*(5), 21-22.

United Nations (UN) & American Society for Public Administration (ASPA). (2001). *Benchmarking e-government: A global perspective. Assessing the UN member states.* Retrieved from http://www.unpan.org/egovernment2.asp

West, D. M. (2000). *Assessing e-government: The Internet, democracy, and service delivery by state and federal government.* Taubman Center for Public Policy at Brown University. Retrieved from http://www.brown.edu/Departments/Taubman_Center/polreports/egovtreport00.html

Chapter VII

E-Participation Models

Suzanne Beaumaster, University of La Verne, USA

Abstract

Participation is the cornerstone of our governance process and has manifested in a variety of constructs since the inception of our government. As public agencies and political leaders discover the possibilities offered by technological mediums, the question becomes, what kind of participation should we be fostering and what do we hope to gain through participative processes? The opportunities to enhance our understanding and approach to democracy have grown tremendously in the past decade. Through technological means, more individuals can gain access to public dialogue and discourse. Given the nature of recent technological opportunities, public leaders are considering the possibilities of e-governance and within that framework, e-participation. This chapter provides a definition and discussion of three e-participation models: information exchange, general discourse, and deliberation. In addition, the chapter will address the issues, characterizations, and criteria that are closely related to the development of electronic participation in the governance process.

Introduction

The advent of the information age has brought with it a host of new monikers for our public-administration lexicon. These terms—e-democracy, e-government, and e-participation—express the direction we are heading with regard to government processes and citizen access. Today no discussion of issues surrounding government or administration can exclude concerns of our new, virtual world, where we engage in business dealings, administration, personal activities, discourse, and government electronically. The issues we have concerning all of these matters must now be discussed with an eye for the implications of what it means to engage virtually. Virtual communities are developing around us, outside of physical space; as these virtual spaces evolve and take shape, individuals, institutions, and governments must reframe their perspectives to meet challenges posed by emerging interactive environments. Old challenges malinger while entirely new challenges must be addressed outside of our conventional perspectives. E-government, and within it e-participation, is one such issue.

The Participation and E-Participation Relationship

E-participation is an approach, a technique, and a tool simultaneously, representing a new way of addressing old problems or issues, specifically regarding citizen access to government processes. In addition, e-participation structures a particular set of activities through which the public can gain information and services, and deliver input. Finally, e-participation is a specific tool for the development and enhancement of public participation as well as a means for access to elected officials and administrators. In other words, e-participation is not just a new way of doing the same old thing, but has the potential to transform citizen access.

Our current participatory structure has grown out of the problematic nature of representative democracy. Since the inception of our government, an ongoing discussion persists regarding the desirability and practicality of direct citizen participation in public decision making (Abilock, 2005; Morse, 2006; Noam, 2005; Timney, 1998). Given the representative nature of democracy in the United States, matters of equity and access are at the forefront of any discussion regarding citizen access and input into the governance process whether done so electronically or via more traditional systems.

In an ideal sense, citizen participation is a hallmark of the democratic process in the United States. However, in reality the transaction costs to obtain information and impart influence are borne differentially by citizens. The desire to interact as a significant member of the governance process necessitates being knowledgeable on the issues to a degree that has proven difficult to achieve for a variety of reasons. Government decision making is comprised of a complex network of political officials, technical experts, interest groups, organizations, administrators, and citizens. Effective input in these decisions demands some degree of sophistication in navigating the complex features of a given issue or problem. With increasing institutional interdependence comes increasing complexity, posing challenges for access and influence that even the well-heeled citizen may find daunting. Regardless of the intricacies involved, it is the citizen who must acquire the status of being informed in order to achieve legitimacy within the system. As the information age brings about new opportunities for participation, governance must address some significant issues including the following. How important is citizen interaction for effective decision making? What level of input is appropriate? To what degree should government facilitate the process of informing the citizenry?

Dealing with the Questions of Participation

There is a great deal of consensus in the literature to support the argument that citizen participation is not only a hallmark of our governance process, but fundamentally necessary for effective decision making. The spirit of this perspective finds its roots in the belief that citizens are interested in public decisions and the issues facing our society; furthermore, it is one of the duties of government to ensure that the citizens are provided the information necessary to be substantively part of the governance process. In the words of Thomas Jefferson,

I know of no safe depository of the ultimate powers of the society but the people themselves, and if we think them not enlightened enough to exercise their control with a wholesome discretion, the remedy is not to take it from them, but to inform their discretion. (as cited in Dumbauld, 1955, p. 93)

From this perspective, in any medium, electronic or traditional, citizen participation is a primary goal. There is a great number of arguments to support the position that participation is fundamental to our governance process. Kathi and Cooper (2005) posit that participation may be viewed either morally or instrumentally. The former suggests that it is a citizen's right in a democratic society to participate in the process of government. The latter perspective provides that participation, at the very least, is necessary for stability within the political community. In addition, the instrumental participation promotes efficiency and effectiveness as well as laying the groundwork for social change. While democracy by its very nature demands the participation of its citizens, practical application of the ideal is much more problematic, especially given the institutional stature conferred on public representatives at all levels of government.

Participation and the Democratic Process

Fundamentally, involvement in decision making by the public improves the democratic process through the enhancement of the regulatory process. In essence, it supports the individual's right to be heard and to influence government activities (Holman & McGregor, 2001). In many ways, public participation is as much about the democratic process as it is about outcomes. There is something to be said for being involved regardless of whether or not your input becomes part of the ultimate decision, policy, or agency plan. As stated by Charles White (1997, p. 24), "Democratic theory envisions a system of government organized as much to foster deliberation as to guarantee participation." In this way, citizens take ownership of the process as well as the outcomes. In essence, the activity of participation sets in motion a chain of events as it were. If people participate in public decision making, then the decisions have a greater likelihood of reflecting what is of concern to the specific community. It follows that higher levels of individual participation lead to a greater feeling of ownership in the process, which may lead to an increased level of interest for other future civic activities. Finally, the greater the level of overall participation, the harder it is for public officials to ignore the concerns, ideas, and input of the citizenry (Al-Kodmany, 2000).

Of course there has always been some dispute with regard to the desirability and viability of public participation. The dissenting opinion fears that too much democracy is as problematic as too little. This has been of concern since the very outset of our constitutional beginnings; a prominent example is found in Federalist 10 where Madison speaks to the fears surrounding the judgment of the general public. His argument states,

to refine and enlarge the public views, by passing them through the medium of a chosen body of citizens, whose wisdom may best discern the true interest of their country, and whose patriotism and love of justice, will be least likely to sacrifice it to temporary or partial considerations. (Cooke, 1982, p. 62)

The opposition to more participation has always maintained that an informed or elite citizenry are more suitable for the task of public decision making than the general, often uninformed public. This perspective argues that direct participation is not an especially effective or efficient way to inform decisions (Innes & Booher, 2004). This viewpoint claims that the methods of public participation, which are legally required, do not achieve that which they are formally bound to accomplish—that is, to review, comment, and provide input on public decisions. Typically, participants in this process are quickly disillusioned; citizens and public officials alike find themselves in an untenable situation. Citizens feel disenfranchised and administrators are beset with all manner of input, which they often find erroneous, cumbersome, and irrelevant. The supporters of this argument speak to the question of what is required to really achieve actual participation in the governance process. The general feeling on the issue suggests that the current system does not satisfy public participants' desire to truly be part of the working dialogue involved in shaping decisions. The worst-case scenario is that current practices actually polarize and antagonize those who are attempting to engage and contribute to the process (Innes & Booher).

As Kathi and Cooper (2005) describe, the commitment to citizen participation is conflicted. On the one hand, such civic engagement is encouraged based on our traditional approach to democracy, yet the framework of the system is designed to "protect political and administrative processes from an active citizenry" (p. 560). Essentially, the technical elite maintain control over decisions within the public sphere and the lack of responsiveness to citizens continues to plague the process.

As the discussion over the best practices for democracy continues, it appears that we are achieving somewhat of a middle ground. The prevailing perspective supports indirect forms of participation such as voting while at the same time acquiescing to the fact that some more direct methods of public participation are needed. This certainly leaves us with agreement that citizen participation should be supported as a fundamental component for decision making. Unfortunately, this agreement is often only grudgingly supported or paid lip service to as a necessary evil. In the 1990s, the "reinventing government" movement (Osborne & Gaebler, 1992) brought this discussion into bold relief. The businesslike approach to government advocated by the reinventing trend canonized a passive customer or user of services over and above an active citizen or direct participant in governance (Innes & Booher, 2004). At best, we are left with some sense of ambivalence about public participation; in the worst case, it may be completely ignored as a part of the governance process. Given the apparent discord surrounding the general issue of citizen participation, it would seem that any further discussion of adding a new ingredient to the mix might muddy the issue even further. What is not in doubt is the fact that government in general and participation constitutively are now moving in the direction of electronic formats. Questions remain as to what role e-participation should assume. Is e-participation more functional as a supplement to the traditional format or should it serve as a replacement for the existing system? What are the overarching purposes of e-participation?

Purposes of E-Participation

Given the problematic reality of participation in general, it would seem that the viability of participating in an electronic format would fare no better. Certainly citizen apathy is not just a result of inconvenience. Thankfully, e-participation is not just a tool that makes access to the government process more convenient. There are a host of purposes that e-participation serves. Following are some of the current objectives that may be fulfilled.

- Participating electronically is convenient. The Internet as a medium offers 24/7 access and there is no need to go to the location of the service in order to acquire information, access services, or conduct business. The anytime and continuous format that e-participation allows can foster more thoughtful engagement over and above the immediate-response format demanded of most real-time venues.
- E-participation allows access to a broader audience (Macintosh, 2004). The range is extensive and essentially borderless.
- The digital format and the use of the Internet allow for many different communication mediums. Participation is supported using multiple technologies that can mitigate numerous circumstances: language barriers, audio and visual deficiencies, transportation issues, individual skill levels and educational differences, communication differences, and technological capacity (Carver, Evans, Kingston, & Turton, 2000).
- Individuals who are unable or unwilling to engage in discourse face to face have access to a medium that allows them to engage at their own pace, over time, with an opportunity to review and revise. This may serve to reduce anxiety and produce more active participation. In addition, e-participation allows for individualized (personal) expression.
- E-participation is informative from multiple perspectives. It provides policy and decision makers with a perspective on public preferences as well as enhancing decisions through the incorporation of the public's knowledge base. From a second perspec-

tive, e-participation can "provide relevant information in a format that is both more accessible and more understandable to the target audience to enable more informed contributions" (Macintosh, 2004, p. 2). Information can be tailored for multiple levels of understanding. An individual reviewing an issue can "drill down" through increasing levels of complexity. In this way, complex relationships can be presented in a user-friendlier format. A user might explore the data and explanations of a given issue through multiple levels of hyperlinks and subsequently engage in dialogue more informed and better prepared.

- E-participation has the capacity to draw in a broader audience and through various methods, promote deeper, more deliberative participation (Macintosh, 2004).
- Finally, through the use of a variety of information technologies, e-participation allows public officials to archive information as well as citizen input on any given issue. In this way, the electronic medium supports case building and the tracking of trends.

E-participation is one of the ways in which democracy is being reshaped. The increase in digital information as well as communication technologies opens up a multitude of possibilities for enhancing public-service delivery, discourse, and citizen-involved decision making. The key is to determine how e-governments should operate and what the nexus is of their functionality. It is also important to ascertain both the desirability and feasibility of e-participation with specific concerns for realistic returns on the initial investments. Specific to this discussion is the question of how e-participation should be addressed by those responsible for creating the commitments and armature that e-participation requires.

E-Participation Models

In its current state, e-participation can be broken down into three types, which are not mutually exclusive. They are information exchange, interactive services, and deliberative participation. Figure 1 shows a three-zone model representing these types and their relationships. Each component of the model functions in conjunction with the others.

The exchange of information as one-way communication and interactive participation as a two-way interchange function closely together. Individuals who access a government Web site typically begin the process because they are looking for specific information; this level of engagement constitutes participation as information exchange. The users then find themselves accessing particular government services or engaging in indirect participatory behavior such as voting, paying utility bills, or reviewing commentary. This commitment level provides a base from which the evolution of e-participation can begin. Accessing a city Web site in order to find information about area parks and recreational services may lead to reserving the park and paying associated fees using an online form. The progressive nature of e-participation can further evolve as the public continues to make use of online information and services; they become more interested in issues regarding specific services and the municipality at large. This is where the process leads to e-participation at the deliberative level. Deliberative e-participation reflects a more interactive version of the citizens interfac-

Figure 1.

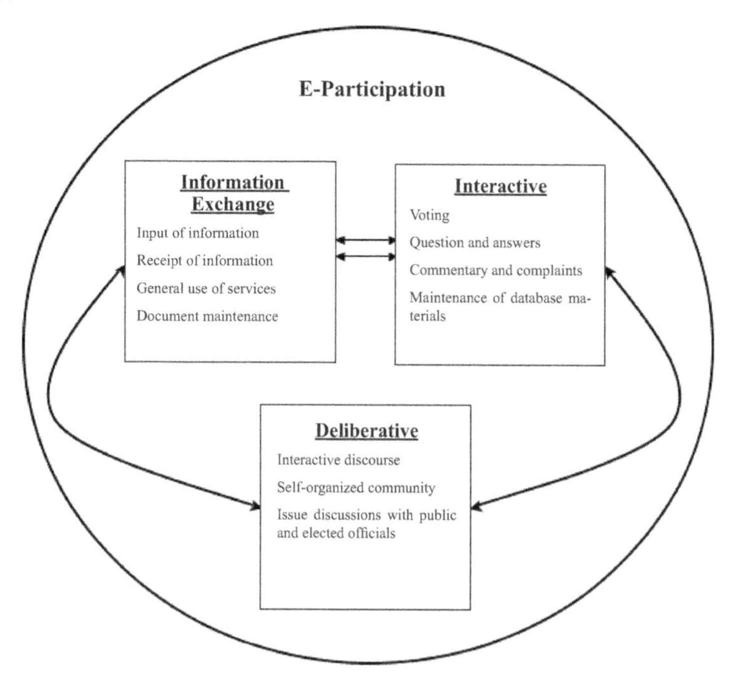

ing with government. This fully engaged zone of participation encompasses direct forms of participation representing two-way, interactive discourse between citizens and citizens, and between citizens and public officials. Such dialogue becomes a stepping-stone for the ongoing development of public interest in online communities.

Information Exchange

As mentioned above, the model of e-participation presented here encompasses a range of participatory approaches. At one end of the spectrum is information exchange. This portion of the model incorporates what is often described as the general delivery of public information. Although certainly not an exhaustive list, the following are included: municipality or agency data, community information, frequently asked questions (FAQs), contacts, available services, and even information about private-sector services related to the given organization. Almost all government agencies and municipalities have some kind of informational Web site. These range from the simple to the sublime. The greater majority of public institutions have designed their electronic portals to include a full range of government services, including everything from information about upcoming events and paying traffic tickets to reviewing building permits as well as transcripts of council sessions.

Developing a framework for e-participation necessitates a discussion of where participation actually starts; as is the case with so many other aspects of life, half the battle is just showing up. What does this mean electronically? When a citizen accesses information or public services in an online format, is that participation? A good argument can be made that information exchange is the place where e-participation truly begins (Hudson-Smith, Evans, & Batty, 2005)—the first step toward the evolution of an informed public. While this kind of basic information does not necessarily constitute a deeply informed populous, it is a start. The convention of providing access to public information or delivering services online is in many ways a factor of convenience. At this level of e-participation, the public is engaging in a unidirectional relationship with government as the producer and citizens as the consumers (Macintosh, 2004). In this way, information exchange is a medium for citizen involvement at its most base level. In essence, bringing the public to a government Web site is a more convenient and efficient way for many citizens to conduct business on a day-to-day basis. Over the last 10 years, most generalized government services have moved to the Web and are now accessible in electronic format. The functionality of the governance process is enhanced by convenience, accessibility, and availability 24/7. Through this medium, citizens have become accustomed to the electronic format; moreover, the more people use online systems, the more comfortable they become, thus paving the way for further electronic interactions.

Information Exchange as Participation

Information exchange constitutes the most fundamental of functional activities with regard to governance, and while it represents the most minimal form of civic engagement, it is, after all, participation. One of the fundamental components of democratic theory is the principal of political equality, which supports the equal opportunity for each citizen to influence political activity and public decision making should they choose to do so (Rosenbaum, 1978). The key component here is choice. In this sense, the individual citizen's choice to access information electronically represents the occurrence of participation. The facilitation of public participation through the delivery of information further supports the participative process. In this day and age, the electronic posting of municipal information is a given: In this way, e-government assumes at least this level of participation no matter how minimal it may appear in the grand scheme of things.

Even at this most basic level, individual ownership of the process can take shape. A sense that an exchange is possible—along with the discovery that not only is information readily available but that access is relatively easy—can open the door for increased demand of both information and services. Information exchange is a critical part of the e-participation model primarily because it lays the foundation for the effectiveness of both the interactive and deliberative components. Holman and McGregor (2001) describe three factors directly related to the effectiveness of e-participation: "information and information flow, level of preparedness, and timeliness" (p. 165). The authors argue that public participation increases the demand for information and that the knowledge gained in turn prompts further, more informed participation. This, it would seem, is a self-perpetuating cycle. This process has the added benefit of raising the level of confidence that public officials have in citizen input. At the information-exchange level, information is typically "process information"

(Holman & McGregor), which is one-way communication from government to the citizen. The accessibility of process information in a timely manner provides two of the necessary components for the preparation of the public as an informed citizenry. Whether it is citizens or, instead, public agencies who drive the need to obtain information, the fact remains that it is a critical resource in the decision-making process.

Interactive E-Participation

The second component to the model presented here is interactive e-participation. On the one hand, citizen interaction is fairly functionalist, providing day-to-day access and usage of government services, electronic correspondence with public officials, or information review and commentary. This particular element is characterized by a two-way relationship. Macintosh (2004) refers to this type of participation as "consultation" in which "citizens provide feedback to government" (p. 2). In this case, it is the agency or public officials who define the parameters of the discussion while citizens provide input and opinions as needed or requested.

Delivering services virtually is at the core of any organizational presence online. The idea began its development in the 1980s as businesses and municipalities considered the possibilities of creating the virtual delivery of services; at the outset, this was primarily entertainment options and online shopping (Hudson-Smith et al., 2005). As information technology evolved, becoming more prevalent and user friendly, the medium became a central part of the design of e-government (Hampton, 2003).

At this level of interaction, participation starts when a citizen chooses to access a service electronically. While this may not seem to have the magnitude necessary to represent true e-participation, it does provide a gateway to civic opportunities. Participation at this level means that the necessary groundwork has been laid technologically to promote further aspects of e-government. In addition, it indicates a desire by the public to make use of electronic or technologically based approaches to conventional activities. It has been argued that technologies can "stimulate participation in democratic processes," which enhances an individual's sense of civic belonging (Hudson-Smith et al., 2005, p. 68). In addition to allowing greater access, ease of use, and convenience, online interaction provides an antecedent to creating a sense of involvement in a virtual community. A strong foundation for continued citizen engagement may be fostered through the facilitation of an online community at this level. After all, no participation happens over night; it is a developmental process.

Deliberative E-Participation

While both information exchange and interactive e-participation are important components for a fully functioning participation model, more direct participation is necessary in order to truly achieve effective governance. The element that brings this model together is deliberative e-participation. This particular component is characterized by advanced levels of communication, which go beyond two-way interactions. The argument here is for the development of collaborative discourse between government and citizens. This arrangement calls for active public engagement in all aspects of decision making, entailing public involvement in the complete process including defining the issues and problems, general commentary and input toward the knowledge base, and finally agenda setting and evaluation. Similar conceptualizations have referred to this level of e-participation as "active participation" (Macintosh, 2004, p. 2), "e-regulation" (Garson, 2006, p. 83), and, from a traditional perspective, "active," "transition" (Timney, 1998, pp. 93-94), and "deliberative democracy" (Weeks, 2000, p. 360).

An excellent framework is presented by Mary Timney (1998), which raises questions for the development of e-participation design. The two possibilities she presents that have applicability for this discussion are (a) government with the people or (b) government by the people. On the first account, Timney describes a "hybrid or transition" form of participation in which citizens share control with decision makers. The problem definitions are developed by the agency, but the ensuing process is open, promoting the desired outcome of consensus. In this case, the public administrators both participate and facilitate. In the second case or "active" model, the author describes a scenario in which the citizens are in control of the entire process. In this case, the agency and administrators act as functional support and consultants. This is government by the people and represents full ownership by citizens—the extreme end of the direct participation spectrum. Certainly, either of these approaches can be achieved in electronic participation formats, with technology allowing for the development of either scenario. The question becomes, given the nature of representative democracy and participation in general, which framework offers the greatest possibility of achieving effective e-participation?

Conceptualizing Deliberative Participation

This conceptualization of e-participation necessitates a set of criteria that is the benchmark of public deliberation. First, citizens must be provided balanced, accessible, and credible information. Deliberation requires that an individual be reasonably informed on the particulars of the issue. While the provision of information does not mean that each individual will make use of it in the same way, or even at all, efforts must still be made to supply adequate information. As discussed with regard to information exchange, the use of technologies allow for a variety of possibilities. Data can be tailored for individual needs, circumstances, and expertise levels. A broader range of services can be offered electronically to more locations and to accommodate a variety of circumstances.

Second, citizens must be allotted adequate time in which to deliberate. The majority of public issues are not simple by any stretch of the imagination. Even with the best information and adequate understanding, the average citizen will need the time to think through the issue. An additional concern in this regard is the openness of the parameters by which each citizen is asked to deliberate. This level of public participation necessitates much greater degrees of depth and tends to be much broader in scope. The discussion must be open to expansion and revision as individuals engage. One of the most ideal characteristics of electronic delivery methods is their transcendence of time and geography. Information and opportunities for engagement may be made available 24/7 and to the farthest reaches of a community.

It is easy to imagine that open and expansive dialogue might be in danger of devolving into something of a free-for-all, especially considering the relative anonymity and accessibility of the electronic medium. Given this possibility, a third criterion is in order: that of a regulated format. As in any discussion, rules of conduct must be set in order to maintain a viable atmosphere for dialogue. The electronic format allows for rules to be agreed upon and set, as in any other environment. However, technology allows for much stricter maintenance of these rules. Time frames can be easily upheld; discussions may be shut down at the end of the allotted period. Users in violation of protocol can be blocked. A wide variety of solutions may be employed to maintain the viability of online discussions.

Finally, deliberative participation requires that the sample of citizens be inclusive, interactive, and representative of the broader population. In any participation model, this means that efforts must be made to include participants who represent those concerned with or affected by the issue at hand. In an electronic format, this typically means addressing the digital divide. Online access must be available for those who typically do not have equal representation as well as those who are less literate or apprehensive of technology, and individuals who are socially marginalized. Information technologies make it possible for more and more citizens to become involved in public discourse. Access points can be made available at a variety of public locations and training modules can be built into most systems. With the rapid advancement of technology come more and more options for delivery and increased user friendliness. In addition, the electronic medium helps strip away many designations that have tended to increase the likelihood of status-based bias or influence. Certainly, the differences between citizens are less obvious, especially typical identifiers such as race, gender, and wealth.

Criteria for Deliberative E-Participation

It is when e-participation becomes deliberative that it can begin to meet the criteria of real effectiveness. While measures of effectiveness can be difficult to come by, Innes and Booher (2004) provide a very usable set of criteria for participatory effectiveness: "supports interaction, dialogue, and collaboration; inclusive; defines future actions; self-organized; raises difficult questions; builds shared knowledge; builds social capacity; and finally, produces innovative responses" (p. 422). The electronic environment presents a medium in which these criteria may be addressed or fostered. E-participation allows public officials and citizens to engage in discourse in a controlled and evolving atmosphere. In essence, each instance of participation in an electronic format shapes future arrangements as the online community develops.

One of the primary goals for deliberative e-participation is breadth. In other words, it is necessary to support a wide participatory base in order to achieve the appropriate level of interaction, dialogue, and collaboration. Too few participants and too narrow of an interest spectrum leads to views and outcomes that may not be representative of the public base. Weeks (2000) argues that wider participation is a necessary feature if the development of a civic culture is one of the desired outcomes. It is equally important for the generation of public will and improving general discourse.

In order for deliberative e-participation to be effective, it must be credible as well as representative. It is not enough that a large number of citizens become part of the public discourse; in addition, the opinions and commentary provided must tenable. If they are not, then the legitimacy of the governance process and its outcomes are undermined. This supports the desirability and viability of well-planned and well-executed information exchange. The better informed the public is, the more credible their input.

One of the more interesting features of deliberative e-participation is the ability it has to inform future actions and decision making. Over time, repeated interaction and extensive dialogue will create a usable knowledge base that can be shared and drawn upon in a multitude of governance scenarios. In this way, social capacity is increased and decision making may improve with regard to both timeliness and innovation.

Facilitating E-Participation

It is clear that public participation is an important and desirable component of the governance process; the difficulty of creating and supporting e-participation is another matter. Civic participation as it exists today in the United States is rife with apathy. The disinterest in public engagement is further degraded by distrust of public officials, estrangement from public institutions, the complexities of life, and overall pessimism regarding government. While this would seem to be an untenable situation, there is reasonable optimism that the development of e-participation may offer new opportunities for enhancement of the existing system. However, one of the key challenges facing supporters of e-participation is the difficulty surrounding the creation of a welcoming and worthwhile environment in which the deliberative process can take place (Weeks, 2000). At its core, facilitating e-participation is a function of three things: technology, usage, and support. The information technologies must be viable and user friendly in order to promote usage, while concurrently, usage must be supported in order to foster expanded access.

Providing information and government services in an electronic format necessitates well-planned information systems. Public institutions must be prepared to provide data and services in an understandable and user-friendly format. This may include access by individuals with varying degrees of expertise, hardware that ranges anywhere from obsolete to cutting edge, multiple platforms, and possibly a multitude of languages. The goal is to use technology to reach a wider and more diverse audience. In order to achieve this, an array of technologies will be necessary.

Information Technology and E-Participation

A plethora of technologies is currently available for use as tools to facilitate e-participation. These technologies allow multiple levels of information delivery and communications, making it possible for citizens to interact with others and to engage in civic dialogue. Any discussion of technological tools must begin with the Internet and what have become our primary service networks: the World Wide Web (WWW) and e-mail. The Internet as a medium provides a backbone of connectivity for both organizations and individuals. Via the Internet, municipalities create accessible and user-friendly Web sites, which allow information to be channeled to citizens as well as providing a conduit for service delivery, discourse, and civic engagement. Once an Internet presence has been established, it is possible to develop a wide array of e-government opportunities using one or more technological means. Following is a discussion of a few of these options.

E-mail is arguably the most extensively used application by individuals and organizations. It has become fundamental to doing business and communicating in our society. While it tends to be a one-to-one communications channel, it is a very versatile and user-friendly option for information delivery and general interaction. Mailing lists or list services offer another option for general delivery of information and provide similar applicability to that of e-mail while allowing for many-to-many communications. These lists are extremely stable forms of communication technologies and have been extensively utilized for over 10 years. Citizens may subscribe to a list, essentially self-selecting their participation channels.

While information delivery and general communication are necessary and fundamental to e-participation, further access is required to achieve interactive participation. Online databases and the forms associated with them are one such technology. Web forms come in a variety of types; they range from simple data input points to structured surveys. Through this medium, government can solicit various kinds of information and input from citizens regarding upcoming situations, decision points, or issues. Web forms also make it possible to archive responses and information into searchable databases. From an e-participation perspective, these are useful for future reference both by public officials and citizens alike.

Finally, information technology must support deliberative e-participation through the facilitation of active discourse. Chat rooms and discussion boards are the most common conduit for such engagement. Chat rooms allow for real-time interactions in all three formats: one to one, one to many, and many to many. These interactions may be scheduled by public officials or citizens in order to raise or discuss issues, converse about general topics of interest or concern, and simply chat about items that are of interest to the community at hand. Online chats have gained in popularity because they allow for communication at any time a group chooses and they approximate a conventional face-to-face interaction. This is one technology that is currently experiencing further advancements through the addition of video and audio capabilities, thus alleviating the need for advanced typing skills. It stands to reason that as the technology gets better, the communications capacity will improve right along with it.

Discussion boards or online forums are more linear, providing 24/7 engagement through threaded asynchronous communication. All discussions appear as posts to a particular topic or thread. Threads may be created and managed by either the public official or the citizen. Discussion boards are very viable for e-participation as they allow the user to come and go in the discussion, offer the opportunity to read through and ruminate on other's posts, allow for adequate time for individual consideration, and finally, the opportunity to modify input or remove it completely. Using this medium, the progression of the discourse may be tracked and the discussion can evolve based on the postings of the participants.

While these technologies are a limited representation of what is available to facilitate e-participation, they do offer a vision of what is possible. New opportunities for enhancing citizen involvement arise regularly as technological advancements are always on the horizon.

Beyond Technologies

E-participation is certainly not just a function of using the most viable information technologies; it is also a matter of providing the necessary support for the systems as well as promoting citizen usage. When developing e-participation strategies, it is important not to overlook the importance of access and user friendliness. Facilitating e-participation requires that the information systems and user applications function appropriately and are maintained so that access is not interrupted or problematic. It is important that citizens are not merely invited to be part of e-governance—they must also be enabled (Ward, 2005). These concerns go beyond the digital divide, shifting focus to designs that accommodate all users. Obviously it is not realistic to expect that any given system can include everyone, but care must be taken to make sure that there is no exclusion of any particular group.

System administrators must be aware of the diverse capabilities of the populous and design the systems to be interactive and user friendly. For many users, technology is still somewhat confounding. It is the job of government to help remove some of the mystery of the electronic medium, especially for those who typically find themselves behind the technological learning curve. As we progress in the information age, this becomes less of a problem. The demystification of the electronic medium is further advanced through user-friendly applications, effective training, as well as reliable and secure systems.

E-participation, especially deliberative e-participation, is relatively new and underutilized. At this juncture, it is very important that the public be supported, particularly as they become more familiar with electronic interactions. No one approach to the facilitation of e-participation will be sufficient. The needs of the populous are too diverse to allow for a one-size-fits-all method. To that end, public discourse and feelings of connectedness may be enhanced through customized applications.

Finally, usage becomes a significant concern for the development of e-participation. Part of facilitating e-participation must be a thorough reading of the needs and desires of the public. It is vital to try to include as many individuals as possible; it is in this way that a more representative group is developed. People who do not participate due to circumstance, physical challenges, and personal choice should be encouraged to access government through electronic means. Citizens must have a sense that they are a necessary and desired part of the public decision-making process—that they have a significant role to play in the development of virtual communities and government processes. Instead of a top-down approach, e-participation must support a bottom-up approach, which Mary Timney (1998) describes as government with the people, or in some cases even government by the people. This necessitates an environment where decision-making power is not centralized but distributed throughout the affected community. Allowing the public to inform governance is important to achieving an effective participation model.

Conclusion

All forms of participation can be valuable in the governance process because they bring individuals together. In essence, participation allows citizens to find out about diverse interests and how each is related to a common good. It redirects a portion of civic responsibility to the citizen: Individuals become part of the process, thereby allowing people to see themselves in a broader context. In an early work that discusses the impacts of technology on democracy, Abramson, Arterton, and Orren (1988) describe the value of direct participation in this way: "... by allowing citizens to act like citizens, participation acquaints each with the interest of all. Participation gives the citizen education in, and responsibility for, the common good" (p. 179). They go on to argue that a crucial consideration in this regard is the need for discourse: "Participation can deliver civic education and civic virtue only when it takes the form of deliberation and dialogue, persuasion and debate" (p. 179). The authors suggest that this requires activity beyond voting, which can be described mainly as the individual's assent of predetermined interests.

One of the tenants of improved participation in this country is the need to facilitate an engaged citizenry, that is, to get the public more fully involved in the process of decision making within their communities. Weeks (2000) argues that this approach necessitates that citizens have an opportunity to fully evaluate the issues at hand. This means that they must have access to reliable and legitimate information, a full view of the possible alternatives, and then the chance to make a judgment about what they believe to be the best option given all of the information.

E-participation has the potential to play multiple roles in the promotion of civic engagement. One possible outcome is the development of diverse virtual communities. E-participation has the ability to bring together diverse populations, previously disenfranchised citizens, and remote voices, thereby stimulating innovative dialogue around complex public issues. In addition, e-participation holds the potential to equalize some of the power most often associated with political power players and technical elites. The average citizen can access information and attain deeper understanding of the issues, allowing for more informed and higher quality civic discourse.

E-participation has the ability to organize like minds, forming a powerful base from which to engage in the decision-making process. Certainly, in a representative democracy there is power in numbers. Consequently, a large group with a cohesive and well-developed message may prove to be doubly powerful. E-participation could conceivably bring these individual voices together, allowing for those with similarity of interest to make their voices heard. No longer must interest demand proximity; electronic mediums allow like-minded individuals to create a community without boundaries. In order for this to happen, certain criteria must be present. E-participation systems must be well designed and supported by public officials. Citizen usage must be encouraged and supported. Also, the information provided must be clear, accurate, and accessible, and the e-participation environment must be open and well managed. Finally, decision makers must agree to use the medium and to respond as well as act based on what is found there.

If citizens are truly the owners of government in a real and active sense, then e-participation must facilitate the direct participation needed to allow this ownership to come to fruition.

References

Abilock, D. (2005). Six promising approaches to civic engagement. *CSLA Journal, 29*(1), 8-11.

Abramson, J. B., Arterton, F. C., & Orren, G. R. (1988). *The electronic commonwealth: The impact of new media technologies on democratic politics.* New York: Basic Books.

Al-Kodmany, K. (2000). Public participation: Technology and democracy. *Journal of Architectural Education, 53*(4), 220-228.

Carver, S., Evans, A., Kingston, R., & Turton, I. (2000). Accessing geographical information systems over the World Wide Web: Improving public participation in environmental decision-making. *Information Infrastructure & Policy, 6*(3), 157-170.

Cook, J. E. (1982). *The federalist.* Middletown, CT: Wesleyan University Press.

Dumbauld, E. (Ed.). (1955). *The political writings of Thomas Jefferson: Representative selections.* New York: Liberal Arts Press.

Garson, G. (2006). *Public information technology and e-governance: Managing the virtual state.* Sudbury, MA: Jones and Bartlett Publishers.

Hampton, K. (2003). Grieving for a lost network: Collective action in a wired suburb. *Information Society, 19*(5), 417-428.

Holman, J., & McGregor, M. (2001). "Thank you for taking the time to read this": Public participation via new communication technologies at the FCC. *Journalism & Communication Monographs, 2*(4), 159-202.

Hudson-Smith, A., Evans, S., & Batty, M. (2005). Building the virtual city: Public participation through e-democracy. *Knowledge, Technology & Policy, 18*(1), 62-85.

Innes, J., & Booher, D. (2004). Reframing public participation: Strategies for the 21st century. *Planning Theory & Practice, 5*(4), 419-436.

Kathi, P. C., & Cooper, T. L. (2005). Democratizing the administrative state: Connecting neighborhood councils and city agencies. *Public Administration Review, 65*(5), 559-568.

Macintosh, A. (2004). Characterizing e-participation in policy making. In *Proceedings of the 37th Hawaii International Conference on System Sciences.*

Morse, R. (2006). Prophet of participation: Mary Parker Follett and public participation in public administration. *Administrative Theory & Praxis, 28*(1), 1-32.

Noam, E. (2005). Why the Internet is bad for democracy. *Communications of the ACM, 48*(10), 57-58.

Osborne, D., & Gaebler, T. (1992). *Reinventing government: How the entrepreneurial spirit is transforming the public sector.* New York: Penguin Group.

Rosenbaum, N. M. (1978). Citizen participation and democratic theory. In S. Langton (Ed.), *Citizen participation in America* (pp. 43-44). Boston: Lexington Books.

Timney, M. M. (1998). Overcoming barriers to citizen participation: Citizens as partners, not adversaries. In C. S. King & C. Stivers (Eds.), *Government is us: Public administration in an anti-government era* (pp. 88-99). Thousand Oaks, CA: Sage Publications.

Ward, J. (2005). An opportunity for engagement in cyberspace: Political youth Web sites during the 2004 European Parliament election campaign. *Information Polity: The International Journal of Government & Democracy in the Information Age, 10*(3/4), 233-246.

Weeks, E. C. (2000). The practice of deliberative democracy: Results from four large-scale trials. *Public Administration Review, 60*(4), 360-372.

White, C. S. (1997). Citizen participation and the Internet: Prospects for civic deliberation in the information age. *Social Studies, 88*, 23-28.

Chapter VIII

E-Government and Creating a Citizen-Centric Government: A Study of Federal Government CIOs

Christopher G. Reddick, The University of Texas at San Antonio, USA

Abstract

This chapter examines the relationship between e-government and the creation of a more citizen-centric government. This study provides a conceptual framework showing a possible relationship among management, resources, security, and privacy issues that would lead to creating a more citizen-centric government with e-government. It explores the opinions of chief information officers (CIOs) on e-government issues and effectiveness. A survey was administered to federal government CIOs in June and July 2005. The survey results revealed that CIOs who have higher management capacity and project-management skills were associated more with creating a more citizen-centric federal government. The contribution of this study to the literature on e-government is that it identifies two key attributes that CIOs can attain in order to reach higher stages of e-government advancement for their department or agency.

Introduction

Electronic government or e-government in this study is defined as the delivery of government information and services to citizens through the Internet 24 hours a day, 7 days a week. This definition has been used in other empirical studies of e-government adoption (Moon & Norris, 2005a). This research adds to this definition with Grant and Chau's (2005) interpretation of e-government as a broad-based transformative initiative, which is consistent with creating more citizen-centric government. Gronlund (2005) reviews the various definitions of e-government. That author has found they share a common theme of the need for organizational transformation through technological implementation. We realize that focusing on the Internet and e-government is a more limited way of examining e-government because of the rise of non-Internet technologies (Gronlund & Horan, 2004).

Citizen-centric government is the delivery of government services continuously to citizens, businesses, and other government agencies through the Internet (Seifert & Relyea, 2004). Citizen-centric government through e-government acts more as a transformation tool that provides a new government model based on being citizen focused (Schelin, 2003). Some scholars have argued that for e-government to fully realize its capabilities, it must transform government from agency centric to citizen centric (Seifert & Relyea).

The term e-government emerged in the late 1990s. It was born out of the Internet boom. The literature of IT use within government is different from e-government because it more often focuses on external use, such as services to citizens and organizational change (Gronlund & Horan, 2004). Definitions of e-government that focus exclusively on service-delivery components fail to capture the more complex aspects of government transformation because of IT (Grant & Chau, 2005). This study attempts to address this issue by focusing on citizen-centric e-government.

Existing Research on CIOs

The existing research on chief information officers (CIOs) or information resource managers (IRMs) has focused on the federal government (Bertot, 1997; Bertot & McClure, 1997; Buehler, 2000; McClure and Bertot, 2000; Westerback, 2000), state governments (Reddick, in press; Ugbah & Umeh, 1993), local governments (Fletcher, 1997), and comparisons between the public and private sectors (Ward & Mitchell, 2004). There have been no scholarly studies, of which we are aware, that examine public-sector CIOs and their opinions on e-government issues and its effectiveness. This is most likely attributed to the Internet being a relatively new research area in the public sector. In general, the management of IT in private-sector organizations has long been a focus of IS research, but the extent of diffusion has not been as extensively explored in public-sector organizations (Fletcher, 1997).

This study empirically focuses on the connection between e-government and creating a more citizen-centric federal government. The existing research has begun to explore the relationship between e-government and increasing citizen-initiated contacts with government; this study fits into that research area (Thomas & Streib, 2003; West, 2004). However, much needs to be done to identify the key attributes of CIOs, which enable them to create more citizen-centric organizations.

Existing Empirical Work on E-Government Adoption

In a survey of state and federal government CIOs and an analysis of their Web sites, West (2004) arrived at the conclusion that e-government has fallen short of its potential to transform government service delivery and trust in government, that is, creating a more citizen-centric government. E-government does have the possibility of enhancing the responsiveness of government and increasing beliefs that government is efficient and effective. There is also evidence that e-government increases citizen-initiated contact with public officials (Thomas & Streib, 2003). The potential of the Internet to improve citizens' access to government and involvement in policy making is well articulated in the literature. However, citizen-centric government is difficult to achieve in the public sector since governments need to provide universal access to their services (Mahler & Regan, 2002).

Citizen-centric e-government is consistent with the four-stage model of e-government adoption in that governments can reach higher levels of adoption if they become more citizen centric. Layne and Lee (2001) proposed a "stages of e-government growth model" that begins with first cataloging online information; second, moving to online transactions; and then third, moving to vertical integration in which local systems are linked to the national systems. The fourth stage of adoption is horizontal integration across different functions leading to one-stop shopping for citizens (e.g., a Web-site portal). Citizen-centric federal government would involve the final stage of the Layne and Lee model of horizontal integration whereby citizens use Web portals to attain services rather than get information from individual departments or agencies.

This study is different from the existing work on e-government adoption since it focuses on one of the highest stages of development, namely, citizen-centric government. Most of the existing empirical work that tests the impact of e-government adoption primarily examines the first two stages of providing online information and municipal e-service delivery (Ho & Ni, 2004; Moon & Norris, 2005b). Additionally, Andersen and Henriksen (2005) argue that the role of government in technological diffusion is studied the least and therefore is the focus of this chapter.

A study of local e-government over 2 years found that as local-government Web sites mature, they will become more sophisticated, transactional, and more integrated vertically and horizontally (Moon & Norris, 2005a). These authors and others found that e-government adoption is progressing rapidly if measured by the deployment of Web sites. However, a movement toward integrated and transactional e-government is progressing much more slowly in more of an incremental fashion (West, 2004). The key question that this study explores is, what can CIOs do to enable their departments or agencies to become more citizen centric to achieve higher levels of e-government adoption?

In order to examine the views of federal government CIOs on e-government and the creation of a government that is more citizen centric, this study is divided into several sections. In the following section, this study articulates the evolution of the roles and responsibilities of federal CIOs. There also is a description of how the public-sector CIO's environment is uniquely different from what can be found in the private sector. We identify what it means to create a more citizen-centric federal government through e-government. A conceptual framework is outlined explaining what factors one would expect to be associated with creating a more citizen-centric federal government. The final sections articulate how these findings can be used to move e-government to higher stages of development, and demonstrate two easily identifiable skills that CIOs can attain through graduate education.

Clinger-Cohen Act and CIOs

The Information Technology Management Reform Act (ITMRA) of 1996 (P.L. 104-106), also known as the Clinger-Cohen Act, established the position of CIO in executive-branch agencies. This act requires agency heads to designate CIOs to lead reforms to help control system-development risks, better manage technology spending, and achieve measurable improvements in agency performance through the management of information resources (General Accounting Office [GAO], 2004b). However, almost a decade after the passage of this act and despite the government's expenditure of billions of dollars annually on IT, GAO's management of these resources has produced only mixed results.

The Clinger-Cohen Act is consistent with the Government Performance and Results Act (GPRA) of 1993 (P.L. 103-62) that requires agencies to establish clear and measurable objectives, to implement a process, to report on the degree to which those objectives are accomplished, and to report regularly to Congress on their progress in establishing and meeting performance objectives (McClure & Bertot, 2000). Together, the Clinger-Cohen Act and the Paperwork Reduction Act (P.L. 10413) of 1995 (which deals with the strategic acquisition and management of information resources by federal agencies) ushered in a new era of IT management practices in the federal government (Relyea, 2000; Westerback, 2000).

With the passage of the Clinger-Cohen Act, federal departments and agencies had the authority and responsibility to make measurable reforms in performance and service delivery to the public through the strategic use of IT (Bertot & McClure, 1997). Prior to this act, the majority of agencies and departments had an IRM official as their top information person who was viewed as an administrative overhead function. IRMs were far removed from the agencies' strategic decision making and program offices that they served with little or no access to senior agency officials. As a solution, the Clinger-Cohen Act states that federal CIOs will report to and work directly with agency directors. As a result, the CIOs were raised to the executive level and were expected to ask the tough questions about strategic planning, outsourcing, and attaining economy and efficiency (Buehler, 2000).

The importance that CIOs place on strategic planning for their department or agency can be found in existing survey research. A survey of senior IT officers and managers of federal departments and agencies revealed that their top priority was aligning IT with strategic goals (AFFIRM, 2004). Existing research on federal CIOs has examined whether proper management of information resources can lead to effective, efficient, and strategic organizations (Bertot, 1997). Evidence was found for a connection between federal agency strategic planning and agency mission attainment, which face different environmental constraints than private-sector organizations.

The Environmental Context of Public-Sector CIOs

In the seminal work by Bozeman and Bretschneider (1986), these authors argued that management information systems (MISs) developed for business administration are not altogether appropriate for public administration. Essentially, the different environmental context of the public organization is an important constraint, which makes public MIS diverge from business. The environment of public MIS differs from that of its private-sector counterpart through greater interdependencies that create increased accountability, procedural delays, and red tape (Bretschneider, 1990). Budget and other constraints on purchasing make it impossible for comprehensive approaches to work well, such as strategic planning (Rocheleau & Wu, 2002).

In a survey of state agencies concerning the ability of public organizations to control and manage information resources, the following was found: (a) Public agencies find their programs and sources of information externally oriented, (b) recruiting and retaining a technically competent workforce in public agencies to manage information resources effectively was found to be a problem, and (c) public agencies are constrained by fiscal crisis and a political climate in which they must operate (Ugbah & Umeh, 1993). These unique differences make public-sector Internet use especially worthy to explore.

Citizen-Centric Federal Government and E-Government

The E-Government Act of 2002 (H.R. 2458) defines electronic government as the use by the government of Web-based Internet applications and other information technologies. This act established the Office of Electronic Government within the Office of Management and Budget (OMB) to oversee implementation of its provisions. The E-Government Act was enacted with the general purpose of promoting better use of the Internet and other information technologies to improve government services for citizens and internal government operations, and provide opportunities for citizen participation in government (GAO, 2004a). According to the General Accounting Office, the OMB and federal agencies have taken many positive steps toward implementing the provisions of the E-Government Act.

Creating a more citizen-centric government can be found in President George W. Bush's management document, the *President's Management Agenda* (PMA) of 2002. This document argues that the "… administration's goal is to champion citizen-centered electronic government that will result in a major improvement in the federal government's value to the citizen" (EOP, 2002b, p. 23). In evaluating the PMA, the GAO showed that the results in terms of e-government implementation were mixed, with many goals only being partially achieved or there being no significant progress made (GAO, 2005b).

Citizen-centric e-government is further elaborated upon in the Bush administration's document *E-Government Strategy* (EOP, 2002a). President Bush has made expanding e-government part of a five-part management agenda for making government more focused on citizens and results. According to the Bush administration, the three main aspects of expanding e-government are to make it easier for citizens to obtain service and interact with the federal government, improve government efficiency and effectiveness, and improve government's responsiveness to citizens. E-government is "... critical to meeting today's citizen and business expectations for interaction with government" (p. 3).

Although the PMA does not specifically define citizen-centric government in its application to the federal government, we can discern from reading the document that it implies a focus on citizen expectations driving government responses rather than the other way around, with an emphasis on performance measures. Citizen-centric government essentially focuses on providing citizens with the services and information they require from their government.

Conceptual Framework

The conceptual framework demonstrated in Figure 1 shows the relationship between six factors that are predicted to create more citizen-centric federal government. Each of these factors is discussed along with its respective hypothesis.

Figure 1. Conceptual framework of factors that predict creating a more citizen-centric federal government

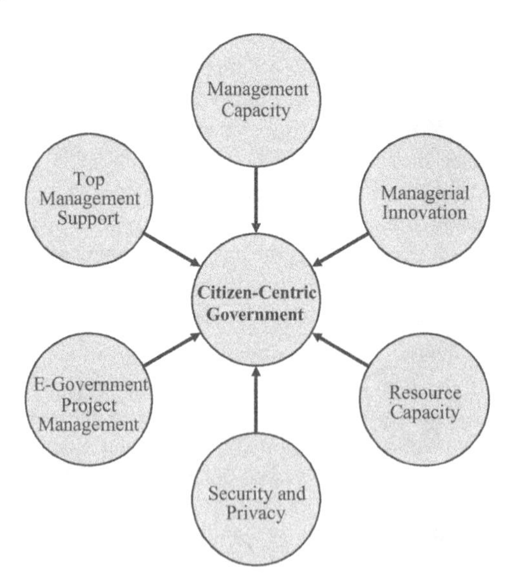

Management Capacity

The literature on public administration in e-government has often argued that effective management is a critical catalyst for its advancement (Brown & Brudney, 1998; Ho & Ni, 2004). The benefits in the public-administration literature of the impacts of IT are often associated with the efficiency and rationality of service provision (Danziger & Andersen, 2002). For instance, putting a strong CIO in place can be done to address the federal government's many information and technology management challenges (GAO, 2005a). The literature on management capacity and its impact on federal e-government initiatives is not as well developed as the local e-government literature. For instance, has e-government made the federal CIO a more effective manager? Empowering employees to make more decisions on their own is a desirable trait according to the total quality management (TQM) literature. Has e-government facilitated the empowerment of employees in the federal government? Finally, the literature on public administration also mentions that performance measures are of critical importance in both public and private sectors. Have federal departments or agencies been able to achieve greater results because of e-government? These three management factors are expected to have an impact on creating a more citizen-centric federal government. Indeed, in a survey of all three levels of government, public-sector IS managers attach more importance to managerial issues than technical ones (Swain, White, & Hubbert, 1995). The following hypothesis can be used to show the impact of management capacity on creating a more citizen-centric federal government.

H_1: *As CIOs believe that e-government management capacity factors are important, this will create a more citizen-centric federal government.*

Security and Privacy

Besides the importance of management capacity outlined in the literature, there also is a growing trend to think about e-government in light of security and privacy concerns. Since the terrorist attacks of September 11, 2001, in the United States, there has been an emphasis on homeland security and emergency preparedness as it relates to information systems (Dawes, Cresswell, & Cahan, 2004). One of the crucial and growing issues for the near future of e-government is the security of information infrastructure and government information applications (Stowers, 2004). This trend is consistent with the security and privacy of digital information. For instance, threats or attacks on information systems could compromise national security. In addition, the privacy of citizens' personally identifiable information (PII) is of paramount importance with increased incidence of identity theft. The federal government must make sure that it has safe and secure information systems. IT security remains a top priority for federal CIOs in President Bush's second term in office (ITAA, 2005). Federal CIOs continue to focus on security and authentication as key building blocks for the advancement of e-government (ITAA, 2004). According to federal government CIOs, in the age of terrorism and identity theft, a clear authentication protocol is necessary for creating a more citizen-centric government (ITAA, 2004).

However, in past studies on differences in the priorities of public- and private-sector IRMs, evidence has shown that public-sector IRMs were much less concerned with protecting information. It was ranked almost last (out of 23 categories) compared to the private sector's relatively high ranking of sixth place (Ward & Mitchell, 2004). This finding is somewhat perplexing given the recent emphasis on privacy and security issues in the federal government. Therefore, we predict that security and privacy issues should have an impact on the ability of the federal government to initiate a more citizen-centric government.

H_2: *As CIOs believe that security and privacy are important issues that must be dealt with in e-government, this will create a more citizen-centric federal government.*

Top-Management Support

The literature also suggests that if top management is supportive of e-government, this provides for greater advancement. For instance, having a champion of e-government, someone who is essentially a cheerleader identifying the benefits of an e-government project and translating it into something of value, is of paramount importance (Ho & Ni, 2004). Additionally, with the increased emphasis on strategic planning in public organizations, leadership is said to be increasingly vital. If a manager is not enthusiastic about e-government and does not see its overall benefits to the organization, this is likely to wear against its advancement. Indeed, existing empirical research shows a connection between top-management support and IT planning at the local level (Ho & Ni; Ho & Smith, 2001). Therefore, support from top management is predicted to have an impact on creating a more citizen-centric federal government.

H_3: *As top management of a CIO's department and agency is more supportive of e-government efforts, this will create a more citizen-centric federal government.*

E-Government Project Management

Along with support from top management, there also is a need for finding and recruiting well-qualified project managers for e-government projects. Project managers are in short supply, especially for government agencies, which must compete for higher paying jobs in the private sector (ITAA, 2004). How widely and quickly have e-government projects been adopted in a department or agency is said to be the barometer of project-management success. Can e-government projects be seen through from start to finish on time and on budget? CIOs were unanimous in the belief that attracting and retaining qualified project-management personnel remains a significant challenge for moving e-government forward (ITAA, 2005).

How have e-government projects changed the interaction of a department or agency with the clients or customers that it serves? Ideally, one would assume that e-government has increased citizen and business interaction with government and has provided for more satisfaction with contacts (Thomas & Streib, 2003). Therefore, we predict that good project management should have an impact on creating a more citizen-centric federal government.

H_4: *As CIOs become more involved in e-government project management, they are more likely to agree that e-government has created a more citizen-centric federal government.*

Managerial Innovation

The transformation agenda of e-government has been promoted under the label of the *new public management,* which calls for reinvention of government as an institutional reform (Grant & Chau, 2005). This managerial innovation has been one of the major thrusts of the theoretical work on e-government. Scholars have argued that e-government is associated with a more decentralized, flexible, efficient, and effective public sector (Ho, 2002; Moon & Norris, 2005b). Research shows that public agencies do indeed face higher levels of formalization and red tape than the private sector (Rainey & Bozeman, 2000).

Existing studies maintain that e-government will break down the silos of information dissemination in government; it will decentralize government, allowing it to run more efficiently and effectively. Agencies will share information more readily and there will be a greater amount of teamwork toward reaching a common goal. Departments and agencies will collaborate more on projects and look at IT not as part of a functional unit, but in terms of serving customers.

An example of the managerial innovation having an influence on e-government is the http://www.firstgov.gov Web portal, where instead of listing departments, the federal government lists services that a citizen most often uses. A number of CIOs also believe that the process of working together across departments, agencies, and in some cases levels of government has resulted in a new model of collaboration through e-government (ITAA, 2005). However, CIOs have started to raise significant concern about the difficulty of changing ingrained cultural attitudes in order to manage change (ITAA, 2004). Existing research suggests a connection between managerial innovation and e-government at the local level (Ho, 2002; Moon & Norris, 2005b), but there is little empirical work in this area on the federal government and the creation of more citizen-centric government.

H_5: *As CIOs believe that they have managerial innovation in their agency, this is likely to have an impact on creating a more citizen-centric federal government.*

Lack of Resource Capacity

One area that should also have an impact on creating a more citizen-centric government is whether the federal government has adequate resources to fulfill e-government mandates. The lack of resource capacity is a perennial problem that federal CIOs face in IT implementation (ITAA, 2005). Is the budget Congress appropriates to an agency adequate to provide for e-government services? Has the government been able to save resources by eliminating manual processes through e-government? Does the department or agency have an adequate amount of IT infrastructure to fulfill its e-government mandates? What kind of outsourcing relationship does the agency have? Does the department or agency fit into the Office of Management and Budget's vision of e-government project management—the OMB being the chief agency responsible for federal e-government? CIOs have expressed frustration with the difficulty of securing budget deliberations from Congress to fund e-government initiatives (ITAA, 2004).

Existing empirical research does not show that resource capacity is a major constraint on public-sector IT planning (Ho & Ni, 2004; Ho & Smith, 2001). However, this factor is included in our model because of its importance identified in focus-group discussions with CIOs (ITAA, 2005). Finally, this chapter predicts that resource capacity should have an impact on creating a more citizen-centric federal government.

H_6: *The more that CIOs believe that resource-capacity issues are a problem for his or her department or agency, the less likely he or she is to agree that e-government has created a more citizen-centric federal government.*

Characteristics of Federal CIOs

This survey of federal government CIOs was administered during June and July of 2005. The contact information for the CIOs was taken from the CIO Council Web site at http://www.cio.gov. This Web site provides the most comprehensive listing of contact information for CIOs employed by the federal government. There were 115 federal departments and agencies that had a designated CIO official. All of them were sent a survey. In total, 38 CIOs responded to the survey, which indicates a response rate of 33%. This is a slightly lower response rate than West's (2004) study of CIOs and e-government service delivery.

The survey protocols were to initially send a cover letter to each of the CIOs indicating that in a few days they would receive a survey. The survey was seeking their opinions on e-government issues and effectiveness. Second, a formal survey and a cover letter with instructions were sent to the CIOs. This was an anonymous survey; therefore, we believe that the responses to the questions are candid. The majority of ideas for questions on the survey were taken from the series of ITAA focus-group discussions with federal CIOs and their views on IT planning and management (ITAA, 2004, 2005).

Table 1 provides the characteristics of CIOs who responded to the survey and their department's or agency's size. The majority of CIOs who responded were from large departments or agencies that employed 5,000 or more full-time equivalent (FTE) employees. Large-sized departments represented 56% of CIOs surveyed. Smaller agency CIOs employing 99 or less FTE composed only 14% of the sample. Therefore, this research is more representative of larger department CIOs than smaller agencies.

Table 1 also indicates that the typical age range of CIOs is between 45 and 54 years, representing approximately half of those surveyed. A third of the respondents were between

Table 1. Descriptive characteristics of CIOs and their departments or agencies

	Frequency	Percent
How many FTE employees are employed in your department/agency?		
99 or less	5	13.9
100 to 499	3	8.3
500 to 999	2	5.6
1,000 to 4,999	6	16.7
5,000 or more	20	55.6
What is your age range?		
35-44	6	16.7
45-54	18	50.0
55-64	12	33.3
What is your gender?		
Female	9	25.0
Male	27	75.0
How many years have you worked for the federal government?		
Less than 5 years	5	13.9
5 to 10 years	4	11.1
11 to 15 years	4	11.1
16 to 20 years	2	5.6
21 to 25 years	5	13.9
26 years or more	16	44.4
How many years have you worked as a CIO?		
Less than 5 years	14	38.9
5 to 10 years	13	36.1
11 to 15 years	3	8.3
16 to 20 years	4	11.1
21 to 25 years	1	2.8
26 years or more	1	2.8
What is your highest level of academic attainment?		
High school diploma	2	5.6
2-year college degree	2	5.6
4-year college degree	9	25.0
Master's degree	19	52.8
Law degree	3	8.3
Doctorate degree	1	2.8

55 and 64 years of age. The smallest number of CIOs surveyed was between the ages of 35 and 44 years, representing around 17% of the sample.

According to Table 1, almost half of the CIOs have worked for the federal government for more than a quarter of a century. Therefore, their tenure in the federal government is substantial. However, those surveyed have not acted as CIOs for long. According to the survey results, 75% of the CIOs have been in that position for 10 years or less. This finding is most likely attributed to the Clinger-Cohen Act of 1996, which established the new position of CIO for most federal government departments and agencies.

The highest level of academic attainment for the CIOs was typically a master's degree, with just over 50% of them holding this advanced degree. Only 25% of CIOs hold a bachelor's degree as their highest level of academic achievement. This finding is what one would expect as the requirement (having an advanced degree) when working at an executive-level position in the federal government.

The characteristics of CIOs and their departments and agencies generally show that the CIOs are from large agencies. They tend to be baby boomers, are male, and have many years of experience in the federal government but fewer years of experience as a CIO. The majority of the CIOs are well educated. The survey results are more representative of large-sized federal departments and agencies than smaller ones. This should be kept in mind when interpreting the findings presented in the following section.

CIOs' Opinions on E-Government

Structured Questions

In this section, the opinions of federal government CIOs on e-government are explored. It examines the views of CIOs on whether they agree that e-government has created a more citizen-centric federal government. We also outline the influences of management capacity, security and privacy, support from top management, project management, managerial innovation, and resource capacity on e-government.

Table 2 presents the impact of e-government on creating a more citizen-centric federal government. Over 60% of CIOs agree that e-government has indeed created a more citizen-centric federal government. However, 14% disagree that e-government has created a more citizen-centric government. A quarter of respondents indicated that they neither agree nor disagree with this statement. However, only 40% of CIOs either disagree or are uncertain of its impact on e-government.

Table 2 also shows the impact of management capacity on e-government. Has e-government made the CIO a more effective manager? First, approximately 40% of CIOs agree that e-government has made them more effective managers. Second, almost 50% of CIOs agree e-government has empowered employees to make more decisions on their own. Third, nearly 66% of CIOs believe that the performance of their agency has improved because of e-government. Overall, the management-capacity findings revealed that e-government has had a major impact on the federal government.

Security and privacy issues are of paramount importance for federal government IT systems. This issue is evident in the CIOs' opinions on security and privacy issues. Almost all CIOs believe that secure storage of citizen and business PII is the most important concern for the future advancement of e-government. Additionally, 92% of the CIOs believe that security and authentication are the key building blocks of e-government advancement. Finally, according to 95% of CIOs, information assurance and security is one of the most pressing concerns for e-government adoption. Not surprisingly, very few CIOs disagreed with these three above-mentioned security and privacy statements.

Another category investigated is the support from top management for e-government adoption, which is one of the critical success factors noted in the literature (Ho & Ni, 2004). Does having a champion of e-government, someone who will spearhead the implementation of e-government efforts, make a difference toward attaining greater levels of adoption? Nearly 90% of CIOs believe that having a champion of e-government is one of the most important, critical success factors. Top management, according to over 70% of CIOs surveyed, has a vision and strategic direction for e-government. Additionally, top management is supportive of CIOs in the e-government decision-making process, according to over 70% of CIOs. There is a general agreement that the OMB has a vision and strategic direction for e-government and has been inclusive in the decision-making process.

This survey outlines the level of adoption of e-government projects in the federal government. Project management has been identified as one of the most critical success factors for a department or agency to possess, according to the ITAA's (2005) survey. E-government projects have been adopted widely and quickly, according to over 50% of CIOs. Over 60% of CIOs believe that e-government projects have increased citizen and business interaction with the federal government. E-government projects are a top priority of departments or agencies, according to almost 60% of CIOs. Seventy percent of CIOs believe that recruitment and retention of qualified e-government project-management staff is of critical importance.

Has e-government been influenced by managerial innovation? The existing literature in this area shows that this has been the case (Ho, 2002; Moon & Norris, 2005b). E-government has allowed for a greater level of information sharing among departments, according to 70% of CIOs. Additionally, e-government has created more teamwork in federal departments and agencies, according to 42% of respondents. Two thirds of CIOs believe that e-government has created a new level of collaboration among departments and agencies. There seems to be support for these three aspects of managerial innovation that have a discernable impact on federal e-government.

A lack of resource capacity in a department or agency is also said to have an impact on e-government advancement. Does the CIO's department or agency lack the necessary IT infrastructure, which would inhibit e-government adoption? Only 20% agree with this statement of not having adequate IT infrastructure. However, approximately 50% of CIOs agree that they do not have an adequate budget to fund e-government in their department or agency. Only 28% of CIOs agree that they have not seen manual processes being eliminated because of e-government. Finally, there is a movement in the federal government to holistically and competitively outsource IT. Approximately 22% of CIOs agree that they do not take a holistic view when it comes to outsourcing e-government projects. The resource-capacity findings provide evidence that key e-government infrastructure is available, but the greater issue is not having adequate budgetary resources to fund e-government. This issue is also addressed in the open-ended question.

Table 2. Summary of CIO opinions on e-government

Category	CIOs' level of agreement and disagreement with the following statements about e-government	Strongly Agree (%)	Agree (%)	Neither Agree/ Disagree (%)	Disagree (%)	Strongly Disagree (%)
Citizen-Centric Government	Have created a more citizen-centric federal government	13.9	47.2	25.0	11.1	2.8
Management Capacity	Has made me a more effective manager	13.2	28.9	47.4	10.5	0.0
	Has empowered employees to make more decisions on their own	5.3	42.1	23.7	26.3	2.6
	Has enabled us to achieve greater performance milestones and results	10.5	55.3	23.7	7.9	2.6
Security and Privacy	Secure storage of citizen and business PII is one of the most pressing concerns for e-government advancement.	45.9	48.6	2.7	2.7	0.0
	Information assurance/security is one of the most important concerns for e-government adoption.	40.5	54.1	2.7	2.7	0.0
	Security and authentication are the key building blocks for the advancement of e-government.	35.1	56.8	5.4	2.7	0.0
Top-Management Support	Having a champion of e-government is one of the most important critical success factors for e-government advancement.	52.8	36.1	11.1	0.0	0.0
	Top management has a vision and strategic direction for e-government for my department/agency.	22.2	50.0	13.9	11.1	2.8
	Top management is very supportive of my department's/agency's participation in the decision-making process for e-government.	27.8	44.4	16.7	8.3	2.8

Table 2. continued

E-Government Project Management	Has been adopted quickly and widely	13.2	42.1	28.9	13.2	2.6
	Has increased citizen and business interaction with my department/agency	22.2	38.9	22.2	13.9	2.8
	Is a top-priority of my department/agency	13.9	44.4	19.4	19.4	2.8
	Recruitment and retention of qualified e-government project-management and -support staff is one of the most critical issues.	41.7	27.8	25.0	2.8	2.8
Managerial Innovation	E-government has allowed a greater degree of information sharing among departments/agencies.	19.4	50.0	19.4	11.1	0.0
	E-government has fostered greater teamwork in employees.	5.6	36.1	38.9	19.4	0.0
	E-government has developed a new level of collaboration among departments/agencies.	13.9	52.8	16.7	16.7	0.0
Lack of Resource Capacity	Lacks IT infrastructure, which inhibits e-government adoption	2.8	16.7	16.7	44.4	19.4
	Does not have adequate budgetary resources to fund e-government projects	13.9	33.3	22.2	22.2	8.3
	Has not seen manual processes being eliminated as a result of e-government	5.6	22.2	19.4	47.2	5.6
	Does not take a holistic view of e-government when competitively outsourcing e-government projects	5.6	16.7	33.3	27.8	16.7

Open-Ended Question

An open-ended question was also asked of federal government CIOs on their opinions concerning e-government issues and effectiveness. The most common responses were that CIOs had issues with a lack of budgetary resources, the OMB dictating e-government projects, and no role for smaller agencies in federal e-government initiatives.

For example, in the budgetary-resources issue, one CIO believed that the most significant hindrance to implementing e-government is OMB's practice of mandating implementation schedules that are shorter than the budget cycle. This CIO argues that agency budgets are developed 2 years in advance. Consequently, the OMB needs to either provide planning guidance that announces e-government initiatives or mandates 2 years in advance of their issuance, or allow at least 2 years for the implementation of initiatives. Another CIO stated that e-government project managers spend far too much time begging for money from reluctant agency partners. Finally, a CIO said that a lack of funding is the biggest inhibitor to e-government advancement.

A second common theme was reaction from some CIOs about the role of top management being either the chief executive of the department or agency or the OMB. A CIO stated that he or she personally has not seen much value in the e-government policies promulgated by the OMB. According to another CIO, $60 billion in federal IT spending is simply too large to be managed on a top-down basis. According to another CIO, e-government is very important to my organization but has sometimes been stymied by the OMB. A CIO commented that executive leadership must really understand e-government opportunities. Thus, they must want and know how to leverage technology for successful e-government outcomes.

A third common response was the role of small agencies in federal e-government initiatives. The scalability of e-government initiatives presents problems for small and microagencies, according to one CIO: One size does not fit all agencies. E-government needs to address the needs of the very small agencies (10 employees or less) as well as the larger agencies, according to another CIO.

Less common responses indicated by CIOs were that too many silos of information and information technologies exist in the federal government. According to a CIO, departments and agencies are finding it difficult to keep up with new government regulations (privacy, security, etc.). A CIO commented that governments must close the digital divide for e-government to reach its full potential. One response that is consistent with creating a more citizen-centric federal government was the comment by a CIO that e-government initiatives are very worthwhile in giving citizens a participatory role in government and providing easy access to information.

The results in Table 2 support many facets of e-government advancement and effectiveness. There is ample evidence that management capacity is an important catalyst for e-government adoption. Not surprisingly, security and privacy were the priorities for e-government advancement according to the vast majority of CIOs. Top-management support and direction is also crucial for e-government development. Project-management skills and support were noted as being critical success factors. Finally, the lack of resources of the department and agency was also found to have a discernable impact on e-government.

The structured questions are also supported by some of the responses to the open-ended questions of a greater need for e-government budgetary resources, top-management support through the OMB of e-government initiatives, and the role of smaller agencies in e-government initiatives. There is some initial evidence that all of these categories are important factors that explain e-government advancement. The next section of this study investigates what impact they have as a group on creating a more citizen-centric federal government.

Descriptive Statistics of Dependent and Predictor Variables

In order to model these hypotheses on creating a more citizen-centric federal government, we composed indexes representing each of these categories. The indexes are presented in Table 3 along with their summary statistics. The dependent variable that is modeled creating a more citizen-centric federal government is shown in this table. All of the statement variables outlined in Table 2 were coded in the following manner. A 2 was recorded for a response of *strongly agree*, a 1 for *agree*, 0 for *neither agree nor disagree*, -1 for *disagree*, and -2 for *strongly disagree*. For the dependent variable, creating a more citizen-centric federal government, the mean score was 0.58, indicating that most of the CIOs agreed with the statement. However, they did not strongly agree with this statement since the mean was above 0 and less than 1.

The responses for the six categories of independent variables are outlined in Table 2. We simply added the responses for each of the categories to get an index score for each respondent. For example, in order to compose the management-capacity index, we added up the responses to the three statements in this category of e-government making the CIO a better manager, empowering employees, and creating greater performance results (see Table 2). For management capacity, there was a minimum score of -5 and a maximum score of 6,

Table 3. Descriptive statistics of dependent and predictor variables

	Observations	Minimum	Maximum	Mean
Have created a more citizen-centric federal government	36	-2	2	0.58
Management Capacity	38	-5	6	1.29
Security and Privacy	37	-1	6	3.95
Top-Management Support	36	-3	6	3.06
E-Government Project Management	36	-8	8	2.67
Managerial Innovation	36	-3	6	1.69
Lack of Resource Capacity	36	-6	5	-0.97

Note: Dependent variable shaded

which means that there was a substantial range of responses when adding the three categories together. However, the mean score was 1.29, which implies that there was more of a tendency for CIOs to agree that e-government affected management capacity.

The remaining five predictor variables are also reported in Table 3. All of the mean values are positive, with the exception of a lack of resource capacity, which was negative. In this variable, a majority of respondents disagreed that resources were not much of a factor. The strongest level of agreement was for the issues of privacy and security, with a mean score of almost 4. Top-management support for e-government registered the second highest mean score of just over 3.

Results of OLS Regression Model

Using the six predictor variables alluded to in the previous section, which of these factors best explains creating a more citizen-centric federal government? Ordinary least squares (OLS) regression was used since we are modeling a dependent variable ranging from -2 to 2.

Table 4 presents two statistically significant impacts of the predictor variables management capacity and e-government project management on creating a more citizen-centric federal government. These variables are both statistically significant at the 0.05% level.

The management-capacity variable implies that as CIOs agree that management has been affected as a result of e-government, they are approximately 33% of a point more likely to

Table 4. E-government factors that predict citizen-centric federal government

OLS Regression Model			
	Beta Coeff.	*t-Statistic*	*Prob. Sign.*
Management Capacity	0.34	(2.20)*	0.04
Security and Privacy	0.04	(0.29)	0.77
Top-Management Support	-0.02	(-0.10)	0.92
E-Government Project Management	0.38	(2.31)*	0.03
Managerial Innovation	0.22	(1.39)	0.17
Lack of Resource Capacity	-0.07	(-0.49)	0.63
Constant		(-0.55)	0.58
Model Diagnostics			
F-Statistic	(10.65)**		
Adjusted-R^2	0.62		
Number of Observations	36.00		

*Note: * significant at the 0.05% level; ** significant at the 0.01% level*

increase their level of agreement that e-government has created a more citizen-centric federal government (beta coefficient=0.34). Additionally, CIOs who agree that project management has been affected are over 33% of a point (beta coefficient=0.38) more likely to increase their level of agreement with the statement that e-government has created a more citizen-centric federal government.

These results essentially imply that the statements that e-government has made the CIO a more effective manager, has empowered employees, and has enabled the CIO to achieve greater performance milestones are correlated with creating a more citizen-centric federal government. If e-government projects have been adopted widely and quickly within a department or agency, the CIO believes e-government projects have increased citizen and business interaction with his or her agency, e-government projects are a top priority of the agency, and recruitment and retention of project-management staff are important (the four project-management statements). These statements were correlated with creating a more citizen-centric government.

The six predictor variables in Table 4 explained just over 60% of the variance of the OLS regression model. The F-statistic of 10.65 indicates that the overall significance of the model is strong, being significant at the 0.01% level. However, since there were only 36 observations for the regression model, this limits the interpretation of the results.

Discussion of Hypotheses

The six hypotheses mentioned previously should be reviewed in order to find out whether the evidence found in this study refutes or confirms them. Hypothesis 1 inquired whether CIOs agreed that manage-capacity factors related to creating a more citizen-centric federal government. The evidence found in this study indeed supported this hypothesis since management capacity was correlated with citizen-centric government. This confirms existing literature that argues managers are more effective if they set performance targets and empower employees to make more decisions on their own. This is consistent with the GPRA initiated in the federal government since 1993.

Hypothesis 4 was also supported in the survey of CIOs. There was evidence that CIOs who had favorable views of project management and its impact on their department or agency were more likely to be of the opinion that e-government has created a more citizen-centric federal government. The ITAA (2005) has also found through extensive interviews with federal CIOs that project management is one of the most important critical success factors for IT advancement. This study lends support that good project management is also applicable to e-government as well.

There is no evidence found in this study that top-management support is correlated with e-government. However, the open-ended question indicates that one of the issues that CIOs face is the OMB dictating e-government projects. Issues of security and privacy are critical to advancing e-government, yet they had no relationship to creating a more citizen-centric federal government. Security and privacy would override issues of citizen-centric government since it has dominated Washington agenda setting since September 11, 2001. There was no evidence found for the impact of managerial innovation popularized in the United

States on creating a more citizen-centric federal government. This is surprising since the literature demonstrates such a connection. Finally, the hypothesis that a lack of resources in the federal department or agency decreases the creation of a more citizen-centric government was not supported. This finding coincides with some of the existing literature that resource capacity is not correlated with e-government advancement (Ho & Ni, 2004).

Conclusion

This study has examined some possible factors that might influence the creation of a more citizen-centric federal government. The existing literature on e-government has not provided a connection between e-government and the opinions of CIOs. This is an important area of IS research given that the federal government is the largest purchaser of IT in the United States. Creating a more citizen-centric government is one indication of the advancement of e-government into the highest stage of development (i.e., horizontal integration). This study has identified that having a greater management capacity, and project-management skills and development leads to the creation of a more citizen-centric federal government.

CIOs can use these identified skills in order to reach higher levels of e-government adoption for their departments or agencies. Working more on project management and the ability to manage more effectively are skills that can be easily acquired through education at, for example, the CIO University. This is a virtual consortium of universities for federal CIOs that offers graduate-level programs that directly address executive core competencies. Since just over 33% of federal CIOs do not have a master's degree, this would help in achieving these two important skills. Additionally, more emphasis should be placed on better recruitment and retention of federal project-management personnel. This issue was also mentioned in the ITAA (2005) CIO focus-group discussions.

One of the major limitations of this study is that it examines the opinions of CIOs. Such responses from CIOs are limited because they are based on perceptions, not assessments of actual figures or data. CIOs have a self-interested stake in promoting the view that what they are doing is effective and efficient. As a result, future work might involve independent verification of CIO achievements in their departments or agencies in terms of e-government projects actually being implemented. A comparison across time might provide further evidence as to whether CIOs are achieving results in terms of e-government advancement.

Acknowledgments

The authors would like to thank all of the CIOs who participated in this survey. Without their generous support, this project would not be possible.

References

AFFIRM. (2004). *The federal CIO: Ninth annual CIO challenges survey.* Retrieved July 30, 2005, from http://www.affirm.org/

Andersen, K. V., & Henriksen, H. Z. (2005). The first leg of e-government research: Domains and application areas 1998-2003. *International Journal of Electronic Government Research, 1*(4), 26-44.

Bertot, J. C. (1997). The impact of federal IRM on agency missions: Findings, issues, and recommendations. *Government Information Quarterly, 14*(3), 235-253.

Bertot, J. C., & McClure, C. R. (1997). Key issues affecting the development of federal IRM: A view from the trenches. *Government Information Quarterly, 14*(3), 271-290.

Bozeman, B., & Bretschneider, S. (1986). Public management information systems: Theory and prescription. *Public Administration Review, 46*, 475-487.

Bretschneider, S. (1990). Management information systems in public and private organizations: An empirical test. *Public Administration Review, 50*(5), 536-545.

Brown, M. M., & Brudney, J. L. (1998). Public sector information technology initiatives: Implications for programs of public administration. *Administration & Society, 30*(4), 421-442.

Buehler, M. (2000). U.S. federal government CIOs: Information technology's new managers. Preliminary findings. *Government Information Quarterly, 27*, 29-45.

Danziger, J. N., & Andersen, K. V. (2002). The impacts of information technology on public administration: An analysis of empirical research from the "golden age" of transformation. *International Journal of Public Administration, 25*(5), 591-627.

Dawes, S. S., Cresswell, A. M., & Cahan, B. B. (2004). Learning from crisis: Lessons in human and information infrastructure from the World Trade Center response. *Social Science Computer Review, 22*(1), 52-66.

EOP. (2002a). *Implementing the President's management agenda for e-government: E-government strategy.* Retrieved July 30, 2005, from http://www.whitehouse.gov/omb

EOP. (2002b). *The President's management agenda.* Retrieved July 30, 2005, from http://www.whitehouse.gov/omb/

Fletcher, P. D. (1997). Local governments and IRM: Policy emerging from practice. *Government Information Quarterly, 14*(3), 313-324.

General Accounting Office. (2004a). *Electronic government: Federal agencies have made progress implementing the E-Government Act of 2002* (GAO Publication No. GAO-05-12). Washington, DC: U.S. Government Printing Office.

General Accounting Office. (2004b). *Federal chief information officers: Responsibilities, reporting, relationships, tenure, and challenges* (GAO Publication No. GAO-04-823). Washington, DC: U.S. Government Printing Office.

General Accounting Office. (2005a). *Chief information officers: Responsibilities and information and technology governance at leading private-sector companies* (GAO Publication No. GAO-05-986). Washington, DC: U.S. Government Printing Office.

General Accounting Office. (2005b). *Management reform: Assessing the President's management agenda* (GAO Publication No. GAO-05-574T). Washington, DC: U.S. Government Printing Office.

Grant, G., & Chau, D. (2005). Developing a generic framework for e-government. *Journal of Global Information Management, 13*(1), 1-30.

Gronlund, A. (2005). State of the art in e-gov research: Surveying conference publications. *International Journal of Electronic Government Research, 1*(4), 1-25.

Gronlund, A., & Horan, T. (2004). Introducing e-gov: History, definitions, and issues. *Communications of the Association for Information Systems, 15*, 713-729.

Ho, A. T.-K. (2002). Reinventing local government and the e-government initiative. *Public Administration Review, 62*(4), 434-444.

Ho, A. T.-K., & Ni, A. Y. (2004). Explaining the adoption of e-government features: A case study of Iowa county treasurers' offices. *American Review of Public Administration, 34*(2), 164-180.

Ho, A. T.-K., & Smith, J. F. (2001). Information technology planning and the Y2K problem in local governments. *American Review of Public Administration, 31*(2), 158-180.

ITAA. (2004). *CIO: Catalyst for business transformation. 2004 survey of federal chief information officers*. Retrieved July 30, 2005, from http://www.grantthornton.com

ITAA. (2005). *Issues in leadership: 2005 survey of federal chief information officers*. Retrieved July 30, 2005, from http://www.grantthornton.com

Layne, K., & Lee, J. (2001). Developing fully function e-government: A four stage model. *Government Information Quarterly, 18*(1), 122-136.

Mahler, J., & Regan, P. M. (2002). Learning to govern online: Federal agency Internet use. *American Review of Public Administration, 32*(3), 326-349.

McClure, C. R., & Bertot, J. C. (2000). The chief information officer (CIO): Assessing its impact. *Government Information Quarterly, 17*(1), 7-12.

Moon, M. J., & Norris, D. F. (2005a). Advancing e-government at the grassroots: Tortoise or hare? *Public Administration Review, 65*(1), 64-75.

Moon, M. J., & Norris, D. F. (2005b). Does managerial orientation matter? The adoption of reinventing government and e-government at the municipal level. *Information Systems Journal, 15*, 43-60.

Rainey, H. G., & Bozeman, B. (2000). Comparing public and private organizations: Empirical research and the power of the a priori. *Journal of Public Administration Research and Theory, 10*(2), 447-469.

Reddick, C. G. (in press). Information resource managers and e-government effectiveness: A survey of Texas state agencies. *Government Information Quarterly.*

Relyea, H. C. (2000). Paperwork Reduction Act reauthorization and government information management issues. *Government Information Quarterly, 17*(4), 367-393.

Rocheleau, B., & Wu, L. (2002). Public versus private information systems: Do they differ in important ways? A review and empirical test. *American Review of Public Administration, 32*(4), 379-397.

Schelin, S. H. (2003). E-government: An overview. In G. D. Garson (Ed.), *Public information technology: Policy and management issues.* Hershey, PA: Idea Group Publishing.

Seifert, J. W., & Relyea, H. C. (2004). Considering e-government from the U.S. federal perspective: An evolving concept, a developing practice. *Journal of E-Government, 1*(1), 7-15.

Stowers, G. (2004). Issues in e-commerce and e-government service delivery. In A. Pavlichev & G. D. Garson (Eds.), *Digital government: Principles and best practices.* Hershey, PA: Idea Group Publishing.

Swain, J. W., White, J., & Hubbert, E. D. (1995). Issues in public management information systems. *American Review of Public Administration, 25*(3), 279-296.

Thomas, J. C., & Streib, G. (2003). The new face of government: Citizen-initiated contacts in the era of e-government. *Journal of Public Administration Research and Theory, 13*(1), 83-102.

Ugbah, S. D., & Umeh, O. J. (1993). Information resource management: An examination of individual and organizational attributes in state government agencies. *Information Resources Management Journal, 6*(1), 5-13.

Ward, M. A., & Mitchell, S. (2004). A comparison of the strategic priorities of public and private sector information resource management executives. *Government Information Quarterly, 21*, 284-304.

West, D. M. (2004). E-government and the transformation of service delivery and citizen attitudes. *Public Administration Review, 64*(1), 15-27.

Westerback, L. K. (2000). Toward best practices for strategic information technology management. *Government Information Quarterly, 17*(1), 27-41.

Chapter IX

The Federal Docket Management System and the Prospect for Digital Democracy in U.S. Rulemaking

Stuart W. Shulman, University of Pittsburgh, USA

Abstract

A large interagency group led by the Environmental Protection Agency (EPA) has worked diligently to set up a centralized docket system for all U.S. federal rulemaking agencies. The result, the Federal Docket Management System (FDMS), is still a work in progress, reflecting technical, administrative, financial, and political challenges. A close examination of the effort to design, fund, and shape the architecture of the FDMS suggests many important lessons for practitioners and scholars alike. While both the new technology and the 60-year-old administrative process of rulemaking offer tantalizing glimpses of innovation, increased efficiency, and remarkable democratic potential, the actual progress to date is mixed. Neither the information system nor its users have turned the FDMS into a techno-fix for all or even much of what ails the sprawling U.S. regulatory rulemaking system. In the great American tradition of incrementalism, the FDMS represents a small step toward a number of worthy but perennially elusive goals now routinely linked to the prospect for digital democracy.

Introduction

Agency personnel and the commenting public are adjusting to the increasingly digital landscape of the "notice and comment" process required under the 1946 Administrative Procedure Act (APA). While the APA sets a "floor below which an agency may not go in prescribing procedures for a particular rulemaking" (Lubbers, 2006, p. 6), scholars, activists, and rulemaking practitioners have long recognized the benefits and burdens that accrue when the baseline requirements for gathering public input are exceeded (Furlong & Kerwin, 2004; Kerwin, 2003).

Running parallel to agency and public efforts to adapt, there is a growing body of e-rulemaking research and scholarship focusing on such fundamental issues as whether Internet-enhanced public participation results in better rules (Carlitz & Gunn, 2002) or in a process characterized by informed deliberation (Brandon & Carlitz, 2002; Emery & Emery, 2005). Information technology opens previously unimaginable avenues for engaging the public in meaningful and well-informed public discourse on a national scale. In its latest incarnation, the "new governance" is increasingly defined by the "tool makers and tool users" (Bingham, Nabatchi, & O'Leary, 2005, p. 547) who are building and using the architecture of electronic participation linking citizens and government.

Inside the federal government, best practices continue to target efficient, cost-effective strategies for overworked information managers in budget-strapped agencies facing statutory deadlines and political pressure to promulgate complex rules, allocating billions of dollars of costs and benefits (General Accounting Office [GAO], 2005). Regulatory rulemaking in the United States can be a highly contested, time- and information-intensive process (Coglianese, 2003a; Johnson, 1998; Rakoff, 2000). How and whether outside commenters influence the process and final outcomes is a matter of some debate (Golden, 1998; West, 2004).

Meanwhile, dedicated users of digital communications technologies have started to exploit the potential to flood agencies with vast quantities of public comments (Lubbers, 2006; Shulman, 2006). The outcome of this approach is often a boon for organized interests that instigate the campaigns. Benefits can include regulatory delay (which allows a return to Congress for redress), favorable publicity, and payoffs in terms of membership identification with a group's mission. However, such efforts may also undermine or dilute the voice of the public as agencies face statutory and administrative deadlines to incorporate public input into a legally defensible final rule (Shulman, 2005). Large-volume public comment submissions that bash agency efforts may decrease the overall level of trust of citizens (Yang, 2005) or else hinder agencies' options as they seek rational and coherent approaches to difficult problems (West, 2005). At the very least, they contribute directly to a preexisting "deep ambivalence about citizens directly participating in their government" (Roberts, 2004, p. 315).

This chapter updates earlier reports on the status of electronic rulemaking. First, it briefly introduces the rulemaking process. Second, it documents the origin and progress of the Federal Docket Management System (FDMS) and looks closely at the evolution of the current interface for Regulations.Gov (the FDMS Web portal) in the context of recent literature on public participation and federally funded research into the impact of e-rulemaking. It draws on workshop, interview, and focus group experiences that have fed into a multiyear dialogue between researchers, regulators, and the regulated public.[1] Finally, it concludes, without much surprise, that there is much work to be done to make the FDMS into a hub for enhanced digital democracy and better rulemaking.

The U.S. Regulatory Rulemaking Environment

While the slow pace of regulatory decision making may occasionally provoke controversy, rulemaking is of necessity methodical and time consuming. Precisely because it is a deliberative process, it tends to be associated with some degree of administrative legitimacy. Yet, it is the often unmet challenge of assembling a complete and usable whole record of the multiyear rulemaking process that is cited by courts overturning final rules (Yackee, 2005). Online rulemaking offers solutions and new challenges in this regard. For example, in a rulemaking widely cited as a harbinger of things to come, the United States Department of Agriculture's (USDA) National Organic Program (NOP) conducted a 12-year rulemaking process involving numerous public hearings, national advisory-board meetings, and a massive outpouring of public comments. Two rounds of notice and comment on the proposed and revised rules resulted in over 300,000 public comments.

The challenges and opportunities were quickly apparent. For example, the precise number of comments reported depended on whether form letters were counted as discrete comments, and how and whether to count such letters remain matters of continuing debate for all parties in this process (Shulman, 2005). The evolving definition of "meaningful input" and the uncertain role of digital public participation are unsettled legal and political issues that constitute an important backdrop for the implementation of the FDMS at this early stage (Lubbers, 2006; Stoll, Herz, & Lazarski, 2006).

Participants in the NOP rulemaking submitted just under 21,000 comments via a USDA Web site where they could also read the previously submitted comments of other participants. This introduced an interesting dialogical element to public input. Democratic theorists continue to watch for signs of two-way communication (give and take between officials and citizens or peer-to-peer citizen contact) that run counter to the tradition of one-way comment collection. Professor Lubbers (1998) has noted that public comments "are much more likely to be focused and useful if the commenters have access to the comments of others" (p. 214), and such was the case in the NOP rulemaking. While many stakeholders found cause to complain about substantive and procedural issues along the way, ultimately the organic standard, and the process by which it was finalized, benefited enormously from transparent, Internet-enhanced decision-making structures (Shulman, 2003; "Revisiting the Rules on Organic Food," 1998). During the interviews with key agency personnel, one USDA official involved in the rulemaking noted,

we built an enormous amount of goodwill with the comment process with people who to this day don't believe we got it right. Now the vast majority of the people think we did, but there still are lots of people out there, who think that we've sold everybody down the river, but there's this enormous goodwill because of the comment process, because they really felt like we cared and we tried, we really tried to listen to what they were saying.

Enhanced public participation and the credibility conferred on federal agencies are considered key e-rulemaking drivers (Beierle, 2003). When public values and knowledge infuse decision making, the results are generally positive (Beierle & Crawford, 2002). For example, the Environmental Protection Agency's review of the National Ambient Air Quality Standards (NAAQS) showed that "informed public debate over competing interpretations of cost-benefit analyses and risk assessments can produce broad political support for enhanced protection levels" (Cooney, 1997, p. 11). A senior career official with rulemaking responsibilities at one federal agency remarked,

the public should participate because we the government agencies don't have all the answers, we don't know all the information. Indeed there are situations where we would have made mistakes if we did not have public input—mistakes that could have been very costly. Secondly, I think that when you provide not just an opportunity to participate, but an effective opportunity to participate, those that are affected by the regulation would be more willing to accept it, to live with and comply with it. One of the basic arguments is that even if you don't get everything you want, by participating effectively in the process, you are buying into the result. If we do normal APA type rulemaking in an effective way, bring the public in an effective way, they should feel better about the end product.

To date there has been little systematic documentation of the effect of this digital transformation on either citizens or agencies. Indeed, one workshop report noted "a striking absence of empirical studies that examine the behavior of developers and governmental users of IT" (Fountain, 2002a, p. 40). While the move to Internet-facilitated governance is accelerating, there is a dearth of political science and sociological research on the impact of the Internet on the administrative state (Fountain, 2001). Scholars of public administration have made headway developing a meaningful research agenda focused on the impact of IT (Garson, 2003). However, perhaps one collateral indicator of the still-novel nature of this area is the membership of the IT and politics section of the American Political Science Association, which remains around 300 despite a total association membership of over 15,000.

The interdisciplinary nature of the process may be a barrier, as well as the tendency of academic endeavors to disregard the rulemaking process generally. Political scientists have been particularly reluctant to take up the interdisciplinary challenges inherent in digital government research. As a result, leading disciplinary journals and textbooks do not reflect the important change under way. One author, noting the lag time for well-developed theory and empirical research in the study of public administration, suggests "scholars and practitioners may be ill equipped to face the challenges of the information age" (Holden, 2003, p. 54). A classic text for students of regulatory rulemaking observes that political scientists have done a particularly poor job of studying the rulemaking process (Kerwin, 2003).

Much of the available scholarship lays out a broad and complex research agenda that remains noticeably forward looking (e.g., Lubbers, 2002; Shulman, Schlosberg, Zavestoski, & Courard-Hauri, 2003). Initial attempts at genuinely interdisciplinary work are tentative but promising (Shulman, Callan, Hovy, & Zavestoski, 2004). While the practitioners in federal agencies are moving forward, the concept of e-rulemaking and its implications for democracy remain largely unexamined by academics. Jane Fountain (2002, p. 118) notes, "There is little theory and no coherent research program within the discipline of political science that seeks to account for the potential or likely effects of information processing on the bureaucracy."

For now, much of the e-governance scholarship remains speculative in nature. For example, Joseph Nye (2002, pp. 11-12) sketches a familiar bipolar vision of IT-enhanced government:

In a bleak vision of the future, one can imagine a thin democracy in which deliberation has greatly diminished. Citizens will use the set-top boxes on their Internet televisions to engage in frequent plebiscites that will be poorly understood and easily manipulated behind the scenes. The growth of thin direct democracy will lead to a further weakening of institutions ...

Alternatively, one can envisage a better political process in the future. New virtual communities will cross-cut geographical communities, both supplementing and reinforcing local community. In Madisonian terms the extensive republic of balancing factions will be enhanced. Access to information will be plentiful and cheap for all citizens. Political participation, including voting, can be made easier.

It is against this historical backdrop that a broader e-rulemaking research community is emerging with the support of the National Science Foundation's digital government program (Coglianese, 2003a). Collaboration between computational and social scientists imposes unique challenges, as does work between academic and governmental personnel (Dawes, Bloniarz, Kelly, & Fletcher, 1999; Dawes & Pardo, 2002). Nonetheless, significant inroads have been made over the past 2 years. The prominent role for federal funding and leadership in IT research and development is part of a well-established tradition whereby the public sector supports innovation and research with an eye toward socially beneficial applications (Comedy 2002; President's Information Technology Advisory Committee [PITAC], 2000).

Digital government research requires interdisciplinary collaboration with government partners to ensure that information technology used for citizen-government interaction meets the requirements of democratic institutions and traditions. Program managers in the Directorate for Computer & Information Science & Engineering (CISE) at the NSF have recognized that "insight from the social sciences is needed to build IT systems that are truly user-friendly and that help people work better together" (National Research Council [NRC], 2000, p. 42). To that end, the research into, and practice of, e-rulemaking must continue to be informed by a broad, inclusive, and transparent dialogue about the tools and information systems that can balance competing visions for a democratically legitimate administrative state.

This nascent research domain is likely to generate innovative data that will enable heretofore impossible empirical studies of rulemaking (Dotson, 2003). The lack of solid empirical studies of rulemaking has vexed administrative-law scholars for some time (Kerwin, 2003). In the information age, however, one observer notes that "[f]ed by non-stop, real-time opinion polling, endless market testing of messages and images, and instant and cheap online focus groups, no social scientist need ever go hungry again" (Applbaum, 2002, p. 20).

New automated collection techniques will track Web logs, click-through data, lengths of visits, page counts, and a host of other potential baseline data generated and easily captured by e-rulemaking systems using legally permissible session cookies. At stake in the research process is the potential to shape data collection and mining techniques that may fundamentally redefine the study and the practice of public participation in administrative decision making. The impetus to develop these tools emerges from observations such as the following from a rule writer working at the USDA:

As you know, one of the things that we had out of our 300,000 comments, we had probably 150-175,000 comments that were basically form letters, said the exact same thing over and over and over. Once you've read one, you've read them all. But that still leaves 100,000 individual comments that may have a kernel, a grain of a marvelous idea that you need to somehow bare it out and find it and use it. That either demands sophisticated IT infrastructure, or lots and lots of people.

Computer science research will facilitate the development of e-rulemaking applications for duplicate and near-duplicate detection, stakeholder identification, and clustering issues and themes, and it will summarize content using natural language processing and information-retrieval algorithms. Over the next 5 to 10 years, these data, and the techniques used to harvest and analyze the data, will infuse a challenging yet energetic dialogue between social and computational scientists as they carve out the direction of interdisciplinary digital government-funded research (NRC, 2002; Shulman, Zavestoski, Schlosberg, Courard-Hauri, & Richards, 2001).

Like much of the ongoing e-government rollout (Schelin, 2003), both fundamental and more subtle choices about the architecture of e-rulemaking will shape the new digital and democratic landscape (Lessig, 1999). The transition is fraught with peril for those who would design, approve, or use such a system. For example, information systems that provide seamless access and increased accountability can result in controversy for public-sector officials (Rocheleau, 2003). As the National Research Council (2000, p. 1) has noted, "IT is anything but a mature, stable technology," and the leeway for innovations in democracy that has devolved to technologists and public administration managers is remarkable.

The challenge for researchers in this unsettled context is to conduct interdisciplinary research capable of evaluating and anticipating the shifting terrain. Visions of e-rulemaking need to retain a long-term and evolutionary perspective in the face of demands for technical quick fixes for information management problems that plague the process:

Overwhelmingly, the most important opportunities lie in not simply automating existing applications, but rather in rethinking and remolding the structure and organization of the business process to reflect the best uses of IT and in redesigning and remolding the technology to make it most valuable in its (rethought) application context. (NRC, 2000, p. 146)

Governmental organizations, we are often told, are not prone to seeking business-process reengineering solutions. The new federal impetus to e-rulemaking, however, may do just that on a historically significant scale, comparable to that created by passage of the APA in 1946.

The Origin of the FDMS

After an erratic and uncertain start government-wide in the mid- and late-1990s, e-rulemaking is moving toward a unified structural system for the entire federal government. Clearly, IT-based approaches to rulemaking hold the potential to increase the volume and lower the cost of citizen-to-government (C2G), government-to-citizen (G2C), and citizen-to-citizen (C2C) interactions. The potential for C2C, however, is only beginning to enter the practice of public administration (e.g., for European e-government, see Holmes, 2001), though it flourishes in the voluntary, peer-to-peer, self-organizing spaces on the Internet. The Bush administration situates its online rulemaking initiative in the government-to-business (G2B) category of its 24-point e-government plan (Office of Management and Budget [OMB], 2002a), leaving some doubt about both the immediate prospect for C2C and C2G development.

The transition to e-rulemaking is attributable in part to the impetus from legislative (e.g., the Government Paperwork Elimination Act and the Paperwork Reduction Act) and administrative directives (e.g., the National Performance Review) that seek to make the regulatory process open, transparent, deliberative, efficient, and effective (GAO, 2000, 2001; OMB, 2002b). Citizen demand for electronic access is also an impetus for this transition (Larsen & Rainie, 2002; Schorr & Stolfo, 1997). These various trends culminated in approval by Congress in 2002 of the E-Government Act, which specifically directs agencies "to enhance public participation in Government by electronic means" (Sec. 206[a][2]). Professor Michael Herz (2003) has noted that the act called for enhancements that were already in the pipeline, resulting in "a classic example of Congress leading from the rear" (p. 168). Indeed, many agencies already had begun to develop e-rulemaking systems when the act was signed into law (OMB Watch, 2002). What the act did was to translate a politically popular trend into a bipartisan victory for advocates of a more modern and accessible government.

Despite the move toward a consolidated endeavor under the current eRulemaking Initiative, many federal agencies have been moving ahead with the implementation and refinement of their own in-house solution to the question of how to incorporate Internet-based public participation in rulemaking. More than 100 subagencies have integrated some form of electronic comment process into their notice and comment rulemaking procedures, with varying degrees of success. At the individual agency level, electronic collection of public comments is often ad hoc and conducted via a nonstandardized process (GAO, 2003).

Meanwhile, at another level and of relevance to the impact of e-rulemaking, some are questioning the statutory and constitutional basis for a recent expansion of the OMB's role in rulemaking (Craig, 2003). Professor Lubbers (2006, p. 20) notes, "Perhaps the most significant administrative law development during the last two decades has been the increased presidential involvement in federal agency rulemaking." Interestingly, many of the tools employed by the OMB when it exerts control over federal rulemaking (e.g., monitoring, prompting, or early collaboration in drafting proposals) are likely to be enhanced by seamless IT systems for e-rulemaking.

According to the *President's Management Agenda*, promulgated in 2001,

IT offers opportunities to break down obsolete bureaucratic divisions. Unfortunately, agencies often perceive this more as a threat than as an opportunity, and in response they make

wasteful and redundant investments in an effort to preserve chains of command that lost their purpose years ago. (OMB, 2002c, p. 23)

This theme emerged during many of the e-government interviews conducted over the last 5 years. In reference to a question about institutional obstacles to change, one long-time federal official stated, "We get in a rut. We get jaded. Bureaucracy can kill you sometimes. It truly can." When cross-agency initiatives challenge traditional stovepipe, bureaucratic culture, a number of obstacles quickly emerge. One official at the OMB noted,

E-Government is much more about process transformation and change management than it is about the technology. The technology is the easy, easy, easy stuff. There's an overwhelming amount of technology out there that's applicable. It's hard to drive that change. And you're seeing that in rulemaking. People don't want to lose their system. People don't want to lose the way they manage their docket. People don't want to lose something they identify with.

Early e-rulemaking adopters in the U.S. Department of Transportation (DoT) developed an impressive Docket Management System (DMS) that improved the flow of information across nine subagencies, as well as to and from the public. The DoT's DMS was the first agency-wide e-rulemaking system that assembled entirely electronic dockets for the commenting public to review (Perritt, 1995). In its infancy, it represented a major transformation of the rulemaking work flow. One person involved with the creation of the DMS recalled the numerous technical and organizational hurdles:

They all had their own individual processes, they all champion what they wanted to do, what they felt was best. And now you're going to try and make everybody do it in a uniform fashion. A lot of resistance to that and technical barriers in the early days, the system required open forms based application, so everybody had to buy a big, expensive PC with a 21 inch monitor and do client server all for forms. Couldn't afford it. This was a DoT wide system but they didn't have the money and the technical configurations were horrible. We had 5 different network operating systems throughout the department. With all kinds of different protocols, all kinds of technical hurdles to make a very complicated client install happen.

Originally, the Department of Transportation played the managing partner role as the Online Rulemaking Initiative that emerged in February 2002 (GAO, 2005). The DoT commissioned Excella Consulting to perform an independent study of the seven major e-rulemaking applications that were in existence at federal agencies.[2] The goal was to find an existing system best suited to become the universal Web-based front end for members of the public wishing to read and comment on proposed federal rules. The report (Excella Consulting, 2002) listed several primary areas of interest:

- The public's ability to comment
- The public's ability to search and view proposed rules and their dockets

- The agency's ability to review, report on, summarize, and incorporate comments
- The agency's ability to internally receive, upload, and process paper comments

Based on a complex scoring system, the DoT's Docket Management System finished a close second to the EPA's newer EDOCKET system. While the DoT's many strengths were noted (automated work flow, strong internal reporting, proactive Listserv communication, flexible key-word search, and mature IT procedures), the areas for improvement included the need for better integration of content and comments, content sorting tools, and full-text search. One official at the OMB noted that there is:

a rule in technology that you really never want to be first. Sometimes first to take advantage of something is a good thing, but the best thing to be is second or third, so that you can learn from the mistakes from others.

Other agency personnel were less sanguine about the results of the selection process, commenting that the EPA's Capitol Hill lobbying for the managing partner role was decisive in the end.

As a result, personnel at the EPA assumed the managing partner role developing the OMB-mandated e-rulemaking portal for all federal agencies. The EPA's EDOCKET was praised in the Excella report for its integration of content and commenting, intuitive display, powerful full-text search and content manipulation and sorting tools, as well as an automated work flow based on an open, technically sound architecture. A unified regulatory access point, Regulations.Gov, went live on January 23, 2003 (Skrzycki, 2003). It expanded on the portal precedent set by FirstGov.gov (Fletcher, 2002). Officials at the OMB (2003b) estimated that the new consolidated system resulting from the eRulemaking Initiative could save the federal government close to $100 million by eliminating redundant systems.

The portal Regulations.Gov was always conceived as an interim system, representing the first of three modules planned for the eRulemaking Initiative. In this first module, the Government Printing Office (GPO) hosted the front end (http://www.regulations.gov) and provided user support, while the EPA's National Computing Center hosted the back-end collection and distribution of the comments. One of the few important innovations in Phase One was the ability to "search by keyword across all government agencies to find proposals of interest" (Brandon, 2003, p. 7). This ability lessens, but does not yet eliminate, the challenge of knowing where to look within the labyrinth of federal government Web pages for an open rulemaking on a topic of interest. An EPA official used the example of a motorcycle to illustrate the long-term goal:

The guy that owns a motorcycle in the Midwest and the small business owner who manufactures motorcycles ... anybody that touches a motorcycle can go on line and find out what rules affect ... whatever his relationship with the motorcycle is. Whether he's a parts distributor or the salesman, owner, operator, or whatever, that if they want to say something about these rules that they can readily, easily, in, as fast and cheap, of manner as they can.

A GAO report in late 2003, noted that, despite some difficulties, during its first 3 months in existence, Regulations.Gov provided substantially greater electronic access to proposed rules and commenting than either the DoT or EPA systems. Nonetheless, this interim portal has received very few substantive comments in its short life span. The GAO report found Regulations.Gov was listed in only 2 of 411 rulemakings published over the period of its study. Part of the problem may be that Regulations.Gov is a passive system, "requiring users to take the initiative to find out about recently proposed rules" and the ability to comment on them (GAO, 2003, p. 17). While EPA officials suggested the low rate of use "could be because commenters have become used to filing comments in a particular way," it remains to be seen whether a central portal will ever become popular with the commenting public.

Focus group responses from organized interest groups suggested a number of technical and organizational reasons why Regulations.Gov may remain underused for some time. For example, groups will be unlikely to promote citizen-to-government communication that will steer traffic away from their own information-gathering Web sites. Instead, advocacy organizations will look for creative ways to reverse engineer public comment portals back to their own Web content, maximizing their influence over the message and the valuable respondent contact information. Nonetheless, over a 33-month period ending in September 2005, the initiative reports Regulations.Gov "is one of the most heavily trafficked E-Gov web sites," with an excess of 11 million hits produced by 1.6 million visitors viewing over 8.6 million pages and files (Morales & Moses, 2006).

In September 2005, the eRulemaking Initiative launched the second module, the Federal Docket Management System, which is presented as "a full-fledged content management system" (Morales & Moses, 2006). The FDMS will be the home for all electronic and paper dockets that are on a phased migration to the centralized system. The 2003 GAO report noted that "a number of legal and policy issues still must be resolved," including the issues involved in making electronic documents the official record, as is currently the case at the DoT. However, the rollout of the second module did not hinge on the resolution of these issues. Indeed, the 2005 GAO report praised the initiative for following 28 of 30 "key practices" for managing the initiative. GAO reviewers were particularly happy with the extensive and effective collaboration with other agencies. Administrators of the initiative reported on a successful "cross-agency FDMS development workgroup" involving "more than 100 IT, regulatory, and docket managers and staff across 25 federal agencies" (Morales & Moses, 2006, p. 1).

For the third module, the eRulemaking Initiative will create the Integrated Federal eRulemaking System. This system will create a seamless electronic process for developing, reviewing, and publishing federal regulations and similar documents for internal agency use. This desktop system (the Regulation-Writers Workbench) will provide a host of tools to assist in all phases of the process. Whereas the docket migration to the FDMS is mandatory, participation in this third module by agencies is voluntary, and its future is far from certain (Stoll et al., 2006). Indeed, the initiative has already encountered, and has usually overcome, a number of specific and general challenges (such as bureaucratic resistance, a lack of awareness, and a temporary suspension of funding) stemming from the fact that, as Jeff Seifert (2006) of the Congressional Research Service points out, e-government has no natural domestic constituency.

The Prospect for Digital Democracy in U.S. Rulemaking

For some, the rise of so-called "click-on democracy" eventually will lead to a viable pathway from public indifference to greater civic engagement on a massive scale (Davis, Elin, & Reeher, 2002). The electronic republic has been held out by some (and dismissed by others) as a remedy for inequities that plague democratic and administrative practice. It remains to be seen whether e-rulemaking will significantly alter the adversarial decision-making characteristic of much administrative procedure in the United States, particularly with respect to environmental matters (Dryzek, 2000; Fischer, 2000; Zavestoski & Shulman, 2002). To date, there is little empirical data, no developed theory, and too few studies to support any authoritative statement about the impact of e-rulemaking. Indeed, much of what is available fosters ad hoc speculation rather than substantive insight.

So while the potential for IT enhancements to improve democratic processes is apparent to many, the risks associated with moving administrative decision making and public deliberation online may be less well understood (Coglianese, 2003b; Tesh, 2002). Some perils of online governance have been clearly identified, including social fragmentation, mass manipulation, and increased political and economic inequality due to the digital divide (Sunstein, 2001; Wilhelm, 2000). Yet, even Professor Sunstein's widely noted fragmentation thesis is made suspect by examples, such as the Howard Dean presidential campaign, which uses peer-to-peer organizing to generate impressive levels of face-to-face political activity involving tens of thousands of people across the country.

According to Cornelius Kerwin (2003) of American University, the core elements of rulemaking are information, participation, and accountability. Each of these elements potentially takes on new significance as IT-based applications are introduced and enhanced over time. The use of IT in rulemaking creates the possibility for transparent, low-cost information flows, improved rulemaking management, as well as many-to-many communication. Rulemaking offers the public a chance to participate in a way that is not present in Congress (Kerwin, 2003). It is a process designed to allow participants to sort through and challenge facts derived from numerous sources, experts, and laypersons alike. How individual agencies actually weigh such comments varies widely. One administrator noted,

> *I will say that in general, the most compelling public comments come from the most influential stakeholders and the stakeholders with the largest stakes, if that's a proper way to put it. And I think the system pretty much has identified ... I mean those folks know where they need to play, that sort of thing. Getting another member of the public ... of course there's probably rulemakings where that's a much bigger, there are probably styles and kinds of rulemakings where comments from the individual or the collective opinion of individuals—like I can imagine food labeling or something. You know, food labeling rules by the FDA may be propelled by the aggregate weight individual commentary or something. The rules I have experience aren't influenced by that.*

In theory, IT-enhanced public participation will result in better and more durable rules that stand up to court challenges and better achieve the goals of the authorizing legislation. Ultimately, the official rulemaking record will be more accessible once it is entirely electronic as existing and new tools (e.g., full-text search, self-indexing databases, or stakeholder identification) allow rule writers and the regulated public to drill down into the many sources of relevant data.[3] In the past, judges have been "somewhat perplexed and unhappy about some of the rulemaking records they have been called upon to review" (Lubbers, 1998, p. 216). As one judge noted, in a lengthy rulemaking proceeding, the record "too often resembles a safari through uncharted lands without benefit of a guide" (Judge Wald, as cited in Lubbers, p. 217). Agencies using mature electronic dockets will likely find it easier to compile the record needed to meet the standards under the logical outgrowth doctrine (Kannan, 1996). If rule-writing agencies can more easily show comments were submitted and considered, then a key threshold for rulemaking durability in the courts is crossed.

Future enhancements to e-rulemaking will create innovative methods for advance notice, allowing agencies to target groups and individuals likely to be affected by proposed rules. While Listservs are the most likely method to transition from a passive to an active notification system, we also can expect to see more ubiquitous plain-English translations of regulatory language, as well as Amazon.com-style referral applications (Coglianese, 2003a). For example, if a commenter has submitted concerns about biotechnology in the organic rulemaking, he or she might be prompted via e-mail or during a future visit to a government portal to look at other open or pending rulemakings dealing with genetic engineering. Just as electronic commerce offers users referral and notification services, so too will electronic government, assuming the Office of General Counsel (OGC) can approve such practices.

On occasion, significant rules are promulgated without targeted outreach to interested parties and thus potential commenters. As a result, public input is limited. For example, new treasury-department rules created under the authority of the USA PATRIOT Act to enhance law-enforcement surveillance raise a range of privacy issues for all users of U.S. financial institutions. Despite the widespread discussion of threats to privacy and other civil liberties post September 11th, only 180 public comments were received. While 70% of the comments came from individuals concerned with privacy, the final rule reflected only the "sophisticated statements made by financial institutions and their lawyers" (Cuéller, 2003, p. 4).

In this case, the researcher concluded that layperson comments could have been more meaningful had they directly addressed the possible methods for allaying privacy concerns. We can expect future iterations of e-rulemaking will explicitly guide commenters to make substantive suggestions, perhaps through an optional TurboTax-style interface that leads the public not only to participate, but also to make more useful submissions that will effectively shape the final rule. What will be interesting will be to see where these more structured comment architectures develop. The most likely developers appear to be e-advocacy action-center firms on contract to interest groups seeking to streamline their role in the management of public sentiment. One can imagine a range of human-computer interfaces proliferating in the marketplace as groups seek to offer a menu of options for commenters of varying levels of sophistication and time commitment to the comment process. For many groups, these flexible comment portals may redefine the nature of advocacy and education in the American political system.

Public access to agency procedures and methods, as mandated under the Freedom of Information Act (FOIA) and the Electronic Freedom of Information Act (E-FOIA; Leahy 1998; Strauss, 1989), will be standardized and hence more manageable. The use of IT in rulemaking should be able to ease the twin burdens of delay and cost that plague users of

FOIA and E-FOIA (Grunewald, 1998). In addition, e-rulemaking will greatly enhance the "government in the sunshine" philosophy that "does more than merely create a visitors gallery and convert rulemaking into a spectator sport...[that] empowers the public to question the proposed regulation, and the data and assumptions on which it is based, before it becomes effective" (Kannan, 1996, p. 219). Ultimately, this increased transparency and accessibility will lower the number of FOIA requests and force agencies to better defend their decisions when they release their final rules.

New systems for e-rulemaking can also be expected to result in user-friendly methods for navigation through complex and heterogeneous dockets. As noted by Professor Lubbers (2002), the growth of useful government Web sites in recent years is nothing short of remarkable. With that expansion of e-government services on the Web, both by rule-writing agencies and the National Archives and Records Administration, came a host of navigability challenges, not only for the commenting public, but also for end users and IT managers in the federal agencies. Access to more information is a boon to democracy only when it is easily penetrable with tools that eliminate insignificant documents from a query.

One useful way to describe this capacity is in terms of horizontal and vertical axes (Lubbers, 2002, p. 4) that allow online inspection and full-text search capacity from cradle to grave in a rulemaking life cycle. The horizontal view captures "every meaningful step" in rulemaking, with end users helping to define what counts as meaningful, while the vertical view allows access to all the studies and comments that shaped a final rule. While these tools are emerging rapidly in the private sector, the process of aligning all federal rulemaking to a "global seamless view," as described by Professor Lubbers, remains a slow and cumbersome one. Nonetheless, IT visionaries like Eduard Hovy imagine a perpetual "Super-Google" on every rulemaking that goes backward and forward in time, gathering, sorting, and clustering documents from every imaginable electronic source. Others, such as Robert Carlitz who is very usefully exposing system flaws by unleashing his army of "bots" on the system, call for technology development with a normative bent. According to Carlitz (2006, p. 9),

an electronic docket should be more than an online incarnation of the old paper-based system; it should be a gateway to new forms of public participation, encouraging interaction among participants who can read each other's comments as they are submitted; and it should provide a platform for online dialogue. It should also facilitate multi-media contributions, allow for complex, linked submissions and have the capacity for direct access to computer models used to generate data for specific comments.

Conclusion

A rich and challenging dialogue about the shape of e-rulemaking is under way. While in its infancy, an interdisciplinary research community has formed to assess and inform the development of information technologies that serve the public and rule writers. To date, little is actually known about whether this transition is likely to benefit or degrade the role of public participation, though the murmurs abound about the many perils of moving rulemaking online. As with all policy innovation, particularly technologically determined innovation, the risk of unintended consequences is present. While the Internet may usher in a new era of more inclusive, deliberative, and legally defensible rulemaking, it may be just as likely to reinforce existing inequalities, or worse, create new pitfalls for citizens wishing and entitled to influence the decision-making process.

Still, much of the evidence gathered over the last 5 years of workshops and sustained dialogue with agency personnel suggest that the parties responsible for building and using an integrated federal e-rulemaking system are aware of the high stakes measured in terms of democratic legitimacy, accountability, and regulatory effectiveness. To translate that awareness into a functioning architecture, scholars, federal officials, and many public stakeholders will need to continue to deliberate in a transparent and inclusive manner.

The public rationale for e-government is often couched in the rhetoric of cost savings and other familiar efficiency metrics. Efficient, effective, and responsive e-government, as defined by the OMB, may be at odds with the core principles of participatory democracy as envisioned by advocates of wider and enhanced forms of public commentary. Since federal agencies are neither equipped nor well positioned to examine the impact of IT on democracy, it remains for students of democratic theory, administrative law, and many others to investigate how e-rulemaking actually impacts citizen notions of trust and legitimacy and the nature of the deliberative process (Schlosberg & Dryzek, 2002). A new generation of interdisciplinary scholars is embracing the study of public administration, intrigued by the unpredictable but seemingly powerful impact of information technology on the regulatory process.

Given the importance of regulatory rulemaking and public participation, these efforts will necessarily provoke greater public scrutiny of the architecture of Regulations.Gov and similarly critical citizen-government interfaces. The dialogue under way will help inform those who find themselves steering the juggernaut known as e-government as well as the end-user public, whose practices and demands drive a cat and mouse game that can occasionally become a part of the rulemaking process. For now, all the parties encountered seem willing to support a wider and far-reaching debate about what a better e-rulemaking system might be.

Acknowledgments

This research was made possible with three National Science Foundation grants, EIA-0089892 "Digital Government: SGER: Citizen Agenda-Setting in the Regulatory Process: Electronic Collection and Synthesis of Public Commentary"; EIA-0328914, "SGER Collaborative: A Testbed for eRulemaking Data," and SES-0322662, "Democracy and E-Rulemaking: Comparing Traditional vs. Electronic Comment from a Discursive Democratic Framework." Any opinions, findings and conclusions or recommendations expressed in this material are those of the author, and do not necessarily reflect those of the National Science Foundation.

References

Applbaum, A. I. (2002). Failure in the cybermarketplace of ideas. In E. C. Kamarck & J. S. Nye, Jr. (Eds.), *Governance.com: Democracy in the information age* (pp. 17-31).Washington, DC: Brookings Institution Press.

Beierle, T. C. (1998). Public participation in environmental decisions: An evaluation framework using social goals (Discussion Paper No. 99-06). *Resources for the Future.* Retrieved June 2, 2003, from http://www.rff.org/CFDOCS/disc_papers/PDF_files/9906.pdf

Beierle, T. C. (2003). Discussing the rules: Electronic rulemaking and democratic deliberation (Discussion Paper No. 03-22). *Resources for the Future.* Retrieved June 2, 2003, from http://www.rff.org/disc_papers/PDF_files/0322.pdf

Beierle, T. C., & Crawford, J. (2002). *Democracy in practice: Public participation in environmental decisions.* Washington, DC: Resources for the Future.

Bingham, L. B., Nabatchi, T., & O'Leary, R. (2005). The new governance: Practices and processes for stakeholder and citizen participation in the work of government. *Public Administration Review, 65*(5), 547-558.

Brandon, B. (2003). An update on the e-government act and electronic rulemaking. *Administrative and Regulatory Law News, 29*(1), 7-9.

Brandon, B. H., & Carlitz, R. D. (2002). Online rulemaking and other tools for strengthening civic infrastructure. *Administrative Law Review, 54*(4), 1421-1478.

Carlitz, R. (2006). *Metadata for the FDMS.* Paper presented at the eRulemaking at the Crossroads 2006 Workshop. Retrieved August 25, 2006, from http://erulemaking.ucsur.pitt.edu/doc/Crossroads.pdf

Carlitz, R. D., & Gunn, R. W. (2002). Online rulemaking: A step toward e-governance. *Government Information Quarterly, 19*, 389-405.

Coglianese, C. (2003a). *E-rulemaking: Information technology and regulatory policy* (Report No. RPP-05). Harvard University, John F. Kennedy School of Government, Center for Business and Government, Regulatory Policy Program. Retrieved August 27, 2006, from http://www.ksg.harvard.edu/m-rcbg/research/rpp/reports/RPPREPORT5.pdf

Coglianese, C. (2003b, April). *The Internet and public participation in rulemaking.* Paper presented at the Conference on Democracy in the Digital Age.

Comedy, Y. L. (2002). The federal government: Meeting the needs of society in the information age. In W. J. McIver, Jr. & A. K. Elmagamid (Eds.), *Advances in digital government: Technology, human factors, and policy* (pp. 215-229). Boston: Kluwer.

Cooney, J. F. (1997). Regulatory reform: The long and winding road. *Administrative and Regulatory Law News, 23*(1), 10-12.

Craig, R. K. (2003). The Bush administration's use and abuse of rulemaking, part I: The rise of OIRA. *Administrative and Regulatory Law News, 28*(4), 8-13.

Cuéller, M. (2003). Notice, comment and the regulatory state: A case study from the USA PATRIOT act. *Administrative and Regulatory Law News, 28*(4), 3-4, 16.

Davis, S., Elin, L., & Reeher, G. (2002). *Click on democracy: The Internet's power to change political apathy into civic action.* Boulder, CO: Westview Press.

Dawes, S. S., Bloniarz, P. A., Kelly, K. L., & Fletcher, P. D. (1999). *Some assembly required: Building a digital government for the 21st century.* Retrieved June 2, 2003, from http://demo.ctg.albany.edu/publications/reports/some_assembly

Dawes, S. S., & Pardo, T. A. (2002). Building collaborative digital government systems. In W. J. McIver, Jr. & A. K. Elmagamid (Eds.), *Advances in digital government: Technology, human factors, and policy* (pp. 259-273). Boston: Kluwer.

Dotson, K. (2003). *Regulations.gov: The future of regulatory policy.* Retrieved June 2, 2003, from http://www.ksg.harvard.edu/cbg/news/e-rulemaking_workshop_1-24-03.htm

Dryzek, J. S. (2000). *Deliberative democracy and beyond: Liberals, critics, contestations.* New York: Oxford University Press.

Emery, F., & Emery, A. (2005). A modest proposal: Improve e-rulemaking by improving comments. *Administrative and Regulatory Law News, 31*(1), 8-9.

Excella Consulting. (2002). *Cross-agency edocket assistance.*

Fischer, F. (2000). *Citizens, experts, and the environment: The politics of local knowledge.* Durham, NC: Duke University Press.

Fletcher, P. D. (2002). Policy and portals. In W. J. McIver, Jr. & A. K. Elmagamid (Eds.), *Advances in digital government: Technology, human factors, and policy* (pp. 231-241). Boston: Kluwer.

Fountain, J. (2001). *Building the virtual state: Information technology and institutional change.* Washington, DC: Brookings Institution Press.

Fountain, J. (2002a). *Information, institutions, and governance: Advancing a basic social science research program for digital government.* Cambridge, MA: Harvard University.

Fountain, J. (2002b). A theory of federal bureaucracy. In E. C. Kamarck & J. S. Nye, Jr. (Eds.), *Governance.com: Democracy in the information age* (pp. 117-140). Washington, DC: Brookings Institution Press.

Furlong, S. R., & Kerwin, C. M. (2004). Interest group participation in rule making: A decade of change. *Journal of Public Administration Research and Theory, 15*(3), 353-370.

Garson, G. D. (2003). Toward an information technology research agenda for public administration. In G. D. Garson (Ed.), *Public information technology: Policy and management issues* (pp. 331-357). Hershey, PA: Idea Group Publishing.

General Accounting Office (GAO). (2000). *Federal rulemaking: Agencies' use of information technology to facilitate public participation* (GAO-00-135R). Washington, DC: GPO.

General Accounting Office (GAO). (2001). *Regulatory management: Communication about technology-based innovations can be improved* (GAO-01-232). Washington, DC: GPO.

General Accounting Office (GAO). (2003). *Electronic rulemaking: Efforts to facilitate public participation can be improved* (GAO-03-901). Washington, DC: GPO.

General Accounting Office (GAO). (2005). *Electronic rulemaking: Progress made in developing centralized e-rulemaking system* (GAO-05-777). Washington, DC: GPO.

Golden, M. M. (1998). Interest groups in the rule-making process: Who participates? Whose voices get heard? *Journal of Public Administration Research and Theory, 8*(2), 245-270.

Grunewald, M. H. (1998). E-FOIA and the "mother of all complaints": Information delivery and delay reduction. *Administrative Law Review, 50*(2), 345-369.

Herz, M. (2003). Rulemaking. In J. S. Lubbers (Ed.), *Developments in administrative law and regulatory practice 2001-2002.* Chicago: ABA Publishing. Retrieved August 20, 2006, from http://www.ksg.harvard.edu/cbg/rpp/erulemaking/papers_reports/Herz_E_Rulemaking.pdf

Holden, S. H. (2003). The evolution of information technology management at the federal level: Implications for public administration. In G. D. Garson (Ed.), *Public information technology: Policy and management issues* (pp. 53-73). Hershey, PA: Idea Group Publishing.

Holmes, D. (2001). *eGov: eBusiness strategies for government.* London: Nicholas Brealy Publishing.

Hovy, E. (2003). *eRulemaking: Research problems for IT.* Retrieved from http://www.ksg.harvard.edu/m-rcbg/research/rpp/RPP-2002-04.pdf

Johnson, S. M. (1998). The Internet changes everything: Revolutionizing public participation and access to government information through the Internet. *Administrative Law Review, 50*, 277-337.

Kannan, P. M. (1996). The logical outgrowth doctrine in rulemaking. *Administrative Law Review, 48*, 213-225.

Kerwin, C. M. (2003). *Rulemaking: How government agencies write law and make policy* (3rd ed.). Washington, DC: CQ Press.

Larsen, E., & Rainie, L. (2002). The rise of the e-citizen: How people use government agencies' Web sites. *Pew Internet and American Life Project.* Retrieved May 5, 2002, from http://www.pewinternet.org/reports/pdfs/PIP_Govt_Website_Rpt.pdf

Leahy, P. (1998). The electronic FOIA amendments of 1996: Reformatting the FOIA for on-line access. *Administrative Law Review, 50*(2), 339-344.

Lessig, L. (1999). *Code and other laws of cyberspace.* New York: Basic Books.

Lubbers, J. S. (1998). *A guide to federal agency rulemaking* (3rd ed.). Chicago: ABA Publishing.

Lubbers, J. S. (2002). *The future of electronic rulemaking: A research agenda* (Regulatory policy working paper RPP-2002-04). Cambridge, MA: Center for Business and Government, John F. Kennedy School of Government, Harvard University. Retrieved February 19, 2007, from http://www.ksg.harvard.edu/m-rcbg/research/rpp/RPP-2002-04.pdf

Lubbers, J. S. (2006). *A guide to federal agency rulemaking* (4th ed.). Chicago: ABA Publishing.

Morales, O., & Moses, J. (2006). *eRulemaking's federal docket management system.* Paper presented at the eRulemaking at the Crossroads 2006 Workshop. Retrieved August 25, 2006, from http://erulemaking.ucsur.pitt.edu/doc/Crossroads.pdf

National Research Council. (2000). *Making IT better: Expanding information technology research to meet society's needs.* Washington, DC: National Academy Press.

National Research Council. (2002). *Information technology research, innovation, and e-government.* Washington, DC: National Academy Press.

Nye, Jr., J. N. (2002). Information technology and democratic governance. In E. C. Kamarck & J. S. Nye, Jr. (Eds.), *Governance.com: Democracy in the information age* (pp. 1-16). Washington, DC: Brookings Institution Press.

Office of Management and Budget (OMB). (2002a). *E-government strategy.* Retrieved December 6, 2002, from http://www.whitehouse.gov/omb/inforeg/egovstrategy.pdf

Office of Management and Budget (OMB). (2002b). *OMB accelerates effort to open federal regulatory process to citizens and small businesses* (OMB 2002-27).

Office of Management and Budget (OMB). (2002c). *The President's management agenda: Fiscal year 2002.* Retrieved July 11, 2003, from http://www.whitehouse.gov/omb/budget/fy2002/mgmt.pdf

Office of Management and Budget (OMB). (2003a). *E-rulemaking.* Retrieved June 2, 2003, from http://www.whitehouse.gov/omb/egov/gtob/rulemaking.htm

Office of Management and Budget (OMB). (2003b, March). *Regulations.gov to transform U.S. rulemaking process and save nearly $100 million* [Press release]. Retrieved June 17, 2003, from http://www.whitehouse.gov/omb/pubpress/2003-03.pdf

OMB Watch. (2002). *Administration pushes e-rulemaking.* Retrieved June 2, 2003, from http://www.ombwatch.org/article/articleview/846/1/39/

Perritt, H. H., Jr. (1995). Executive summary. *Electronic Dockets: Use of Information Technology in Rulemaking and Adjudication Report to the Administrative Conference of the United States.*

Retrieved June 2, 2003, from http://www.kentlaw.edu/classes/rstaudt/internetlaw/casebook/Downloadsexecutiv.htm

President's Information Technology Advisory Committee. (2000). *Transforming access to government through information technology.* Retrieved June 2, 2003, from http://www.hpcc.gov/pubs/pitac/pres-transgov-11sep00.pdf

Rakoff, T. D. (2000). The choice between formal and informal modes of administrative regulation. *Administrative Law Review, 52*(1), 159-174.

Revisiting the rules on organic food [Editorial]. (1998, April 13). *New York Times.* Retrieved February 22, 2000, from *LEXIS®-NEXIS® Academic Universe.*

Roberts, N. (2004). Public deliberation in an age of direct citizen participation. *American Review of Public Administration, 34*(4), 315-353.

Rocheleau, B. (2003). Politics, accountability, and governmental information systems. In G. D. Garson (Ed.), *Public information technology: Policy and management issues* (pp. 20-52). Hershey, PA: Idea Group Publishing.

Schelin, S. H. (2003). E-government: An overview. In G. D. Garson (Ed.), *Public information technology: Policy and management Issues* (pp. 120-137). Hershey, PA: Idea Group Publishing.

Schlosberg, D., & Dryzek, J. S. (2002). Digital democracy: Authentic or virtual? *Organization & Environment, 15*(3), 332-335.

Schorr, H., & Stolfo, S. J. (1997). *Toward the digital government of the 21st century.* Retrieved June 2, 2003, from http://www.isi.edu/nsf/final.html

Seifert, J. (2006). *Keeping e-rulemaking on the docket: Analog challenges to digital government.* Paper presented at the eRulemaking at the Crossroads 2006 Workshop. Retrieved August 25, 2006, from http://erulemaking.ucsur.pitt.edu/doc/Crossroads.pdf

Shulman, S. W. (2003). An experiment in digital government at the United States National Organic Program. *Agriculture and Human Values, 20*(3), 253-265.

Shulman, S. W. (2005). The Internet still might (but probably won't) change everything. *I/S Journal, 1*(1), 111-145.

Shulman, S. W. (2006). Whither deliberation: Mass e-mail campaigns and U.S. regulatory rulemaking. *Journal of E-Government, 3*(3), 41-64.

Shulman, S. W., Schlosberg, D., Zavestoski, S., & Courard-Hauri, D. (2003). Electronic rulemaking: New frontiers in public participation. *Social Science Computer Review, 21*(2), 162-178.

Shulman, S. W., Callan, J., Hovy, E., & Zavestoski, S. (2004). SGER: A testbed for erulemaking. *Journal of E-Government, 1*(1), 123-127.

Shulman, S., Zavestoski, S., Schlosberg, D., Courard-Hauri, D., & Richards, D. (2001). Citizen agenda-setting: The electronic collection and synthesis of public commentary in the regulatory rulemaking process. In *Proceedings of the First National Conference on Digital Government Research.*

Skrzycki, C. (2003, January 23). US opens online portal to rulemaking. *Washington Post*, p. E01.

Strauss, P. (1989). *An introduction to administrative justice in the United States.* Durham, NC: Carolina Academic Press.

Stoll, R. G., Herz, M., & Lazarski, K. E. (2006). Rulemaking. In J. S. Lubbers (Ed.), *Developments in administrative law and regulatory practice 2004-2005.* Chicago: ABA Publishing.

Sunstein, C. (2001). *Republic.com.* Princeton, NJ: Princeton University Press.

Tesh, S. (2002). The Internet and the grassroots. *Organization & Environment, 15*(3), 336-339.

West, W. (2005). Administrative rulemaking: An old and emerging literature. *Public Administration Review, 65*(6), 655-668.

West, W. F. (2004). Formal procedures, informal processes, accountability, and responsiveness in bureaucratic policy making: An institutional analysis. *Public Administration Review, 64*(1), 66-80.

Wilhelm, A. G. (2000). *Democracy in the digital age: Challenges to political life in cyberspace.* New York: Routledge.

Yackee, S. W. (2005). Sweet-talking the fourth branch: The influence of interest group comments on federal agency rulemaking. *Journal of Public Administration Research and Theory, 16*, 103-124.

Yang, K. (2005). Public administrators' trust in citizens: A missing link in citizen involvement efforts. *Public Administration Review, 65*(3), 273-285.

Young, I. M. (2000). *Inclusion and democracy.* Oxford: Oxford University Press.

Zavestoski, S., & Shulman, S. W. (2002). The Internet and environmental decision-making. *Organization & Environment, 15*(3), 323-327.

Endnotes

1. Semistructured interviews were conducted by a team of three academic researchers between July 21 and July 24, 2003. Most of the 15 interviews were conducted on the condition of anonymity in accordance with Human Subjects guidelines at each of the researchers' universities. Interviews were conducted at the Environmental Protection Agency (EPA), USDA, DOT, IRS, HHS, and GSA.

2. The seven agencies assessed in the Excella report were the EPA, DoT, Occupational Safety & Health Administration (OSHA), Nuclear Regulatory Commision (NRC), Federal Communications Commission (FCC), FDA, and Department of Energy (DoE).

3. For more information on the computer science research on information retrieval, summarization, clustering and other techniques being brought to bear using real-world eRulemaking data, visit the NSF-funded eRulemaking project home page at http://erulemaking.ucsur.pitt.edu/.

Section II

Computer Applications in Public Administration

Chapter X

IT Innovation in Local Government: Theory, Issues, and Strategies

Charles C. Hinnant, U.S. Government Accountability Office, USA

John A. O'Looney, University of Georgia, USA

Abstract

We examine the adoption of information technology within local governments in the United States. The social and technical factors that impact the process of technological innovation are discussed in reference to the adoption of advanced electronic government (e-government) technologies in local government. In particular, we discuss how the adoption of IT, and e-government, is influenced by the local government's motivations to innovate, technology characteristics, available resources, and stakeholder support. We then discuss several strategies that may address these factors. We argue that local governments should seek to formally assess the need to adopt e-government technologies, develop new funding strategies, and develop a mix of in-house and contracted IT services. While local governments have aggregately adopted advanced transaction-based forms of e-government at a lower rate than state and federal governments, it is our contention that local governments are merely reacting to innovation factors within their social and technical environments.

Introduction

Within the governmental framework of the United States, it can legitimately be argued that local governments play a far more prevalent role in the day-to-day existence of both individual citizens and private and nonprofit organizations than do state and federal governments. After all, a simple assessment of the numbers reveals that local governments outnumber state and federal governments 87,586 to 51 (U.S. Census Bureau, 2002). In addition, many citizens and organizations are simultaneously subject to a diverse array of local government jurisdictions and authorities, such as city, county, or special-district authorities. Furthermore, local governments play a crucial role in the provision of key public services, such as education, community development, public health activities, public utilities, solid-waste removal, law enforcement, and public safety.

Given the critical services that local governments have traditionally provided, it is not surprising that, like other public institutions, they have long employed information technology as a means to improve internal operations in the production of those services. If we take a systems perspective, we can define IT broadly as we might any "computer-based information system [which is an] information system that requires hardware, software, databases, telecommunications, procedures, and people to accomplish goals" (Stair, 1992, p. 27). Employing this broad definition, it is obvious that the use of IT is necessary for even the most ordinary of activities undertaken by local governments. For example, examinations of local governments in the United States indicate that there has been a growing trend toward adoption of IT within local government over the past 25 years and that today virtually all make use of IT to one extent or another (Kraemer & Norris, 1994; Norris, 2003). In fact, a 1997 survey of city and county governments carried out by the International City/County Management Association (ICMA, 1997) indicated that only 3% of respondents did not use computers of some kind to support operations.

As new forms of IT became available and adopted within society, local governments learned new ways of employing the technology to achieve their own institutional goals. For instance, during the 1970s and 1980s, much attention was given to the adoption and impact of mainframe and then personal computers on the operations and internal environment of local governments (Kraemer, Dutton, & Northrop, 1981; Kraemer & Norris, 1994; Norris & Kraemer, 1996). By the mid-1990s, the increasingly widespread use of distributed networks, such as the Internet and World Wide Web (WWW), began to shift the focus toward how public-sector organizations could harness this new form of IT to deliver information and services directly to the public. This new focus on electronic government, or e-government, reoriented the focus on how IT could be used by broader government reform initiatives to have agencies provide programmatic information and services to citizens and other stakeholders (Kraemer & King, 2003; Watson & Mundy, 2001).[1]

Local governments have quickly adopted at least rudimentary aspects of e-government. As indicated in Table 1, the U.S. Census reported in 2002 that 45.1% of counties, 31.1% of cities, 13.4% of townships, 17.8% of special districts, and 64.3% of school districts responding indicated that they provided information regarding their central activities via a Web site. Similarly, 54.1% of counties, 40.6% cities, 21.2% of townships, 34.6% of special districts, and 73.7% of school districts indicated that they provided a means for the public to communicate or transact business by use of the Internet or another computer-based

Table 1. Local government use of Internet to interact with public (Source: 2002 Census of Governments, Vol. 1, n. 1)

		Central activity information is provided on an Internet Web site that is maintained or controlled by the government.			***Public can communicate or transact business with the government using Internet, e-mail, or other computer-based system.***		
Government Type	**Total Respondents (n)**	**Yes (%)**	**No (%)**	**Response Rate**	**Yes (%)**	**No (%)**	**Response Rate**
County	2,453	45.1	54.9	80.9	54.1	45.9	71.2
Municipal	15,116	31.1	68.9	77.8	40.6	59.4	73.8
Town or Township	10,397	13.4	86.6	63.0	21.2	78.8	61.4
Special Districts	20,337	17.8	82.2	58.0	34.6	65.4	57.9
School Districts	10,880	64.3	35.7	80.6	73.7	26.3	79.8
Total governments	59,183						
Governments by Population Size*							
under 25,000	25,125	19.8	80.2	70.1	30.7	69.3	68.1
25,000 to 49,999	1,329	69.4	30.6	85.5	67.1	32.9	70.8
50,000 to 99,999	784	81.0	19.0	92.9	74.5	25.5	71.0
100,000 to 249,999	458	91.5	8.5	96.0	83.3	16.7	74.2
250,000 to 499,999	149	91.3	8.7	98.7	85.3	14.7	67.5
500,000 or more	121	93.4	6.6	99.2	90.6	9.4	69.7
Total governments	27,966						

*Note: Calculations performed by author; *includes only county, municipal, and township governments*

information system. Furthermore, the size of the government, as measured by population, seemed to have a distinct impact on the adoption of e-government.[2] For governments with populations under 25,000, only 19.8% indicated that they provided information regarding their activities on a government-controlled Web site, and only 30.7% indicated that they provided the ability for the public to communicate and conduct business via the Internet or other computer-based information system. Conversely, 93.4% of local governments with populations over 500,000 indicated that they provided information on a government-controlled Web site, while 90.6% indicated that they provided the ability for the public to communicate and conduct business via the Internet or other computer-based information system. Similarly, a more recent ICMA survey conducted in 2004 reported that almost all (99.4%) of respondents indicated that their local governments had Internet connectivity and 91.1% indicated that their local governments had a Web site.[3] Furthermore, 92.3% indicated that they employed DSL (digital subscriber line), cable, or high-bandwidth connections to facilitate Internet connectivity (ICMA, 2004).

While almost all local governments have adopted at least basic forms of e-government, such as a simple Web site or providing the ability for the public to communicate with the government through the use of e-mail, there is some question as to the extent to which local governments have pursued the adoption of more advanced forms and use of IT that might improve both the production and delivery of public services. Research indicates that many local governments have initiated at least rudimentary attempts to post information and provide basic online services, but few local governments have adopted more advanced forms of ICT that foster high levels of interactivity, communication, and actual political participation (Moon, 2002; Norris & Moon, 2005). Results from a 2004 ICMA survey of cities and counties also indicate that local governments may be relatively late adopters of more recent advanced forms of Web-based technologies. While a majority of respondents indicated that their Web sites provided the ability to download information or forms (council minutes, codes and ordnances, job applications) and participate in online communication with officials, relatively few offered more sophisticated abilities to interact or conduct transactions online.[4] Similarly, the survey indicates that local governments also seem to be slow in adopting Web-based information systems, such as intranets, which are purported to improve internal information exchange and operations. Only about 50% of the local governments who responded indicated that they currently used an intranet. Of those local governments employing an intranet, most used them to facilitate internal communication and few used them for more sophisticated purposes, such as online training (29.4%), online procurement (26.6%), online project team collaboration (33.4%), and the management of time sheets (27.3%; ICMA, 2004). Most local governments seem to be late adopters—adopting e-government technologies only after they have become established technologies and practices. There is little evidence to suggest that there exists high levels of adoption by local governments for relatively advanced IT, when compared to advanced IT's rate of adoption by the private sector or state and federal governments.

A recent study, based on survey responses conducted as part of the Pew Internet and American Life Project (2006), estimated that 73% of adults in the United States were Internet users and that 42% of adults had broadband connections at home. Despite the seemingly high rate of general Internet use by the U.S. public, a recent study of dial-up and broadband users indicates that Internet users still have a slightly less positive view of how much the Internet has improved their interaction with local government when compared with the perceived improvements in interaction with state and federal government (Pew Internet and American Life Project, 2004).[5] This seems to indicate that many local governments are not adopting and implementing advanced forms of IT in a manner that could potentially bring the purported benefits of openness, transparency, and more efficient service delivery often associated with e-government.

This chapter attempts to provide a theoretical lens through which to examine the adoption of more advanced forms IT by local governments. By examining the literature on technological innovation, as well as more recent work regarding public-sector information systems, we hope to highlight the most important factors that influence the adoption of advanced forms of IT by local governments. Furthermore, we propose several strategies that may influence local governments' ability to successfully adopt new forms IT.

IT as Technological Innovation

The adoption and use of IT by local governments is certainly not a new phenomenon, but the adoption of e-government across all levels of government during the past decade has rejuvenated the interest in government's use of IT by practitioners and academics alike. Unfortunately, the growing interest in e-government has not always coincided with an improved understanding of how local governments adopt and implement the more advanced forms of IT. If e-government can be viewed as the latest step in a progression of innovations in government that is predicated on the use of new forms of IT, then the literature associated with the adoption of technology within complex organizations may provide some insight into the social and technical factors that influence how e-government may be adopted by local government. Since technological innovation is typically defined as "the situationally new development and introduction of knowledge-derived tools, artifacts, and devices by which people extend and interact with their environment," it would seem that a more complete understanding of e-government adoption within local governments would be informed by a better understanding of the innovation process (Tornatzky & Fleischer, 1990, p. 11).

The process of technological innovation is often described as a set of stages that includes an awareness of new technology, understanding the match between the technology and organization, adoption of the technology, implementation of the technology, and routinization of its use (Tornatzky & Fleischer, 1990). While the process of technological innovation is often viewed in terms of a linear progression of stages, there is no certainty of a linear progression or ultimately success; the process often has many delays or reverses (Tornatzky & Fleischer, 1990). Similarly, the innovation process for most complex organizations is heavily influenced by the interrelationship of social and technical factors within the organization's internal and external environments. With regard to technological innovation, issues surrounding the technology itself, the availability of required resources, the fit with the organization's primary task, and an organization's structural arrangements all play significant roles in the nature and success of technological innovation.

Research indicates that the characteristics of the technology play an important role in determining whether or not it will be adopted by an organization. Researchers have identified a rather long list of innovation characteristics that can increase or decrease the likelihood of adoption by organizations (Zaltman, Duncan, & Holbek, 1973). With regard to the adoption of technological innovation, a study by Tornatzky and Klein (1982) indicates that three primary characteristics are repeatedly associated with the adoption of new technologies: relative advantage, ease of use, and compatibility. Similarly, the extent to which a particular technology alters existing organizational processes also plays a role in the innovation process. So-called radical innovations usually involve a significant alteration of an organization's processes or outputs, or significantly impact the organization's key stakeholders (Dewar & Dutton, 1986; Ettlie, Bridges, & O'Keefe, 1984). Radical innovations generally experience more risks for failure or setbacks than do technological innovations that involve only incremental changes in an organization's existing technological environment.

The decision to adopt a technological innovation, such as a new form of IT, is predicated on the assumption that organizational decision makers have sufficient awareness of new technologies to understand their potential benefits. This awareness is often discussed as a function of the knowledge that internal stakeholders possess regarding the applicability of new technologies or processes to the organization's operations. In essence, the greater the information and knowledge assets that an organization has at its disposal, the more likely it is to find new technologies to address operational problems, and the more likely it is to understand and implement the technology (Fichman & Kemerer, 1997; Nilakanta & Scamell, 1990). Although all technologies require some learning on the part of the staff participating in the adoption, some technologies place many more demands on adopters for new knowledge and skills. Such technologies are believed to have inherent knowledge barriers because the knowledge required to implement them creates a barrier to diffusion (Attewell, 1992).

In addition to the availability of resources, technologies that have greater levels of congruency with key organizational tasks may be perceived as more useful and, therefore, as a more successful adoption of technology (Cooper & Zmud, 1990). Similarly, studies have shown that internal decision makers, such as managers, search for technological solutions to problems that they face in carrying out their jobs. For example, they may show interest in new forms of IT if the organizational task for which they are responsible requires high levels of communication with external agencies (Bugler & Bretschneider, 1993). Such interest and support by key stakeholders may be important for the successful implementation of new technologies, as studies highlight the importance of managerial support (Beath, 1991). In addition, champions who informally and enthusiastically promote the new technology are also viewed as important during adoption and implementation (Tushman & Nadler, 1986).

While some researchers have focused on user attitudes and perceptions toward IT innovations, there is some debate over the relative importance of the attitudes held by organizational members in comparison to the importance of organizational structure and processes (Hall, 1996). Some organizational-level research indicates that the structural arrangement of the organization plays an extremely important role in the ability to fully accept technological innovations. Organizations that have high levels of structural complexity, low levels of formalization, and low levels of centralization tend to initiate relatively more innovations than do organizations with opposing characteristics (Damapour, 1991; Duncan, 1976). Most of this research assumes that an organization's successful adoption of technological innovations is linked to a perceived need for technology to improve performance or address requirements from external stakeholders. Overall, these findings indicate that the decision to adopt and fully accept IT innovations is a joint result of technology characteristics and the social system in which they are embedded.

Social and Technical Issues to Consider When Adopting E-Government within Local Government

E-government is often associated not only with the adoption of IT, but specifically the adoption of advanced forms of IT, as well as organizational practices that are often employed in the private sector to facilitate electronic commerce, or e-commerce. IT innovations, such as Internet-based applications that provide integration and efficiency in a company's supply chain or more transaction-based services for customers, are often proposed as a means to facilitate similar efficiencies for government. As previously discussed, several studies have highlighted the difficulty for local governments, on a whole, to adopt the more advanced forms of IT that are often associated with e-government (Moon, 2002; Norris, 2003; Norris & Moon, 2005). The slow adoption rates of many of the advanced e-government interactive technologies, such as transaction-enabled Web sites, personalization of Web sites, and online political forums, seem to highlight the relatively slow diffusion of more advanced forms of IT within many local governments (Hinnant & O'Looney, 2003; O'Looney, 2001a, 2001b; Norris & Moon). However, it is important to understand that the adoption of advanced IT almost always necessitates significant investments in resources, as well as back-end integration of the new technology within the government's technological and social processes. Our discussion of technological innovation within complex organizations provides additional insights into the sociotechnical factors that most likely impact the adoption of IT within local governments. A review of the literature on technological innovation reveals several dimensions that should be considered when evaluating the adoption of innovation by local governments. These include goals and motivations to adopt, technology characteristics, availability of resources, and the support of various stakeholders.

Goals and Motivations to Innovate

By most accounts, the reasons for adopting new forms of IT at the local government level are varied, but not necessarily surprising. The adoption of internal IT systems by local governments during the 1970s and 1980s was notably marked by the desire to increase internal efficiencies (Northrup, Kraemer, Dunkle, & King, 1990). Another motivation for the adoption of e-government is purported to be gains in efficiency within the provision of public services. Still another is the improved interaction and transparency with the public that many of the Internet-based technologies promise to facilitate (Moon, 2002). While most local governments have not yet adopted the most advanced forms of Internet-based technologies, many have implemented some type of Web site to improve communication with the public. Given that one goal of local government adoption of e-government should be to more effectively provide services to external stakeholders, it is somewhat surprising that most local governments do not seem to actively gauge what online services or features are desired by their citizens. A 2004 survey conducted by ICMA demonstrated that, while 67.8% of respondents indicated that they believed e-government has improved communication with the public, only 10% of local governments actively survey citizens or businesses to establish what they actually want online (ICMA, 2004).

Most local governments that have adopted more advanced forms of e-government practices seem to do so in order to facilitate or improve a specific functional area or to achieve specific operational cost reductions. For example, in a few cases, local governments, such as Montgomery County, Maryland, have begun to reduce its costs for energy by consolidating the buying power of 18 county agencies and organizations, as well as conducting online auctions for energy (Robinson, 2006). In other circumstances, local government agencies or departments adopt IT systems that facilitate intergovernmental efforts within a respective functional area or jurisdiction. For example, Pennsylvania's Justice Network (JNET) is a system initiated by a consortium of government agencies, in order to integrate criminal justice database systems throughout the commonwealth. JNET is designed to facilitate criminal justice information sharing and has many local, commonwealth, and federal law-enforcement organizations as stakeholders (Hinnant & Sawyer, 2003). While such case studies are in many ways idiosyncratic, they highlight how local governments do not seem to adopt IT for impromptu reasons. Rather, they seek to adopt IT when it is believed that it will result in the attainment of goals that are desirable to a specific program, department, or to the broader jurisdiction.

Technology Characteristics

The characteristics associated with a specific form of IT also play a role in whether it is adopted within a local government. For example, studies examining the adoption of computer applications by local governments indicate that applications perceived to have greater visibility and less uncertainty with regard to cost were more likely to be adopted (Perry & Danziger, 1979; Perry & Kraemer, 1979). This also seems to be true of the more current forms of IT that are often associated with e-government. For example, this may provide some insight into why most local governments offer a basic Web site, but few have adopted more interactive online technologies. In addition, it is important to understand that some advanced forms of IT may also possess higher levels of risk for local governments. For example, the use of some advanced Web sites employed by private-sector firms, such as Amazon.com, track user behavior in order to personalize the future content that the user views on the Web site. Similar technologies and practices are available for use by local governments and could be used to provide the citizen user with tailored government information or services. However, the collection of such data by local governments may present significant challenges since they may be legally required and/or understandably expected by the public to protect all personal information collected (Hinnant & O'Looney, 2003).

While it is useful to understand a technology's specific characteristics, it is also important to understand that such characteristics cannot be considered outside the societal and organizational contexts within which the technology may be adopted and implemented. At one level, the adoption of IT may be assessed across a set of local governments in order to evaluate overall levels of diffusion. However, it should be clear that the successful adoption of a particular IT, within a specific local government context, has more to do with how well the IT is judged to accomplish a specific task and the degree to which it fits within the existing technical environment.[6] With regard to the adoption of more advanced e-government technologies, local governments or their subunits are probably less likely to adopt IT that is significantly more sophisticated than its existing technical systems. The adoption of more radical forms of e-government technologies would most likely cause too heavy an investment in resources or introduce the need for significant adjustments to internal administrative systems. For example, recent surveys indicate that local governments perceived the greatest potential impacts of e-government to be increased demands on staff, changes in the role of staff, and the need to reengineer business processes (Norris & Moon, 2005).[7]

Availability of Resources

Tied closely to both the motivation to innovate as well as the characteristics of a specific IT, is the local government's available resources. Resources may take several forms, such as financial resources, the number of available staff, or the knowledge assets that are required to adopt and implement a respective technology. The adoption of any IT is known to have potentially significant impacts on the internal processes of any organization and this is also true of IT adoptions by local government organizations. The adoption of many forms of IT associated with e-government potentially requires significant departures from the organization's existing technical and administrative systems. In order to successfully identify, adopt, and implement a new form of IT, the organization must possess, develop, or obtain significant knowledge and expertise regarding not only the technology in question, but also how that technology may be successfully integrated with the existing social and technical structures of the organization. Surveys of local governments consistently indicate that the lack of sufficient financial resources, Web staff, and Web expertise are the top three barriers to e-government initiatives (ICMA, 2004; Norris & Moon, 2005). Furthermore, examinations of advanced online technologies, such as personalization of services, also indicate that the adoption of new online services is often slowed by limitations of technical expertise and budgetary considerations (Hinnant & O'Looney, 2003).

Financial resources may be crucial not only for the initial development or acquisition of IT, but also for its successful integration within an organization's administrative systems. Integration of new IT often requires significant adjustments to existing systems. These adjustments may come in the form of reengineering administrative systems or subsequent changes in the necessary skills possessed by staff who operate the new IT system. More importantly, financial resources may also provide a means for acquiring additional knowledge that the IT may require. This may be in the form of training current staff or in acquiring knowledge from private-sector consultants or vendors. Despite the need for financial resources, 81.7% of local governments had no separate budget item for e-government and only 40.6% of local governments have a separate IT department that is responsible for IT and e-government (ICMA, 2004). The more advanced the IT is relative to the local government's current technical systems, the more knowledge and financial resources may be required since prior work has indicated that a local government's existing level of technological sophistication influences its ability to adopt new forms of IT (Norris & Kraemer, 1996). With regard to the adoption of e-government, this may simply mean that those that are already relatively technologically sophisticated have the resources to draw on in order to adopt and implement even more advanced forms of e-government. Conversely, less sophisticated organizations are less likely to possess the skills and knowledge necessary to sidestep the knowledge barriers that are so often associated with the adoption and implementation of new IT.

Support from Managers, Political Leaders, and Other Stakeholders

While goals, resources, and IT characteristics all play important roles in whether a local government organization will adopt a new form of IT, so will decision makers from both within the organization and in the broader political environment. Obviously program managers, and in some cases political leaders and external stakeholders, should play a role in the development of the organizational goals that precede the actual adoption of any new IT. Their political support throughout the innovation process is also important for the successful adoption of new IT. With regard to the adoption of advanced forms of e-government, program managers, and the operations that they oversee, are more likely to experience the impacts of any changes that the new IT brings. For example, surveys indicate that e-government has the potential to alter the role of staff and the demands on staff, and necessitate the reengineering of processes (ICMA, 2004). Or, the adoption of IT may reinforce the power of existing administrative social structures (Kraemer, King, Dunkle, & Lane, 1989). While e-government, or any IT adopted, may improve operations and introduce efficiencies into internal administrative processes, it may also alter processes in a variety of unforeseen ways. Program managers are in a position to reap the benefits of any positive changes initiated by IT, but they also bear some of the risk for any negative impacts.

While local government administrators and managers play a role in IT adoption, political leaders also play a potentially important role. Although surveys indicate that only 12.5% of local governments cite lack of support from elected officials as a barrier to e-government, e-government can directly impact the nature of the interaction elected politicians have with the public. Over 65% of local governments now provide direct communication with elected officials through their Web sites. Furthermore, the same survey indicates that 39.9% of respondents indicate that e-government did increase citizen contact with elected or appointed officials (ICMA, 2004). So, the potential exists for e-government to directly impact the political lives of elected officials. Some scholars (e.g., Ihrke, Rabidoux, & Gabris, 2000; Svara 1990) suggest that council-staff relations can also play a role in how innovations are perceived. Essentially, they argue that a positive relationship between council and staff will lead to innovations being perceived as successful by other local government actors. Such a positive relationship between council and staff may also signal additional buy-in and support from elected officials in the case of e-government adoption by local governments. In addition, a key role played by elected officials has to do with funding. A 2004 ICMA e-government survey indicated that 92.6% of local governments funded e-government initiatives through general revenues (ICMA, 2004). Therefore, regarding IT, the most important role elected officials could play is to provide the financial support, through general revenues, for local government organizations to be innovative.

Strategies for Adopting IT in Local Government

Considering the social and technical factors that influence the technological innovation process and introduce risk into such a public context, it is important to highlight some strategic activities that may assist local governments in successfully adopting and implementing new forms of IT. Such activities should include an organizational assessment of needs and planning, as well as the development of funding sources and acquisition strategies. While prescriptive actions can be derived and discussed for the adoption of new forms of IT, it should be noted that the sheer variety of local governments, along with the intraorganizational variety with regard to knowledge and resources, makes a prescriptive discussion about the adoption of new IT less than systematic and certainly limits its generalizability. However, prescriptive discussions may be used to elicit new ideas and considerations for both future research and practice.

Formally Assess the Need for IT and Plan for Innovation

Prior to undertaking the adoption of an IT innovation that may require a local government to invest significant amounts of resources, it is important for the adopting organization to fully analyze the goals it intends to address with the IT, what tasks the IT is designed to accomplish, and the resource allocations required to implement the IT over its life of operation. In the case of e-government technologies that are designed to facilitate interaction with citizens, businesses, or other external stakeholders, an assessment of goals should include an assessment of whether the IT is actually desired by the potential user. With regard to the adoption of advanced forms of e-government, such as the use of personalization of Web sites, Hinnant and O'Looney (2003) proposed that governments should pay close attention to the level of sophistication of the citizens within their jurisdiction, in order to gauge the real demand for a particular online innovation. However, an ICMA study in 2004 indicated that only about 10% of local governments conduct surveys to determine what online services their citizens want. As for other e-government technologies, such as online procurement systems, successful adoption would depend on whether private-sector vendors either possess, or are willing to acquire, the knowledge and systems that may be required to use the online innovation.

Similarly, any acquisition of new IT or IT-management practices should also include an assessment of the level of fit between the IT and the institutional goals it is attempting to address. This may require the organization to undertake formal evaluations of the administrative processes that the IT seeks to improve, as well as of its specific technology characteristics. This assessment may also be completed as a precursor to the development of a formal technology plan that addresses how the IT will deal with specific organizational goals, as well as analyzing the return on investment associated with acquiring the IT. While such formal assessments and justifications of new IT systems have become prevalent in large private firms, state governments, and federal government agencies, it is less likely that most local governments have developed the internal capacity to complete such tasks (Hinnant & Sawyer, 2003; U.S. General Accounting Office [GAO], 2004). Furthermore, it is important to understand that the extent of any assessment and planning activities should be considered with regard to the adoption process and the IT being adopted. IT innovations that may require significant resources during adoption or that have significant inherent risks may give reason for more formalized, thorough, and rigorous planning efforts.

In addition, an early assessment of need and IT planning also serves a broader purpose to inform and, in some instances, educate other stakeholders. For example, such preparatory efforts may illuminate the benefits of potential adoption efforts to managers and political leaders, therefore assisting in garnering their political and financial support. Likewise, such efforts can also highlight the potential costs and risks inherent in IT adoption. Overall, such planning activities may not only enable local governments to better understand how the adoption of an IT innovation might address organizational goals or administrative processes, but they also provide a means to mitigate some of the risk associated with adopting IT within such a public environment.

Develop Alternative Funding Sources

The ability to acquire new IT is also predicated on the availability of the financial resources necessary not only to acquire the IT system, but implement the IT system and operate the IT system in the future. The availability of financial resources is consistently cited by local governments as one of the most significant barriers to e-government adoption (ICMA, 2004). Since the management of local government IT systems is often the responsibility of functional subunits that are often funded through general revenues, it would appear that IT innovation is often hampered by both the lack of a central IT department, as well as specific funding sources within local government. This situation is often exacerbated if the local government has limited sources of revenue with which to invest in new IT. If financial resources are limited, the local government should seek out alternative sources of revenue or participate in functionally specific IT projects that are funded through state or federal government funding programs.

In assessing how e-government projects are funded, survey data indicate that currently 92.6% of local governments use general revenues to fund e-government. Some other sources of revenue, such as utility revenues (9.8%), federal and state grants (7.5%), enterprise funds (7%), and transaction fees from services (5.2%), are used by local governments to fund part or all of their e-government initiatives (ICMA, 2004). Given the somewhat limited number of financial resources that are readily available from within government administration, it would seem prudent to make use of whatever other mechanisms possible to make IT innovations available. One alternative is the increased use of grants from federal and state governments. While such grants may be very limited for general IT acquisitions, they may be used to foster IT innovation within specific functional domains. For example, Department of Homeland Security grants aimed at enhancing emergency preparedness, law enforcement, or domestic security may be used to fund IT and interoperable communications systems, as well as cybersecurity improvements in local government IT infrastructures (U.S. Department of Homeland Security, 2005).

In addition to grants, financial costs can sometimes be alleviated by seeking partnerships with state and federal governments. For example, state and federal agencies may sometimes take on most of the development costs of some intergovernmental IT systems in order to achieve a specific programmatic outcome. The financial investment for local government may be dramatically reduced, since it may be limited to smaller acquisitions of equipment or personnel training. In other situations, private-sector firms may be willing to partner with local government in order to provide online services to citizens. For example, local governments may contract with private firms to provide online for-fee services, such as the provision of marriage licenses or birth certificates, especially if there are other means of citizens requesting those services in person or by mail. Given the diversity within local governments with regard to the public services provided, existing IT knowledge, and the probability of limited financial resources, adopting innovative e-government solutions may require a mixed-funding approach in order to develop the different components and systems of a broader e-government initiative within local government.

Mix In-House Knowledge and External Expertise

Each local government usually provides a variety of services and, therefore, may possess a wide variety of knowledge assets across functional departments. This, coupled with differences with regard to available financial resources and the sheer variety of IT characteristics, means that local governments are faced with several choices regarding how much of a particular IT initiative to develop in house and how much to acquire through contracts with private-sector firms. According to a recent ICMA survey, only 38.5% of local governments report that they host their own Web sites, while more report that their own staff design their Web sites (55.9%), manage Web-site operations (77.1%), and integrate Web sites into existing databases (68%; ICMA, 2004). However, the mix of in-house vs. contracted IT development and operations may depend on the relative complexity of the IT being considered in relation to the knowledge and expertise present within the local government organization.

It may be reasonable for a local government's in-house IT staff to develop some IT if the technology is already fairly established or if the IT staff have sufficient knowledge and experience working with the specific form of IT. For example, the knowledge required to design and build a basic Web site would not generally be considered a complex undertaking in today's technological environment. In fact, late adopters of any technology can employ the lessons and knowledge developed by early adopters, potentially reducing costs and risks of failure. At the somewhat other extreme, local governments may need to rely heavily on IT vendors for development and operation of IT applications and systems that are far more complex than the local government organization's current level of technical sophistication. In some instances, the high costs associated with updating and maintaining aging IT systems, maintaining knowledgeable staff, and developing new e-government systems in house are viewed as cost prohibitive. Some large cities, such as Minneapolis, Minnesota, have contracted with private-sector companies and outsourced the city's IT operations in an attempt to reduce overall costs (Butterfield, 2006).

In reality, most local governments will likely find some mix of in-house knowledge and external expertise that is the most efficient and effective means to develop, implement, and maintain new IT. Contracting for services that depend on employee knowledge of end users' needs, desires, and behaviors, as well as accepted administrative processes, puts contractors at a particular disadvantage, since they are more likely to have explicit knowledge of the technology, but less knowledge of the local government's administrative processes and culture. Brown and Brudney's (1998) study of the use of geographic information systems (GISs) provides some evidence as to the risks of an outsourcing strategy in an area that is likely to involve a high level of specialized knowledge. Their study indicated that there is a limit to the cost effectiveness of outsourcing core information-system functions. By surveying local government GIS managers, they found that these managers reported that small to moderate levels of contracting out of GIS services yielded clear benefits for their organizations, while higher levels of outsourcing did not appear to produce clear benefits. In fact, in cases where the government moved to near exclusive provision of these services via contractors, the costs and delivery time for the services reached unacceptable levels.

Moreover, contract-only projects tended to produce fewer benefits in terms of operational productivity and performance, organizational decision making, and customer service. The estimated break-even point for the outsourcing of GIS projects in the study was 25% contracted work. That is, projects in which more than 25% of the total budget was spent on outsourced services tended to produce diminished returns.

Findings such as these indicate that a local government's own IT staff and program administrators possess necessary knowledge of administrative processes with which new IT must be integrated. Likewise, external vendors and consultants may be at times necessary to supply knowledge and support more advanced forms of IT. The reliance on contracted IT development and operation services from the private sector does necessitate additional skills within local government focused on overseeing the administration of those services. For example, local governments need to maintain communication with the contractor in order to evaluate, maintain, and assess performance information regarding the contracted IT system or services. Furthermore, in many cases, the local government must possess significant project-management skills in order to provide oversight of the contractor's development of new IT systems (Chen & Perry, 2003).

Conclusion

The diffusion of IT within society during the past decade has refocused attention on how new forms of IT may be used to transform institutions within the private and public sectors. Academics, political observers, and vendors have often advocated e-government initiatives as a means of increasing transparency, openness, and overall efficiency in public-service delivery. However, local governments have sometimes been singled out as a group for being late adopters of IT innovations when compared to state and federal governments or private-sector firms. At first glance, the impression may be that most local governments are IT laggards that fail to serve the citizens and businesses in their jurisdictions by not creating sophisticated IT infrastructures or by not offering a myriad of highly interactive online services. However, is this impression accurate or fair? Any objective assessment would probably indicate the opposite to be more accurate, as well as fair.

In reality, it could be argued that local government organizations are incredibly heterogeneous with regard to their administrative and political goals, as well as the social and technology environments in which they operate. Whether a local government should adopt a new form of IT, such as advanced online services, is a question that should ultimately be based on the government's own specific jurisdictional considerations. A local government's administrators, as well as appointed and elected political leaders, are ultimately responsible for allocating their jurisdiction's resources in a manner that stands up to public scrutiny. Since many local governments are actually quite small in size, have limited resources to allocate, and may not experience significant public expectations about adopting advanced forms of IT innovations such as e-government, it could be argued that remaining somewhat behind the technological learning curve may be, for many governments, the most politically responsible course of action. This course allows other institutions to bear the costs and risks associated with developing and adopting new IT.

Note

The views expressed in this chapter are solely those of the authors and do not necessarily reflect the views of the GAO or the U.S. government.

References

Attewell, P. (1992). Technology diffusion and organizational learning: The case of business computing. *Organization Science, 3*(1), 1-19.

Beath, C. M. (1991). Supporting the information technology champion. *MIS Quarterly, 15*(3), 355-372.

Brown, M. M., & Brudney, J. L. (1998). A "smarter, better, faster, and cheaper" government: Contracting and geographic information systems. *Public Administration Review, 58*(4), 335-345.

Bugler, D., & Bretschneider, S. (1993). Technology push or program pull: Interest in new information technologies within public organizations. In B. Bozeman (Ed.), *Public management: The state of the art* (pp. 275-294). San Francisco: Jossey-Bass.

Butterfield, E. (2006). What a relief. *Washington Technology, 21*(13). Retrieved July 30, 2006, from http://www.washingtontechnology.com/news/21_13/statelocal/28905-1.html

Chen, Y.-C., & Perry, J. L. (2003). Outsourcing for e-government: Managing for success. *Public Performance & Management Review, 26*(4), 404-421.

Cooper, R. B., & Zmud, R. W. (1990). Information technology implementation research: A technological diffusion approach. *Management Science, 36*(2), 123-139.

Damapour, F. (1991). Organizational innovation: A meta-analysis of effects of determinants and moderators. *Academy of Management Journal, 34*(3), 555-590.

Delone, W. (1981). Firm size and the characteristics of computer use. *Management Information Systems Quarterly, 5*, 65-78.

Dewar, R. D., & Dutton, J. E. (1986). The adoption of radical and incremental innovations: An empirical analysis. *Management Science, 32*(11), 1422-1433.

Duncan, R. B. (1976). The ambidextrous organization: Designing dual structures for innovation. In R. H. Kilman, L. R. Pondy, & D. P. Slevin (Eds.), *The management of organizations: Strategy and implementation* (pp. 167-188). New York: North-Holland.

Ettlie, J. E., Bridges, W. P., & O'Keefe, R. D. (1984). Organizational strategy and structural differences for radical versus incremental innovation. *Management Science, 30*(6), 682-695.

Fichman, R. G., & Kemerer, C. F. (1997). The assimilation of software process innovations: An organizational learning perspective. *Management Science, 43*(10), 1345-1363.

Hall, R. H. (1996). *Organizations: Structures, processes, and outcomes* (6th ed.). Englewood Cliffs, NJ: Prentice Hall.

Hinnant, C. C., & O'Looney, J. (2003). Examining pre-adoption interest in online innovations: An exploratory study of e-service personalization in the public sector. *IEEE Transactions on Engineering Management, 50*(4), 436-447.

Hinnant, C. C., & Sawyer, S. (2003, October). *From keystone to e-stone: Assessing architectural innovation in state government.* Paper presented at the meeting of the National Public Management Research Conference, Washington, DC.

Howell, J. M., & Higgins, C. A. (1990). Champions of technological innovation. *Administrative Science Quarterly, 35*(2), 317-330.

Ihrke, D. M., Rabidoux, G., & Gabris, G. T. (2000). *Managerial innovation in local government: Some effects of administrative leadership and policy board behavior*. Poster presented at the Midwest Political Science Association National Conference, Chicago.

International City/County Management Association. (1997). *Technology in local government survey.* Washington, DC: Author. Retrieved July 30, 2006, from http://www.icma.org/upload/bc/attach/{F69C870D-81B7-4DFB-AEFB-0CCD99F13611}tech1997web.pdf

International City/County Management Association. (2004). *Electronic government 2004.* Washington, DC: Author. Retrieved July 30, 2006, from http://www.icma.org/upload/bc/attach/{9BA2A963-DDCC-40B7-836D-F1CFC17DCD98}egov2004web.pdf

Kraemer, K. L., Dutton, W. H., & Northrup, A. (1981). *The management of information systems.* New York: Columbia University Press.

Kraemer, K. L., & King, J. L. (2003, September). *Information technology and administrative reform: Will the time after e-government be different?* Paper presented at the Heinrich Reinermann Schrift Fest, Post Graduate School of Administration, Speyer, Germany. Retrieved July 30, 2006, from http://www.crito.uci.edu/publications/pdf/egovernment.pdf

Kraemer, K. L., King, J. L., Dunkle, D., & Lane, J. P. (1989). *Managing information systems: Change and control in organizational computing.* San Francisco: Jossey-Bass.

Kraemer, K. L., & Norris, D. F. (1994). Computers in local government. In *1994 municipal yearbook.* Washington, DC: International City/County Management Association.

Moon, M. J. (2002). The evolution of e-government among municipalities: Rhetoric or reality? *Public Administration Review, 62*(4), 424-433.

Nilakanta, S., & Scamell, R. W. (1990). The effect of information sources and communication channels on the diffusion of an innovation in a data base environment. *Management Science, 36*(1), 24-40.

Norris, D. F. (2003). Leading-edge information technologies and American local governments. In G. D. Garson (Ed.), *Public information technology: Policy and management issues* (pp. 139-169). Hershey, PA: Idea Group Publishing.

Norris, D. F., & Kraemer, K. L. (1996). Mainframe and PC computing in American local governments: Myths and realities. *Public Administration Review, 56*(6).

Norris, D. F., & Moon, M. J. (2005). Advancing e-government at the grassroots: Tortoise or hare? *Public Administration Review, 65*(1), 64-75.

Northrup, A., Kraemer, K. L., Dunkle, D., & King, J. L. (1990). Payoffs from computerization: Lessons over time. *Public Administration Review, 50*(5), 505-514.

O'Looney, J. (2001a). *The future of public-sector Internet services: Pt. I. Government technology.* Retrieved July 20, 2006, from http://www.govtech.net/magazine/story.php?id=5806&story_pg=5

O'Looney, J. (2001b). *The future of public-sector Internet services: Pt. II. Government technology.* Retrieved July 20, 2006, from http://www.govtech.net/magazine/story.php?id=6097&issue=10:2001

Perry, J. L., & Danziger, J. (1979). The adoptability of innovations: An empirical assessment of computer applications in local government. *Administration and Society, 11*(4), 460-492.

Perry, J. L., & Kraemer, K. L. (1979). *Technological innovation in American local governments: The case of computing.* New York: Pergamon Press.

Pew Internet and American Life Project. (2004). *How Americans get in touch with government.* Washington, DC: Author. Retrieved July 21, 2006, from http://www.pewinternet.org/pdfs/PIP_E-Gov_Report_0504.pdf

Pew Internet and American Life Project. (2006). *Data memo: Internet penetration and impact.* Washington, DC: Author. Retrieved July 21, 2006, from http://www.pewinternet.org/pdfs/PIP_Internet_Impact.pdf

Robinson, B. (2006). Energy auctions save county $25M. *FCW.com.* Retrieved May 20, 2006, from http://www.fcw.com/article94542-05-17-06-Web

Stair, R. M. (1992). *Principles of information systems: A managerial approach.* Boston: Boyd & Fraser Publishing.

Svara, J. (1990). *Official leadership in the city.* New York: Oxford University Press.

Tornatzky, L. G., & Fleischer, M. (1990). *The process of technological innovation.* Lexington, MA: Lexington Books.

Tornatzky, L. G., & Klein, K. G. (1982). Innovation characteristics and innovation adoption-implementation: A meta-analysis of findings. *IEEE Transactions on Engineering Management, 29*(1), 28-45.

Tushman, M., & Nadler, D. (1986). Organizing for innovation. *California Management Review, 28*(3), 74-92.

U.S. Census Bureau. (2002). *2002 census of governments* (GC02(1)-1). Washington, DC: U.S. Government Printing Office.

U.S. Department of Homeland Security. (2005). *FY 2006 homeland security grant program: Program guidance and application kit.* Washington, DC: Author. Retrieved July 30, 2006, from http://www.ojp.usdoj.gov/odp/docs/fy2006hsgp.pdf

U.S. General Accounting Office. (2004). *Information technology investment management: A framework for assessing and improving process maturity* (GAO-04-394G). Washington DC: Author.

Watson, R. T., & Mundy, B. (2001). A strategic perspective of electronic democracy. *Communications of the ACM, 44*(1), 27-31.

Zaltman, G. R., Duncan, R., & Holbek, J. (1973). *Innovations and organizations.* New York: Wiley Interscience.

Endnotes

[1] While e-government is most often touted as a means to facilitate interaction with citizens, the same underlying technologies can also be configured to interact with a variety of other stakeholders such as other local governments, businesses, state governments, or the federal government.

[2] Organizational or institutional size is commonly found to be related to the adoption and use of IT (Delone, 1981). Population size is essentially a proxy measure for the amount of resources that an organization may have available to facilitate the adoption, development, and implementation of IT.

[3] The 2004 ICMA e-government survey was mailed to chief administrative officers in cities with populations of 2,500 and over and to county governments with council-administrator or council-elected executive forms of government. The survey achieved an overall response rate of 42.9% with 3,410 respondents.

[4] Fewer than 10% of respondents indicated that the local government Web site provided means to pay for taxes, utilities, bills, fines, fees, or business-license applications online. Only 10.2% of respondents indicated that they provided a means to complete permit applications online (ICMA, 2004).

[5] Dial-up users indicated that the Internet has improved interactions with both state (47%) and federal (49%) governments a lot or somewhat. Only 35% of dial-up users indicated that the Internet had improved their interaction with local governments. Similarly, broadband users indicated that the Internet has improved interactions with both state (59%) and federal (61%) governments a lot or somewhat. Only 45% of broadband users indicated that the Internet had improved their interaction with local governments (Pew Internet and American Life Project, 2004).

[6] It is important to note that there is often considerable variation in the goals, needs, and technical sophistication across the departments within a single local government. Therefore, an IT system may fit and be perceived as beneficial by some departments and not by others.

[7] Norris and Moon (2005) also report that the same ICMA survey indicates that fewer local governments mention potentially positive impacts of e-government such as reduced time demands for staff, reduced administrative costs, a reduction in the number of staff, or increases in nontax revenue.

Chapter XI

Information Technology as a Facilitator of Results-Based Management

James E. Swiss, North Carolina State University, USA

Abstract

The most widely accepted normative model of good public and nonprofit management is often called results-based management. It encourages planning and target setting to make the organization more proactive, an emphasis on outcomes to make the organization better focused on its mission, quick performance feedback to make the organization more responsive, and continuous process improvements to make the organization better able to serve its clients. These changes are possible only with supporting information technology. This chapter discusses ways that IT, including GIS, dashboards, and data mining, can support the new management model. However, IT can increase management effectiveness only if its role has been carefully designed. Before implementing major IT changes, top public and nonprofit managers must begin by determining what information would best guide upcoming major decisions. They must also decide how they wish to balance system integration vs. costs, disintermediation efficiencies vs. client guidance, internal information accessibility vs. security, and frontline worker empowerment vs. organizational uniformity.

Introduction

The role of information technology in government and nonprofit management has constantly evolved over the years. This evolution has generally been driven by continuing improvements in IT. However, IT's management role has recently been changing more rapidly, and this time it is because public management, at least as much as IT, has substantially changed. This chapter will discuss some of the ramifications of this new role for IT within public and nonprofit management. In the first of three sections, it will briefly describe the new normative model of best management practices. The second section will look at ways that IT can advance this new management model. The final section will examine some policy choices that must be confronted when incorporating IT into public and nonprofit management systems.

The New Model of Public and Nonprofit Management

The Changing Definition of Good Management

During the past decade, consultants, texts, academics, and professional societies have all combined to change the accepted definition of good management. Twenty years ago, public or nonprofit managers who listened to their workers, kept their operations moving smoothly, and calmly handled occasional crises would usually be considered outstanding managers by peers and by outsiders. Today such managers would likely be viewed as competent, but disappointingly passive. To be seen as outstanding, today's managers must proactively move their agencies to higher levels of performance each year. The current normative model of management calls upon managers to continuously improve their agencies' performance by scanning their environments purposively and then using the resulting information to set agency targets for improved results, including improved customer satisfaction. The new model then suggests ways of flattening the organizational structure, streamlining its processes, and speeding up its feedback systems in order to help produce these continuously improving results.[1]

Confusingly, this single new model has been given at least three different names: results-based management, strategic management, and performance management. We will use the term results-based management in this chapter. (Adding to the confusion, a fourth term—new public management, or NPM—is also sometimes treated as a synonym. However, NPM is a broader concept that includes results-based management, but then also adds a strong market emphasis, calling for more public entrepreneurship and increased privatization of government programs. NPM's market-oriented policy prescriptions lie outside this chapter's focus on internal management improvements, and we will not consider NPM further.)

Results-based management combines several streams of management changes that developed separately, primarily in the business world. Total quality management (TQM) was first. Total quality management has a long history, but its current influence stems from the late1980s, when many major U.S. corporations adopted such TQM approaches as frontline worker empowerment, process benchmarking, and continuous measures of customer satisfaction. Reengineering, with its tools for process improvement, was implemented by many companies a few years later.[2] Government and nonprofit agencies soon began to adopt modified versions of these business techniques, and later added performance (outcome) measurement to the mix. By the start of this century, these tools had been combined and synthesized into a model of best management practices that looked far different from those of 20 years earlier.

This results-based model has achieved widespread acceptance. At the federal level, it was endorsed by the Clinton administration as well as by the George W. Bush administration, and by Congress through the Government Performance and Results Act of 1993. The model is also widely accepted by city managers (Kearney, Feldman, & Scavo, 2000; Moon & de Leon, 2001), and has been promulgated by the major public managers' professional societies.[3] Actual governmental implementation has substantially lagged in the symbolic acceptance of these new models, but it too shows steady increases (Kettl, 2005; Moon & de Leon, 2004; Moynihan, 2006; Poister & Streib, 2005).

The Primary Characteristics of Results-Based Management

The differences between the older and newer models of good management can be summed up in three terms. The new model attempts to make government and nonprofit agencies more proactive, more agile, and more results oriented. To produce these characteristics, an organization undertakes a number of steps. Four of the most important are:

- As part of its strategic planning process, the agency scans its environment to spot relevant trends (whether opportunities or threats) before they affect the agency.
- The agency proactively sets results-based targets, both long range and short range. Results, also called outcomes, are measurable effects on people outside the organization. For a school, the results (outcomes) might include increases in learning, or in later earnings. For a police department, results could include increases in convictions, decreases in crimes, and greater feelings of safety. For a health department, results include fewer days out from work, fewer days of hospitalization, or fewer deaths. Because customer satisfaction is always one important goal for any public or nonprofit agency, customer ratings are always among the results tracked.[4]
- The agency then gathers continuing information on how well it is reaching those outcome targets and takes remedial action if there are shortfalls.
- The agency attempts to increase customer satisfaction by offering quick, one-stop shopping that allows the client to receive coordinated services from one location, and in many cases, from one person. Such coordinated service delivery is promoted by empowering frontline workers to handle more of a customer's needs, and by mechanisms for coordinating (and in some cases permanently cutting across) functional silos.

Empirical Evidence for the Efficacy of Results-Based Management

Both business and government managers have enthusiastically embraced many new management approaches over the past half century. Some of these new management tools and approaches have left lasting imprints, but far more have quickly died and are remembered today only as fads (Light, 2006). Does empirical evidence indicate that results-based management is more than another short-lived fad?

The record is promising but mixed, with strong evidence for the efficacy of some aspects of results-based management, but slim evidence for other aspects. For example, there is as yet little evidence for the efficacy of strategic planning in the public or nonprofit sector, although private-sector studies suggests that strategic planning can, under some circumstances, produce improvements (Boyne, 2001; Miller & Cardinal, 1994). There is good evidence that process-improvement techniques drawn from reengineering and TQM sometimes produce positive changes in internal governmental activities, including fewer errors, lower costs, and shorter cycle times. However, their effect on outcomes is far less clear (Mani, 1995; Poister & Harris, 2000). The evidence is best, and the results most dramatic, for management approaches based on target setting and feedback. A large number of studies show that goal setting and results tracking often produce major gains in organizational outcomes (Locke & Latham, 1990; O'Leary-Kelly, Martocchio, & Frink, 1994; Perry, Mesch, & Paarlberg, 2006; Rogers & Hunter, 1992). An overall evaluation, then, might conclude that the efficacy of some aspects of results-based management is unclear, but that many important components do have empirical backing.

Results-Based Management's Intense Appetite for Information

Results-based management makes rigorous demands of government and nonprofit agencies. A results-oriented agency proactively determines a long-range direction and then advances in that direction through a series of goals and objectives expressed in terms of desired results, that is, the specific, measurable outcomes expected. In order to pursue these outcomes nimbly, it avidly seeks and digests current information about changes in the environment, about its own progress toward outcomes, and about customer satisfaction. It ensures that it is able to act quickly upon this information by restructuring its operations so that its organizational structure is flatter, its frontline workers are more empowered, and its processes are streamlined to reduce cycle time and handoffs.

To meet the demanding expectations of this management model, an agency requires a great deal of complex data: timely, complete information about both processes and outcomes. The model also requires that the organization have the ability to act promptly on that information. These requirements can be met only by agencies that take advantage of recent developments in information technology, and this management role for IT is the subject of the next section.

Information Technology and Results-Based Management

IT in Public and Nonprofit Agencies: The Historic Emphasis on Operations Over Management

Management texts distinguish operations from management. Operations, usually carried out by frontline workers, are the day-to-day activities of public and nonprofit agencies. Operations would include such activities as a teacher teaching fifth-grade students, a doctor treating patients, or a sanitation crew picking up garbage. Routine supporting services for these frontline activities, such as purchasing or inventory control, are also classified as operations.

Management, on the other hand, is the process of coordinating and directing many such operational activities to produce an effective agency. Management has commonly been the mission of the middle and upper levels of organizational workers.[5] For a school district, for example, keeping the records of one student is a concern of operations. However, when the student records are aggregated and used to determine how one school has performed compared to another school (or compared to a target), the information is now a concern of management because this aggregated information can be used to direct the organization.

For most government and nonprofit agencies, IT has historically been much more important for operations—running the inventory reorder system, the accounting system, and the payroll system, or maintaining records for each client transaction—than it has been for management. (This suggests that many of the governmental and nonprofit IT systems over the years that were termed management information systems [MISs] might more accurately have been termed operations information systems.)

However, IT's management role in these organizations is now rapidly expanding. We will look at IT's role in results-based management by examining its impact on each management step that was discussed in the first section.

IT's Role in Scanning the Environment and Setting Long-Range Goals

As part of strategic planning, agencies scan the organizational environment to spot early trends (opportunities and threats) so they can move proactively. IT, in the form of data mining, can help agencies in such scanning by showing agencies relationships and trends that they never expected. At the federal level in particular, as Cahlink (2000) reports, data mining is often used for long-range planning. For example,

the Justice Department has used it to find crime patterns so it can focus its money and resources on the most pressing issues. The Veterans Affairs Department has used it to predict the demographic changes among its 3.6 million patients so it can prepare more accurate budgets ... [and] the Internal Revenue Service [IRS] uses the technology at its customer service center to tracks calls in order to pinpoint the most common customer needs.

The IRS can then address these needs through its long-range plans (see also U.S. Government Accountability Office, 2004a).

Geographical information systems (GISs) also play an important role in strategic planning, especially for local governments. By using the GIS data to track physical and demographic changes block by block and neighborhood by neighborhood, a city or county gains the insights it needs to understand trends and to determine the best long-range approaches.

IT's Role in Promoting Agility Through Quicker Feedback

Results-based management emphasizes making the organization more nimble so it can adjust quickly to change. An agency's agility is enhanced by the early warnings of outside changes provided by strategic planning and by the quick reaction times of empowered frontline workers. However, an agency's agility depends most of all on its ability to quickly secure current information about the agency's ongoing performance. Such real-time performance data ensures that good performance can be immediately reinforced and that performance shortfalls can be promptly remedied.

The steady increase in IT portability has been an important contributor to more timely feedback. One good example is provided by the management of a very large parks maintenance program:

For a decade, New York City's parks and recreation department sent inspectors into parks to rate the performance of its maintenance teams. The inspection crews compiled massive reports covering all 1,500 city parks. The task was so overwhelming that the report could only come out three times a year, and by the time it came out, the data was stale and too mountainous to be useful.

Three years ago, the department overhauled the inspection program and bought four handheld computers. The computers fit into the palm of one hand, and show a menu of the twelve parks features, from lawn condition to graffiti, that inspectors rate as either acceptable or unacceptable. Inspectors enter information through a touch-screen keypad and can add additional comments through a handwriting feature. At the end of the day, the data is uploaded straight into the database that the department uses to churn out performance reports.

... By speeding the data-collection process, the technology allowed ... [the department] to move from voluminous triennial reports to biweekly performance reports on roughly 115 parks at a time. (Swope, 1998, p. 61)

Because of new forms of IT, the top managers of the parks department had fresher, more usable data to establish accountability. The department became more agile, with far quicker remedial responses. The overall conditions of the parks rapidly improved, with the percentage of parks graded *acceptable* almost doubling from 44 to 83% (Swope, 1998).

IT's Role in Promoting a Focus on Overall Results

Two IT-based tools, GIS and dashboards, have played particularly strong roles in helping organizations to focus on overall results.

The Role of GIS

Geographic information systems are a particularly useful managerial tool because they allow managers to see overall patterns. Most large police departments, for example, now use GIS to display crime results district by district. Inspections departments, planning departments, fire departments, and other agencies also gather, analyze, and display their performance measures through GIS. New York's Comstat is the premier example of this approach. As one observer notes,

In police circles the tense and theatrical NYPD Comstat meetings are the stuff of legend. All the top cops pack into a "war room" around a large rectangular table. Colorful computer-generated maps flash up on a screen, showing recent crime trends and statistics. A precinct commander takes the hot seat, and questions start flying: "Why are robberies up in this neighborhood? What are you doing about it?"

... [In all cities with similar systems,] geographic information systems and database technology bring the statistics to life. And commanders are held responsible for reducing crime in their areas. For many departments, this process—especially the accountability part—is nothing short of revolutionary. (Swope, 1999)

As it developed, Comstat (also commonly spelled Compstat) broadened the range of results it tracked. For example, citizen groups feared that too much attention to crime rates alone led to goal displacement, and so Comstat also began to hold commanders accountable for such outcomes as the number of citizen complaints about police behavior and similar measures of police civility. The resulting range of measures helps produce a good overall picture of performance.

Other cities have taken the same IT-enabled feedback model and applied it to all city services, not just police. Baltimore, for example, holds Citistat agency meetings every 2 weeks, with "the projection screens awash with spreadsheets, graphs and maps, visual representations of how well each city agency is doing its job." The head of the Baltimore Parks and Recreation Department says, "Citistat has redefined what our jobs are...We need to constantly evaluate what we're doing and see it makes a difference" (Swope, 2001). Different versions of Citistat-style performance-monitoring meetings are now found in many cities throughout the United States (Behn, 2006).

The Role of Dashboards (and Scorecards)

Beyond GIS, other IT-based tools also help agencies track and examine their performance. As illustrated by Comstat and Citistat, the pace of feedback has quickened in many organizations. In the past, agency personnel would often receive written summaries of their unit's performance measures infrequently, with quarterly reports most common. This undercut the rapid feedback that is one of the main advantages of performance measurement. Now many agencies post performance information on intranets, making weekly and even daily information available for decision making at all levels.

Because performance-reporting systems are still evolving, different public and nonprofit agencies use different names for their online performance reports. For example, some give them agency-specific names, and others term them *balanced scorecards*.[6] However, the most common term for online reporting systems, especially ones that emphasize graphics and interactivity, is *dashboards*.[7]

Dashboards integrate the information that is often now scattered among incompatible legacy systems so the agency can gain a complete overall view of its ongoing performance. A dashboard will show 6 to 12 key performance indicators, usually including such measures as the number of clients processed, customer-satisfaction ratings, error rates, cost data, progress in ongoing projects, and client outcomes.

Dashboards display much of their data graphically, through bar charts, pie charts, and trend lines. Graphical presentations help managers to more quickly see patterns that might otherwise be hidden in long columns of figures. When managers look at a dashboard, the graphical presentation immediately alerts them to changes or problems. A few dashboards are less traditional and move beyond bar charts to display many key measures as dials, with worrisome "red zones" marked. When the arrow for, say, this month's error rate moves into a red zone, all managers recognize that their error targets have been missed, and that they should quickly begin to implement changes to improve accuracy.

Because dashboards tend to report relatively short-term data, with some categories updated weekly or even daily, managers get the quick feedback necessary for results-based management. Because the dashboards are interactive, managers can drill down to learn more about how each category applies to particular geographic areas, or particular kinds of clients.

Timely performance information, such as that provided by dashboards, is also an important facilitator of organizational learning and continuous improvement. The best organizations constantly experiment with new approaches. Such experiments provide important trial-and-error-based learning, but only if timely performance results are readily available so that promising new approaches can be recognized and then institutionalized, and disappointing experiments can be terminated promptly.

In short, the best performance information systems—whether called online performance reports, balanced scorecards, or dashboards—provide all the information that managers would need in order to set short-term and long-term performance targets, to gather process and outcome information about progress toward those targets, and to conduct research on ways to perform better (Dover, 2004).

Dashboards are built, in part, from enterprise resource planning (ERP) systems. Therefore organizations without an ERP system will usually find that developing an effective ERP system is a major step toward securing useful performance information. However, dashboards extend well beyond ERPs. As Liang and Miranda (2001, p. 15) point out, "The very data that executives need for strategic and tactical decision making often requires the combination of data from ERP and non-ERP application sources (e.g., cost accounting information, efficiency indicators, customer information, performance measures, historic data for forecasting)."

Dashboards are still far less common in government and nonprofits than they are in businesses. Even many public and nonprofit agencies that do use online reporting systems such as dashboards still have a long way to go to automatically integrate the data from what were once stand-alone systems. However, substantial progress has been made in many cases, and the pace seems to be accelerating.

IT's Role in Coordinating and Streamlining Services to Provide One-Stop Shopping

Fostering Coordination Through Information Integration

Results-based management promotes one-stop shopping, which brings together all related services in one location. For example, a low-income client could receive help with housing, food stamps, income maintenance, health care, and job training from one location and possibly from just one case manager. One-stop shopping is usually advocated as a way of helping clients save time and avoid confusion, and these in fact are major benefits. However, far more importantly, such cross-functional coordination can help government achieve its most important long-range program results because attacks on any social problem—poverty, crime, or substance dependency—are more likely to succeed if they are coordinated rather than delivered piecemeal by different organizations at different times under different rules. Such a coordinated service system requires coordinated information streams. Integrated information systems, then, not only increase client satisfaction through one-stop shopping, but more importantly increase the agency's ability to produce meaningful results.

Fostering Streamlined Services Through IT-Driven Disintermediation

Information systems, particularly Internet-based systems, can also foster greater efficiency and quicker services through disintermediation: in plainer English, by eliminating the middleman or -woman (Jallat & Capek, 2001). In the private sector, for example, online travel sites allow consumers to order tickets directly from airlines, eliminating the time, information loss, and cost of travel agents who once mediated between consumers and airlines. Disintermediation leads to greater efficiency because the service is provided with fewer handoffs (i.e., steps involving different actors).

In government and nonprofits, information systems provide a faster, more seamless experience for clients by enabling two types of organizational disintermediation: decreasing the number of handoffs to middle managers, and decreasing the number of handoffs to frontline information processors and advisors.[8] IT's potential to decrease layers of middle management has been widely discussed and analyzed. IT, particularly expert systems and online help lines, can take over some of the roles that middle managers once played as resident experts and coordinators for less-experienced frontline workers. Moreover, middle managers traditionally carried information from upper management to frontline workers, and now this role, too, is diminished because IT allows direct, immediate connections.

IT also allows agencies to decrease or even eliminate many frontline advisors and coordinators, such as job-placement brokers or college course advisors, because clients can now directly access job or course listings, and can use online systems to register for job interviews or for courses. This form of disintermediation saves money and empowers clients, but it also raises policy choices that we will discuss in a later section.

Summarizing IT as a Management Facilitator

Results-based management aspires to be proactive, agile, and results oriented. IT in the form of data mining helps agencies proactively scan the environment. IT in the form of handheld computers helps the agency to agilely monitor results, and client-focused interactive Web sites helps the agency to quickly and agilely deliver services. Finally, IT in the form of GIS and integrated performance information systems (i.e., dashboards) helps the agency focus on the overall results. At every step, IT provides capabilities that are crucial to results-based management.

Policy Choices in Using IT to Advance Results-Based Management

When managers first consider ways that they can use IT to facilitate results-based management, they must begin by making a series of policy decisions. Among the most common choices facing top managers are deciding where the balance should lie between costs and the level of systems integration, between the efficiency of disintermediation and the need to guide clients, between internal data accessibility and privacy concerns, between efficiency and political openness, and between frontline empowerment and organizational uniformity. This section will briefly discuss each of these choices in turn.

Balancing the Extent of System Integration with Costs

Looking into the future, an ideal information system would move beyond a dashboard for a single agency. One of the most important management trends of recent years is that many services are often delivered not by one agency, but by networks of agencies. For example, often federal, state, local, nonprofit, and contracted for-profit agencies all work together in an area like job training and placement. To take another example, an even larger network of agencies works together to help at-risk families, concurrently providing psychological counseling, job training, food stamps, housing aid, health care, and other services. The most important desired result—a satisfactorily functioning family—is produced by all these agencies together. When dealing with such common policy areas, then, an ideal performance-reporting system would foster the exchange and integration of management data across many cooperating agencies, including both governmental and nonprofit ones.

However, even within a single large agency with multiple legacy systems, it is often extremely costly and time consuming to integrate existing systems in order to produce the most useful management information. Moreover, the more ambitious the integration effort, the higher the possibilities of failure. Says a report on integrated governmental ERP systems, "From Arkansas to Iowa to San Antonio and beyond, ERP implementations have been tripped up and nearly deep-sixed by problems …" (Perlman, 2004). Gartner, an independent research firm, has estimated government's failure rate at close to 60% for outsourced systems-integration projects (Perlman, 2002).

Therefore, managers taking their first steps toward using IT in results-based management must make difficult decisions about how to balance the advantages of integrating a large number of information systems on one hand, and implementation cost and risk on the other.

Balancing the Efficiency of Disintermediation vs. the Need to Guide Clients

As discussed earlier, IT allows agencies to eliminate intermediaries, allowing customers to directly appraise and choose their own services. This disintermediation has great advantages in reducing costs and empowering clients. At the same time, though, it can have disadvantages as well.

The type of disintermediation that raises the most complex management questions is the potential elimination of many frontline client advisors. A generation ago, a job seeker at a government employment agency might well work entirely through a job counselor who would screen job opportunities and help contact potential employers. Now, job seekers usually interact much less with such frontline advisors because they are often quickly directed to Internet-based job listings that they screen on their own; they can then use IT to send applications and otherwise communicate directly with employers. Agencies have similarly used IT to eliminate many frontline advisors for such functions as helping government employees to invest their retirement funds, or helping community-college students choose and register for courses. This type of disintermediation is often termed *coproduction* because it replaces the agency's intermediary with the efforts of the clients who help coproduce their own results.

Such IT-driven efficiencies speed the process and give the client more direct control, but some agencies have discovered that their systems assumed too high a level of client information. Some clients may want the advice and reassurance provided by human intermediaries, such as job counselors who suggest job-seeking strategies, retirement counselors who suggest investment choices, or college advisors who help students to evaluate possible course selections. Other clients may be pleased to be able to avoid such advisors, yet they may make consistently poor choices without them, thereby diminishing the program's results. System designers, then, must balance the efficiency gains of disintermediation with the clients' need for personalized guidance.

Balancing Information Integration and Accessibility vs. Privacy Risks

When all the information that applies to a single client is scattered among various unintegrated databases, management is greatly impeded, but privacy is less threatened. Results-based management calls for integrating information, but gathering all the information about a client in one place and making that information accessible to any worker that the client may contact (one-stop shopping) greatly increases the possibility of privacy infractions: More information is available from any one leak, and there are also more places from which it can leak (Ghere & Young, 2001; U.S. Government Accountability Office, 2005, 2006). Similarly, defense, foreign affairs, and police agencies often need to safeguard sensitive information about national security or ongoing police investigations, and therefore they face a trade-off between making internal information quickly accessible to their workers vs. information security.[9]

Management information services are increasingly outsourced for efficiency reasons (Thibodeau, 2001) and this, too, may increase the risks of information leakage because it increases the number of individuals with access to the information while decreasing the ability of government leaders to directly monitor them. In making decisions about IT and management, then, top managers must first make a policy decision about the desired balance between efficiency, usually promoted by making complete information easily accessible to employees, and information security.

Balancing Political Openness vs. Organizational Efficiency

Just as there are trade-offs involved in determining the extent of employee access to some types of organizational information, there are sometimes trade-offs in determining the extent of public access to ongoing governmental management information.

Some organizations have begun to publicly post their weekly and monthly performance information on their Web sites to demonstrate that they are open and accountable. For most agencies, there is no downside to these short-term public reports because they discover that the (unfortunately) general public has an unwavering lack of interest in their performance measures.

However, such short-term postings present a greater trade-off for health, environmental, regulatory, and other agencies with politically charged missions and intense interest groups because interest groups are able to use this information to slow the managerial decision process and to move the ultimate decisions away from efficiency (Ghere & Young, 1998; Musso, Weare, & Hale, 2000).

For example, a state forestry agency may find that its weekly performance information about controlled burns or acres planted will enable both environmental groups and industry groups to intervene with agency leaders and legislative overseers every time a short-term target they approve of is missed or one they disapprove of is exceeded. Managers must now spend time responding to these interventions. Openness is gained, but efficiency is lost, and sometimes responsiveness to the wider public is lost because of the need to respond to intense narrow groups. If top agency administrators decide these costs are too great, they may decide to distribute short-term performance information only internally through intranets, and otherwise report only broad outcome information to the public. That has been the path taken, for example, by the Atlanta dashboard designers (Edwards & Thomas, 2005). Whatever their choice, top managers in such controversial policy areas must decide on the appropriate balance of openness vs. efficiency as they plan their use of IT in management.

Balancing Frontline Worker Empowerment vs. Organizational Uniformity

Even if frontline workers have access to agency or client information, a distinct although related question is what power they are given to act on that information. The more decisions are set by the top, the more the organization will treat each client uniformly, and this uniformity is often viewed as fairness. However, different clients bring subtly different problems and concerns to organizations, and if frontline workers are empowered (within some limits) to use their own discretion, the organization will be more flexible, and the morale of frontline workers will usually be enhanced as well.

Bovens and Zouridis (2002) point out that IT advances have greatly changed the roles of a certain type of "street level bureaucrat," those frontline government workers who make decisions about whether clients are eligible for certain levels of health care benefits, public housing, food stamps, or job training.[10] In the past, such workers had to apply a great deal of individual judgment (Lipsky, 1980). Now they may be better seen as "screen-level bureaucrats" with much less discretion because software will often suggest and sometimes even determine levels of eligibility with no human intervention.

For most large agencies, the decision about how much power to devolve to frontline workers will greatly affect the information system. Ironically, the same IT that enables frontline workers to be empowered, because it puts complete information at their fingertips, allows top management to micromanage intensely because they can instantly call up every bit of information about what frontline workers are doing and deciding. A decision about the level of frontline empowerment that best meets society's needs, then, must precede many decisions about the information system (Bovens & Zouridis, 2002; Ghere & Young, 1998).

An Overview of the Managerial Balances: Why Management Choices Must Precede IT Choices

Many discussions of planning for IT systems focus on relatively narrow IT issues. For example. They often recommend that organizations plan for scalability so the system can grow as the organization grows, break major IT projects into pieces so they can be implemented and monitored better, and set clear project-management expectations (U.S. Government Accountability Office, 2004b). These are important guidelines, but they must be preceded by more fundamental decisions about the agency's mission and about what facets of management are most important in advancing that mission. Only then can meaningful decisions be made about how any new IT system can support management.

When managers are trying to determine the right organizational structure, Chandler (1969) once famously advised that they should recognize that "structure follows strategy." In other words, whether any given structure is right for an organization cannot be determined in the abstract. The organization must first choose its strategic direction, and then the right structure is the one that best fits that managerial strategy (Chandler, 1966).

This simple concept applies to information systems as well, but it is sometimes ignored by top government managers when they are making (or quite often, delegating) IT decisions. Decisions about the information systems, particularly managerial information systems, should not come before decisions about the organization's managerial direction. Managers must first decide what managerial balances—between efficiency and openness, for example—they wish to create; only then will some IT systems be correct and others ill-suited for that chosen strategic direction. We have already discussed five decisions about balances or trade-offs that must precede many IT decisions. Beyond those five broad policy decisions, more specific managerial decisions that precede IT decisions would include the following:

- What are the most important decisions facing our agency, and what specific information would be most useful in helping us make those decisions?
- For each specific item of information gathered, who in the organization needs it to make better decisions, and in what form will they find it most useful?
- Who outside the organization needs to know it, and again in what form?
- What incentives do managers now have to actually use the information they are given, and how can those incentives be strengthened?

If IT experts make all the major choices about an agency's IT systems, they are in fact determining much of the agency's strategic and management direction. As Zuurmond and Snellen (1997, p. 205) have said, "Managing the information architecture means managing the organization ... [because] these architectures control the decision premises of all organizational members." An organization's strategic decisions should not be a by-product of IT choices made by technical experts on technical grounds. Instead, they should reflect the long-range plans, management goals, and policy balances that have been thoughtfully determined by agency managers. IT choices then naturally follow.

Because it forces managers to face basic questions about the agency's policies and management, the process of designing information systems can be a powerful form of strategic analysis that can create a much more effective, results-driven agency.

References

Behn, R. D. (2006). The varieties of Citistat. *Public Administration Review, 66*(3), 332-340.

Bovens, M., & Zouridis, S. (2002). From street-level to system-level bureaucracies: How information and communication technology is transforming administrative discretion and constitutional control. *Public Administration Review, 62*(2), 174-184.

Boyne, G. (2001). Planning, performance and public services. *Public Administration, 79*(1), 73-88.

Cahlink, G. (2000). Data mining taps the trends. *Government Executive, 32*(12), 85.

Chandler, A. D. (1969). *Strategy and structure: Chapters in the history of the American industrial enterprise.* Cambridge, MA: MIT Press.

Dover, C. (2004). How dashboards can change your culture. *Strategic Finance, 84*(4), 43-48.

Edwards, D., & Thomas, J. C. (2005). Developing a municipal performance measurement system: Reflections on the Atlanta dashboard. *Public Administration Review, 65*(3), 369-376.

Ghere, R. K., & Young, B. A. (1998). The cyber-management environment: Where technology and ingenuity meet public purpose and accountability. *Public Administration & Management: An Interactive Journal, 3*(1).

Jallat, F., & Capek, M. J. (2001). Disintermediation in question: New economy, new networks, new middlemen. *Business Horizons*, *44*(2), 55-60.

Kearney, R. C., Feldman, B. M., & Scavo, C. P. F. (2000). Reinventing government: City manager attitudes and actions. *Public Administration Review, 60*(6), 535-548.

Kettl, D. F. (2005). *The global public management revolution* (2nd ed.). Washington, DC: Brookings.

Liang, L. Y., & Miranda, R. (2001). Dashboards and scorecards: Executive information systems for the public sector. *Government Finance Review, 17*(6), 14-19.

Light, P. C. (2006). The tides of reform: Patterns in making government work, 1945-2002. *Public Administration Review, 66*(1), 6-19.

Linden, R. M. (1994). *Seamless government: A practical guide to reengineering in the public sector.* San Francisco: Jossey-Bass.

Lipsky, M. (1980). *Street-level bureaucracy.* New York: Russell Sage Foundation.

Locke, E. A., & Latham, G. P. (1990). *A theory of goal setting and task performance.* Englewood Cliffs, NJ: Prentice Hall.

Mani, B. G. (1995). Old wine in new bottles tastes better: A case study of TQM implementation in the IRS. *Public Administration Review, 55*(2), 147-158.

Miller, C. C., & Cardinal, L. B. (1994). Strategic planning and firm performance: A synthesis of more than two decades of research. *Academy of Management Journal, 37*(6), 1649-1665.

Moon, M. J., & de Leon, P. (2001). Municipal reinvention: Managerial values and diffusion among municipalities. *Journal of Public Administration Research & Theory, 11*(3), 327-351.

Moynihan, D. P. (2006). Managing for results in state government: Evaluating a decade of reform. *Public Administration Review, 66*(1), 77-89.

Musso, J., Weare, C., & Hale, M. (2000). Designing Web technologies for local governance reform: Good management or good democracy? *Political Communications, 17*(1), 1-19.

O'Leary-Kelly, A. M., Martocchio, J. J., & Frink, D. D. (1994). A review of the influence of group goals on group performance. *Academy of Management Journal, 37*(5), 1285-1301.

Perlman, E. (2002). Contract hoops and loopholes. *Governing, 15*(10), 53.

Perlman, E. (2004). The ERPworks. *Governing, 17*(5), 55-60.

Perry, J. L., Mesch, D., & Paarlberg, L. (2006). Motivating employees in a new governance era: The performance paradigm revisited. *Public Administration Review, 66*(4), 505-514.

Poister, T. H., & Harris, R. H. (2000). Building quality improvement over the long run: Approaches, results and lessons learned from the PennDOT experience. *Public Performance & Management Review, 24*(2), 161-176.

Poister, T. H., & Streib, G. (2005). Elements of strategic planning and management in municipal government: Status after two decades. *Public Administration Review, 65*(1), 45-56.

Rogers, R., & Hunter, J. E. (1992). A foundation of good management practice in government: Management by objectives. *Public Administration Review, 52*(1), 27-39.

Swope, C. (1998). Performance measurement's essential ingredient. *Governing, 11*(7), 61.

Swope, C. (1999). The Comstat craze. *Governing, 12*(12), 40-43.

Swope, C. (2001). Restless for results. *Governing, 14*(7), 20-22.

Thibodeau, P. (2001). Government eyes IT outsourcing. *Computerworld, 35*(17).

U.S. Government Accountability Office. (2004a). *Data mining: Federal efforts cover a wide range of uses.* Washington, DC: Author.

U.S. Government Accountability Office. (2004b). *Information technology management: Improvements needed in strategic planning, performance measurement, and investment management governmentwide.* Washington, DC: Author.

U.S. Government Accountability Office. (2005). *Data mining: Agencies have taken key steps to protect privacy in selected efforts, but significant compliance issues remain.* Washington, DC: Author.

U.S. Government Accountability Office. (2006). *Privacy: Key challenges facing federal agencies.* Washington, DC: Author.

Zuurmond, A., & Snellen, I. T. M. (1997). From bureaucracy to infocracy. In J. A. Taylor (Eds.), *Beyond BPR in public administration: Institutional transformation in an information age* (p. 205). Amsterdam: IOS Press.

Endnotes

1. Results-based management also encourages workplace teams, and suggests ways of tying results to group incentives and to budget decisions. However, these are areas where IT has a smaller role, and so they will not be covered here.

2. Reengineering included both a philosophy, which was rapidly jettisoned by most government agencies, and a series of tools, which were widely adopted and continue to be used. The philosophy held that the most effective process improvements required very radical changes that were usually conceived and implemented in a top-down manner. This philosophy proved strikingly impractical for government. The tools included approaches for decreasing cycle times and making services more seamless for customers. Among these approaches were ways of capturing information just once, parallel processing, job enlargement, and case management. These process tools and approaches proved quite useful for producing incremental government improvements. The best public-sector introduction is Linden (1994).

3. Of the major professional managers' societies, the American Society for Public Administration (ASPA) has most explicitly endorsed the approach. In 1996, it established the Center for Account-

ability and Performance in order, according to the ASPA Web site, "to address the requirement for all levels of government to move to performance-based, results driven management." The two other professional societies, International City/County Managers Association (ICMA) and the National Association of Schools of Public Affairs and Administration (NASPAA), have also published official books and monographs on results-based government.

4 In the public sector, increased customer satisfaction is never the only goal and often it is not the major goal. Regulatory agencies, such as bank examiners, restaurant inspectors, highway patrol officers, will often find that doing a good job will rarely delight their most attentive customers such as bank owners, restaurant managers, and speeders. Meanwhile, their other customers (the general public) are inattentive. Other agencies, like the U.S. State Department and health research agencies, will often find that most of their customers do not have enough information to evaluate their performance. Therefore, even though customer ratings are always useful in government, they play a less central role than they do for business, and they are accordingly only one piece of information within a range of results.

5 Although management is traditionally the purview of middle and upper organizational levels, we will later discuss how this distinction is breaking down because both results-based management theories and IT advances have made it possible to move more management decisions to frontline organizational workers.

6 A balanced scorecard is a particular type of performance report that usually focuses on four organizational sectors: processes, human-resource development, results, and customer ratings. Unlike the term dashboards, the term scorecards does not indicate that the reports are necessarily online.

7 Of course, very small organizations will sometimes track and report performance measures using only simple spreadsheets to total and compare monthly figures. For most organizations, however, performance data are usually tied to more sophisticated information-processing capacities, as discussed in this section.

8 In government, many types of Internet-based disintermediation can also produce operations-level effects, such as allowing government to buy supplies, screen bids, and sell bonds much more efficiently. Because these are not managerial-level effects, they lay outside our scope in this chapter.

9 Some sophisticated systems do tailor the information shown by the need to know of each user. However, such systems are expensive and difficult to keep up to date as users, information, and clearances change month to month.

10 Bovens and Zouridis (2002) point out that this IT-driven loss of discretion and autonomy does not apply to all of Lipsky's (1980) street-level bureaucrats. Those who deliver particularistic services to a small group of clients at a time—health care providers or teachers, for example—have not lost much discretion to IT. Parenthetically, the authors also point to a level beyond "screen level" bureaucrats where all human discretion has been removed and government decisions are made by IT systems with no face-to-face contact. They call this level "system level bureaucracies" and provide a number of European examples.

Chapter XII

Managing IT Employee Retention: Challenges for State Governments

Deborah J. Armstrong, University of Arkansas, USA

Margaret F. Reid, University of Arkansas, USA

Myria W. Allen, University of Arkansas, USA

Cynthia K. Riemenschneider, University of Arkansas, USA

Abstract

IT employees are critical to the successful functioning of contemporary governmental agencies. Researchers and practitioners have long sought to identify workplace factors that influence employee retention. In this chapter we review the existing literature on factors that may reduce the voluntary turnover of public-sector IT professionals. Examples are presented that illustrate what states have been doing to improve their ability to retain their technology workforces. We conclude with an in-depth review of two studies addressing workplace factors that may influence state government IT personnel retention. The first is a study from a single state, designed to test factors potentially influencing affective commitment (a precursor of turnover intention) for state IT workforces. The second is a study from two states, designed to examine factors potentially influencing retention of state IT workforces. The message for decision makers is clear: When it comes to the retention of IT personnel, workplace and job characteristics matter. The insights from this chapter should aid public agencies in their ongoing efforts to retain quality IT professionals.

Introduction

All levels of government face serious staffing challenges. These challenges are especially salient when it comes to IT employees who are critical to the successful functioning of contemporary governmental agencies. Estimates of unfilled public- and private-sector IT jobs range from 342,000 to 600,000 (Catlette & Hadden, 2000; Passori, 2000). Almost 87% of state governments lack the IT personnel they need (Newcombe, 2002) at a time when the federal government is devolving more responsibilities for public-policy implementation and quality service delivery to the state level, many current public-sector employees are contemplating retirement, and the public and private sectors of the economy are competing for a limited number of skilled IT employees. Unfilled public-sector IT positions present a significant threat to government's ability to serve the public, making it vital that governments identify ways to attract and retain IT personnel.

Researchers have long sought to identify workplace factors that influence employee retention and commitment. In the most comprehensive study to date on voluntary turnover in state governments, Selden and Moynihan (2000) found retention was more likely when state governments provided better pay, more internal mobility opportunities, and employee-friendly human-resource practices. In their study of state government personnel, Thatcher, Stepina, and Boyle (2002-2003) found employees expressed considerable willingness to quit when they perceived better opportunities elsewhere. Thus, turnover in IT environments is affected by the state of the economy, pay rates, and attractive opportunities in other organizations.

This chapter provides insights that may be utilized to influence the retention of IT personnel in the public sector. Initially, the chapter describes how public-sector managers face private-sector competition for qualified IT employees. Sometimes staffing shortages and budgetary concerns are managed through outsourcing or offshoring. For states that do not select those options, it becomes important to identify how to create attractive working environments. Therefore, the scholarly literature discussing issues related to voluntary turnover and affective organizational commitment is reviewed in order to identify what organizations can do to create more attractive working environments for IT employees. This is followed by a discussion of what a sample of states that have been recognized for their excellence in IT personnel initiatives have done. An in-depth review of two public-sector IT workforce studies follows. The first study identifies factors associated with affective commitment and job satisfaction, and the second identifies factors associated with turnover intention (retention). The chapter concludes with ideas on what public-sector managers can do to retain valued IT employees.

Public-Sector Employers Face IT Staffing Issues

Public-sector employers face two related staffing issues. First, they face recruitment and retention issues since private-sector business can offer IT employees better salaries and incentives. Second, in order to deal with labor shortages and/or control costs, many federal, state, and local agencies are considering outsourcing or offshoring some IT functions.

Private-Sector Competition

As stated previously, the number of unfilled public- and private-sector IT jobs is sizable and expected to grow. Thus, developing ways to retain current employees and attract future ones is important. Frequent turnover of valued employees can be a costly challenge in organizations and even more so in an area such as IT that is facing a renewed labor shortage. High turnover often is more damaging to an organization than the inability to recruit new employees (U.S. Merit Systems Protection Board, 2002).

With regard to hiring and retaining IT personnel, research has found that public-sector agencies have historically had limited ability to compete with the private sector. Bozeman and Bretschneider (1986) report that public-sector agencies experience difficulties in hiring IT personnel because of a real or perceived lack of flexibility in their hiring practices and limited support for in-house training. Lan, Riley, and Cayer (2005) found that public-sector agencies had trouble recruiting and retaining IT personnel because of uncompetitive compensation packages, inflexible recruitment and human-resource practices, tight technology budgets, and an unfavorable public image.

One factor contributing to turnover intentions of IT personnel in the public sector may be the weak relationship between performance and pay and between performance and promotion compared to the private sector (Bozeman & Bretschneider, 1986; Bretschneider & Wittmer, 1993; Mohan, Holstein, & Adams, 1990). Another contributing factor may be the array of job options open to skilled IT personnel outside the public sector. In a study of state government employees, Kim (2005) found turnover intentions of IT personnel are affected by attractive job opportunities in other organizations. The limited public-sector literature focused on retention and turnover among IT professionals (e.g., Kim; R. W. Perry, 2004; Reid, Riemenschneider, Allen, & Armstrong, 2006; Selden & Moynihan, 2000; Thatcher et al., 2002-2003) suggests that maintaining a supportive work environment as well as providing a challenging task environment could provide measures to counterbalance the compensation issues and attraction of jobs outside the public sector.

Outsourcing and Offshoring

An organization outsources by contracting all or part of its information-technology functions to another company (Nam, Rajagopalan, Rao, & Chaudhury, 1996; Sinensky & Wasch, 1992). In an onshore outsourcing operation, the outsourced IT work takes place in the same country as that of the outsourcing firm's IT department, whereas with offshoring, the outsourced IT work takes place in a country different from that of the outsourcing firm's IT department (Fish & Seydel, 2006). As in the private sector, state and federal governments are increasingly considering outsourcing and offshoring as viable options to deal with labor shortages in the IT field (Garson, 2003). According to Edelman, Reynolds, and Holle (1997, p. 2), "Outsourcing has increasingly been examined by various governmental units as a potential solution to declining growth in agency revenues while attempting to enhance program service delivery in an era of devolution."

Government's reliance on nongovernmental businesses and workforces is of course not new, nor are the perennial problems with service integration across levels of government and coordination among multiple public and private providers. Light (1999), in a study for the Brookings Institute, estimated that approximately 8 million contract workers complement a federal workforce of about 2 million. The Bush administration's competitive sourcing initiative essentially sees public and private workforces competing for work that traditionally was considered to be the domain of the public sector (Light, 2003). Imposed head-count constraints force agencies to consider contracting and outsourcing even if they are philosophically opposed to it, or if the true cost savings or other organizational benefits cannot be clearly ascertained (Chen & Perry, 2003; White & Korosec, 2005). Thus, in many instances, governments, because of their complex technology needs, are increasingly using the services of large private-sector companies. Chen and Perry (p. 5) report that "government IT outsourcing is expected to be the fastest-growing segment of the overall federal IT market. The growth rate is estimated to be about 16 percent per year between fiscal year 2001 and 2006, reaching $13.2 billion." According to these authors, the forms of outsourcing most widely used by federal agencies include network services, data-center services, call centers, Web hosting, and other types of application services.

Coordinating the activities of literally thousands of public and private IT providers offers daunting technical and accountability challenges for public-sector managers. Service integration across multiple platforms and applications and across geographical space are only the most obvious challenges. Avoiding service interruptions using largely proprietary applications, keeping up with rapidly advancing technologies, and justifying and communicating increasing costs to the general citizenry and legislative bodies remain ever-present challenges. From the perspective of the public-sector manager, preserving transparency, accountability, and privacy protection can clash with private businesses' intent to maintain their competitive position. Contracting, and more recently outsourcing and offshoring, reflects an ongoing constitutional struggle over the continued blurring of the public and private spheres within government since the potential for the intermingling of interests is a serious concern (Chen & Perry, 2003). A core tension within this approach to governing is to keep government operations small while holding private contractors accountable for their performance (for an extensive discussion see Guttman, 2003).

At the state level the picture is a similar one but more murky with some states embracing outsourcing or offshoring and others resisting the trend. IT remains one of those areas where state governments see opportunities to achieve cost savings or to compensate for lack of available expertise. In states with outsourced services, the majority of such services are for technology or administrative support. State policy makers are getting better in discerning where true cost savings can be achieved while avoiding some of the pitfalls of outsourcing (Chi, Arnold, & Perkins, 2003; Government Accountability Office, 2006). A recent Government Accountability Office report of state-administered social-services programs reported that in the 15 states surveyed, actual savings associated with outsourcing and/or offshoring parts of the service, mostly in customer service and software development, ranged from less than 1% to over 20%. For example, Florida awarded a private outsourcing firm a multiyear contract to take over the state's benefits and payroll administration as well as the training and recruiting of state workers. The state is expected to save millions of dollars in managing its antiquated and disparate personnel systems (Overman, 2003).

Other states have been less receptive to outsourcing and/or offshoring. In 2004 more than 200 bills were introduced in state legislatures to prevent outsourcing and/or offshoring (Sakar,

2005). Two states, New Jersey and Arizona, prohibit outsourcing (Government Accountability Office, 2006). In 2005, 19 state legislatures were presented with bills that addressed outsourcing issues ("2005 Outsourcing Bills," 2005). Many of the bills included prohibitions against outsourcing, included measures to tightly monitor contracted services, or prevented outsourcing if, as a consequence, state jobs were lost. Pressures to resist outsourcing come from several sources: unions attempting to protect relatively high-paying jobs in light of continued manufacturing job losses, security or privacy concerns over protecting sensitive data, and concerns over hidden costs in managing mixed public and private systems.

In summary, states face pressure to provide IT services as the federal government devolves more responsibilities to the state level. This is taking place within a labor environment where there is competition between public- and private-sector employers for a limited pool of qualified IT personnel within the United States. Some states are deciding to manage the lack of personnel by outsourcing or offshoring IT functions, whereas other states are resisting those trends. For states that resist outsourcing or offshoring, deciding how to attract and retain IT personnel capable of managing the various phases of IT resource acquisition and managing complex contracts or outsourcing arrangements, while assuring contract compliance, becomes vital. The next section discusses literature identifying workplace conditions associated with voluntary employee turnover in order to help public-sector managers create a working environment conducive to IT personnel retention.

Voluntary Turnover of Public-Sector IT Professionals

Employee turnover has been associated with significant economic and noneconomic costs. Economic costs include separation, replacement, and training costs. The annual economic per-person cost of turnover can range from $1,200 to $20,000 depending on the position (Hatcher, 1999), and the replacement cost of an experienced IT computer programmer may reach well over $20,000 (Griffeth & Hom, 2001). Noneconomic costs include losses in terms of leadership, organizational knowledge, and innovation capacities. Additional losses may include disrupted work processes, inattention to different employee groups, increased stress on the remaining workforce, and diminished attention to external changes and stakeholder needs (Pinkovitz, Moskal, & Green, 2001).

Given such costs, the study of voluntary employee turnover, largely conducted in private-sector organizations, has been an area of interest for almost 30 years (e.g., Bluedorn, 1982; Mobley, Griffeth, Hand, & Meglino, 1979; Steers & Mowday, 1981). Researchers have identified numerous antecedents to voluntary employee turnover covering categories such as leadership, coworkers, stress, pay, job content, and the external environment to name a few (Griffeth, Hom, & Gaertner, 2000; Mobley et al., 1979; Porter & Steers, 1973). Griffeth et al.'s meta-analysis is the "most wide-ranging quantitative review to date of the predictive strength of numerous turnover antecedents" (p. 463). Two of the most important proximal precursors to the withdrawal process that they identified were organizational commitment and job satisfaction.

While organizational commitment has been found significant in the private sector, it is important to investigate the extent to which it is vital to public-sector employee retention because some public-administration scholars have emphasized the distinct motivations, ethical obligations, and service commitments of public-sector agency employees (e.g., Frank & Lewis, 2004; Mosher, 1982; Nyhan, 1999; Perry & Porter, 1982; Rainey, 1982; Romzek, 1990; Staats, 1988; Wright, 2001). Relevant differences between private- and public-sector employees have been reported in terms of reactions to intrinsic and economic rewards (e.g., Crewson, 1997; Kim, 2005), the value of extrinsic and intrinsic employee motivations (e.g., Frank & Lewis), and affective organizational commitment (e.g., Nyhan; Zeffane, 1994). As organizational commitment emerged as a significant antecedent of turnover in the Griffeth et al. (2000) meta-analysis, we focus on it in the next segment and look at the role of organizational commitment in the retention of public-sector IT employees.

Affective Organizational Commitment

Organizational commitment involves an individual's level of commitment to his or her organization. Affective organizational commitment is a strong belief in and acceptance of the organization's goals and values, and a willingness to expend effort on behalf of the organization (Mowday, Steers, & Porter, 1979). Several studies have confirmed the strong (negative) relationship between affective organizational commitment and voluntary turnover (e.g., Farrell & Stamm, 1988; Griffeth et al., 2000; Mathieu & Zajac, 1990; Michaels & Spector, 1982; Mowday et al., 1979; Tett & Meyer, 1993).

Few studies of affective organizational commitment have occurred in public-sector organizations (for exceptions, see Balfour & Wechsler, 1996; Irving & Coleman, 2003; Kim, 2005; Liou & Nyhan, 1994; Maranto & Skelley, 2003; Nyhan, 1999; Steinhaus & Perry, 1996; Thatcher et al., 2002-2003). According to Reid et al. (2006, p. 322), "This lack is surprising due to the concept's prominence in the organizational behavior literature, and the critical need to identify factors that might favorably influence the retention of public sector IT employees." In one of the few studies investigating affective commitment in public agencies, Liou and Nyhan (p. 111) concluded that "affective commitment not only explains more than 27% of the variation in public employee commitment to their organization but also correlates significantly with employee tenure."

Researchers (e.g., Steers, 1977) have consistently identified personal characteristics (e.g., tenure, sex, education, and need for achievement), job characteristics (e.g., challenges, opportunity for interaction, feedback from supervisor), and work experiences (e.g., organizational support, rewards for performance) as key antecedents of affective organizational commitment (Reid et al., 2006). While Meyer, Stanley, Herscovitch, and Topolnytsky (2002, p. 38) found that demographic variables tend to "play a minor role in the development of organizational commitment" when looking across employment sectors, IT employees share a common set of personal characteristics that may be especially relevant to their affective organizational commitment and eventual retention. They have a high need for challenging work, have a lower need for social interaction (Couger & Zawacki, 1978), and are more ambitious, logical, and conservative than the norm (Wynekoop & Walz, 1998).

Hackman and Oldham (1976) introduced the concept of job characteristics. Job-characteristics variables that have demonstrated a positive relationship with affective organizational commitment include goal setting, work schedules, and task variety (e.g., Hackman & Oldham, 1980; Locke, Latham, & Erez, 1988; Thatcher et al., 2002-2003). Job-characteristics variables that have demonstrated a negative relationship with affective organizational commitment include role ambiguity, subjective stress, and intergroup conflict (e.g., Cammann, Fichman, Jenkins, & Klesh, 1978; Chan, 1989; Greaves & Sorenson, 1999; Irving & Coleman, 2003; Jamal, 1990). We highlight these variables because of their prominence in and relevance to IT work settings.

Work-experiences variables that are commonly associated with voluntary turnover and/or retention transcend immediate job contexts and reflect more generally on organization-wide policies such as interrole conflict (i.e., work-family conflict; Eisenberger, Armeli, Rexwinkel, Lynch, & Rhoades, 2001; Thompson, Jahn, Kopelman, & Prottas, 2004; Thompson & Mastracci, 2005), perceived support from the organization (Eisenberger, Huntington, Hutchison, & Sowa, 1986; Hutchison, Sowa, Eisenberger, & Huntington, 1986; Meyer et al., 2002), leader-member exchange (e.g., Albrecht, 2005; Albrecht & Travaglione, 2003; Harris, Kacmar, & Witt, 2005; Somech, 2003), and pay for performance (Meyer et al.). Although the role of pay for performance in enhancing affective organizational commitment has received mixed support in public-sector settings (Thatcher et al., 2002-2003), state and federal governments are adjusting their pay scales and benefit packages in hopes of reducing employee turnover. Interrole conflict has been discussed as a stressor for IT employees (Ahuja, 2002; Greenhaus, Collins, Singh, & Parasuraman, 1997).

The scholarly literature has identified key variables influencing an employee's affective organizational commitment and has verified a strong link between affective organizational commitment and voluntary turnover. Many state government organizations interested in retaining valuable IT employees have considered implementing policies and practices, based on findings in the scholarly literature, that influence the affective organizational commitment of their employees. In the following section, we will illustrate with a few examples what state governments are currently doing to address the employee-retention pressures that they are facing.

State Government Experiences

The following provides a few illustrations of what states have been doing to improve their ability to retain their technology workforces. We conclude this segment with a review of two studies. The first study explored factors influencing affective organizational commitment for state IT workforces, and the second examined factors potentially influencing the retention of state IT workforces.

State Initiatives

The following states were selected based on scores calculated by *Governing Magazine* and reported in the magazine's 2005 report titled *Government Performance Project: Grading the States*, which recognized states for their excellence in IT initiatives. The Web sites of all the states that received ratings of B+ or better were read by the authors, along with the Web sites of those states' IT departments. The following states provided the most information regarding their IT initiatives.

Missouri

In February of 2005, the state of Missouri conducted an IT environmental scan including input from 23 agencies to identify the trends and issues related to Missouri state government (*Missouri IT Environmental Scan Summary: February*, 2005). Of particular importance was the area of IT staffing—identified as one of four major subject areas along with citizens, IT consolidation, and agency business partners, management, and users. The lack of organizational commitment, lack of advancement and salary increases, increased workload, lack of training opportunities, and uncertainty regarding the IT consolidation initiatives were identified as IT employee dissatisfiers and demotivators. The proposed initiatives for IT staffing included the creation of a strategy for managing potential staff losses due to budget cuts, recruitment by the private sector, and retirements. Other initiatives included the identification of training needs and the development of training plans, and the provision of flexible work schedules, recognition, performance-based incentives, and raises when possible. With the improvement of the economy in Missouri and in general, concern was expressed over losing IT workers to the private sector.

Washington

In the Washington State Department of Information Services (DIS) 2005-2007 biennium strategic plan, one of the core values states, "Our employees are our most valued assets and their well-being is crucial to our success" (Washington State Department of Information Services, 2005, p. i). A part of the 2005-2007 strategic initiative is to conduct surveys every 2 years of all DIS employees, to report the results, and to follow up with responsive action plans. Another strategic objective is to move to a performance-based culture. DIS proposes to establish key competencies for all employment positions, ensure that each DIS employee has an annual evaluation, train employees to know how to use the performance development plan, and assure that proper training and policies are in place for each employee to meet the goals of his or her performance management plan. Additionally, the state has been working to develop a mechanism to be able to capture every DIS employee's training profile in order to ensure the employee had taken all of the mandatory training. Internal communications between all DIS employees are enhanced via Webcasting and video on demand. This allows all employees to be connected regardless of the time of day or location. Finally, the state of Washington is implementing a succession planning process to help employees plan for advancement by knowing what key competencies are needed for each position. If the employee needs training to develop a particular competency for a position, he or she can get the training before applying for a vacated position. The goal of DIS is to increase employee retention and maximize return on training investments. Therefore, in attempting to retain its IT employees, Washington appears to be paying attention to characteristics specific to IT employees including their ambition and their desire for professional development.

Delaware

The state of Delaware's strategic annual report sets the goal to become the employer of choice with a workforce that is empowered, capable, supportive, and accountable (Department of Technology & Information, 2004). Some initiatives to meet this goal include providing IT employees with the opportunity to grow professionally and personally, improving communication throughout the organization, and improving the performance management and compensation plans. Another initiative is to deploy and improve an employee recognition plan. Delaware computes an employee satisfaction index that is based on employee satisfaction surveys. Like Washington, Delaware appears to be targeting some of its retention initiatives toward those personal characteristics of IT employees including ambition and the need for challenging work, as well as work-experiences variables such as pay for performance.

Michigan

In Michigan, close to 60% of the IT workers are age 45 or older, and this state is faced with the challenge of replacing the IT knowledge base as these workers begin to retire (From Vision to Action, 2005). One strategy is to increase the diversity of the IT workforce by creating interview panels that are diverse in terms of gender and race. Michigan is also recruiting from college job fairs throughout the state. The state further focuses on improvements in the management of its IT workers in order to maintain a balance between work performed by contract employees and that performed by state IT employees. The Michigan Department of Information Technology (MDIT) hired 140 staff to replace contract employees and saved $19 million annually (From Vision to Action, 2005). Another area where Michigan is making changes is the classification and compensation of its IT workers. The state has recently initiated a special project to review how the private sector and other public organizations compensate and classify the delivery of IT services. The goal is to be able to offer a competitive compensation package designed to attract and retain IT workers. The MDIT has regularly scheduled town-hall meetings and department blood drives to encourage teamwork, and offers telecommuting. Finally, Michigan provides a series of online training programs that cover both technical and behavioral issues. Therefore, in an attempt to retain and replace employees, Michigan is addressing the issues of outsourcing, job characteristics, work environment, and personal growth needs of IT employees previously discussed in this chapter.

Virginia

Like Michigan, the Virginia Information Technologies Agency (VITA) has around 35% of its IT staff being age 50 or older. In the next 5 years, this percentage will increase to 55%. Additionally, the need for workers with specialized IT skills will increase as will the competition for hiring and retaining these workers. VITA recognizes the need to offer specialized training for all of their IT workers (Virginia Information Technologies Agency, 2006b). In November of 2005, the Commonwealth of Virginia entered into an IT infrastructure partnership with Northrop Grumman Corporation (Virginia Information Technologies Agency,

2006a). This partnership is the first of its kind in the United States. The infrastructure partnership includes such tasks as maintaining equipment and services for mainframes, servers and operating systems, and providing e-mail as well as data-center facilities and staff. The current VITA employees who perform infrastructure operations have the option of going to work for Northrop Grumman or staying with the state. Virginia appears to be taking a unique approach to the outsourcing issue discussed earlier in this chapter.

State IT Workforce Studies

The following studies were selected based on their usage of public-sector IT workforces as the sample, and their applicability to the phenomenon under study (IT workforce retention). The first study tests factors influencing affective organizational commitment for state IT workforces, and the second tests factors influencing turnover intentions for state IT workforces. Taken together, these studies demonstrate the importance of job-characteristics and work-environment variables for the retention of public-sector IT employees.

Study 1

The first study by Reid et al. (2006) is intended to provide a backdrop for the challenges that governmental employers face in retaining their IT workforces. The sample for this study was IT employees working for a south-central state with one of the highest voluntary turnover rates of state government employees in the nation. The independent variables in their study included goal setting, role ambiguity, task variety, subjective stress, intergroup conflict, interrole conflict, perceived organizational support, leader-member exchange, and performance-pay contingency, and the dependent variables were affective organizational commitment, and its correlate, job satisfaction.

Reid et al. (2006) found that perceived organizational support, leader-member exchange, role ambiguity, and task variety were the largest contributors to a canonical function that explained 53% of the variance in affective public-sector commitment and job satisfaction. Consistent with previous research (Albrecht, 2005; Albrecht & Travaglione, 2003; Lee, 2004), their findings indicated that in the public-sector IT environment, it is important to have a climate of organizational support and a supportive working relationship with supervisors.

In contrast to research findings in the private sector (e.g., Ahuja, 2002; Igbaria, Parasuraman, & Greenhaus, 1997), in Reid et al.'s (2006) study, interrole conflict and salary issues did not rise to a significant level of concern for employees. Perhaps this is because larger organizations such as state government agencies (Durst, 1999) and organizations with sizable numbers of female employees are more likely to provide family-friendly benefits (Osterman, 1995). In addition, employees may enter the public sector understanding that their salaries are not likely to be comparable with their private-sector counterparts, or feeling other benefits available to public-sector workforces provide sufficient counterbalance so that pay does not become a major issue. However, since the pay for the IT employees in this state was in the bottom 25% of the states, their finding was surprising.

Among the job-characteristics variables, Reid et al. (2006) also found that role ambiguity and task variety strongly contributed to the variance explained in affective public-sector commitment and job satisfaction. Reid et al.'s findings were consistent with past research regarding the negative relationship between role ambiguity and affective organizational commitment (e.g., Igbaria & Greenhaus, 1992), but in contrast to Thatcher et al. (2002-2003). Reid et al. also found support for the link between task variety and affective public-sector commitment in the state government IT workforce:

One explanation for this finding may be that many public sector agencies are severely understaffed so individuals are asked to "pick up the slack" and take on tasks outside of their official job description or serve constituents across many agency units. While these challenges are a source of task variety, they may also adversely affect employees by increasing the potential for role ambiguity. (p. 327)

In contrast to previous private-sector research, Reid et al. (2006) found in this public-sector context other job-characteristic variables (i.e., goal setting, intergroup conflict, and subjective stress) appear to have a lesser influence on affective commitment and job satisfaction. One explanation provided by the authors for this finding may be that "researchers have found that goal setting by individual employees is often limited in public sector settings, and may, therefore, be less central to employee affective public sector commitment and job satisfaction than other factors" (p. 327). Interestingly, no gender differences were found in this sample, which was 53% male.

Study 2

The second study by Kim (2005) is intended to provide a different perspective on the challenges that governmental employers face in retaining their IT workforces. The sample for this study was IT employees working for two states, Nevada and Washington, with voluntary turnover rates of 16% and 6 to 10%, respectively. The independent variables investigated included work exhaustion, role ambiguity, role conflict (job characteristics), participatory management, project resources (work environment), advancement opportunities, training and development, and performance-pay contingency (human-resources management), and the dependent variable was turnover intention.

Among the job-characteristics variables, only work exhaustion was positively associated with turnover intention. This finding is consistent with previous research in the private-sector IT workforce (e.g., Moore, 2000). In response to this finding, Kim (2005, p. 149) asserts, "To address the issue of work exhaustion, managers and supervisors can adjust work expectations and establish more reasonable target dates for the completion of IT projects."

Among the work-environment variables, participatory management was negatively associated with turnover intention. Kim (2005, p. 149) states, "Participatory management practices also facilitate a clear understanding of IT project objectives, policies, and role expectations." Among the human-resources-management variables, consistent with previous research (e.g., Selden & Moynihan, 2000), advancement opportunities were negatively associated with turnover intention. Interestingly, no gender differences were found in this sample, which was 66% male.

From a managerial perspective, the findings from these studies may aid public agencies in their ongoing efforts to attract and retain quality IT professionals. First, whether addressing turnover intention directly or affective organizational commitment as a proximal precursor to turnover intention, these studies indicate that in contrast to the private sector, in the public sector, gender and monetary compensation do not play a role. Second, public-sector managers play a key role in enhancing employees' affective public-sector commitment and job satisfaction, two key constructs previously and strongly linked to turnover in the public sector (Thatcher et al. 2002-2003). Thus, quality supervision may play a significant role in influencing retention by creating an environment where employees feel supported by both their agency and their managers. Participatory management practices and career advancement opportunities may be two mechanisms through which management may demonstrate support. In addition, public-sector managers may be able to increase task variety and/or decrease role ambiguity by shifting tasks and/or workloads in response to individual needs. This may help increase job satisfaction and affective organizational commitment, and decrease work exhaustion and turnover intentions.

What is to be Done?

From our analysis of the states recognized for their excellence in IT initiatives, it appears that there are a few common strategies these states are pursuing. States may want to look to these strategies to increase the retention of quality IT employees. See Table 1 for a summary of strategies being employed.

For example, four of the five states drawn from *Government Performance Project: Grading the States* (2006) appear to believe that training and development efforts are a key to retaining valued IT workers. This is consistent with private-sector IT personnel research (e.g., Couger, Oppermann, & Amoroso, 1994; Couger & Zawacki, 1978; Wynekoop & Walz, 2000) that found the need for achievement, challenge, autonomy, opportunities, and learning are important to IT professionals given the personal characteristics of such employees.

Three of the five states are focusing on performance-based incentives as a means of retaining quality IT personnel. Bozeman and Bretschneider (1986) found that public-sector managers perceive a weaker relationship between performance and pay than private-sector managers. From the study results presented, the pay-for-performance contingency appeared to have a minor impact on affective public-sector commitment, job satisfaction, and turnover intention. Thus, in contrast to the private sector, in a public-sector context, it appears that salary issues may be managed if employees perceive a climate of managerial support and receive recognition in other forms from management. The states of Delaware, Missouri, and Washington understand this aspect and are focused on performance-based pay as a means of achieving employee retention.

Communication seems to be a key focus for the successful states. The Washington state government is using Webcasting and other technologies to communicate with employees, while the Michigan state government uses regularly scheduled town-hall meetings to keep employees informed. Lastly, employee surveys and recognition are being used to keep in

Table 1.

Strategy	State				
	MO	WA	DE	MI	VA
Training and Development	×		×	×	×
Flexible Work Schedules	×				
Telecommuting				×	
Employee Recognition	×		×		
Performance-Based Incentives	×	×	×		
Compensation				×	
Employee Surveys		×	×		
Annual Evaluations and Performance Development Plans		×			
Communication		×	×	×	
Succession Planning	×				
Increase Diversity				×	
Decrease Contract Employees				×	

touch with employee needs and concerns. Communication is a key to reducing role ambiguity and uncertainty, which were negatively related to commitment and job satisfaction in the in-depth case study reported in this chapter. Clear supportive communication from supervisors, as discussed in the scholarly literature on leader-member exchange, also is a key to retaining IT employees.

From the previous discussion, it is apparent that all levels of government are concerned with hiring and retaining vital IT workforces but are not averse to outsourcing some functions to the private sector to keep pace with the technological changes. At the same time, managers must be sensitive to the unique responsibilities of public agencies and the unique political environments to which they must respond. It is clear that blended workforces of contract and permanent employees have become quite common in some public agencies, and some states and federal agencies have engaged in outsourcing and offshoring. Such approaches are being used to deal with the persistent pressures of finding qualified employees or responding to unique circumstances. Yet, even with demonstrable cost savings that might be derived from such activities, these benefits can only be part of the equation. The message for public-sector decision makers is clear: workplace and job characteristics matter. A supportive work environment with clear, reasonable work expectations and identified career opportunities is a key factor in IT personnel retention. IT professionals, who have the option of working almost anywhere in the economy, may be more demanding in this regard. Public-sector managers should take notice of these aspects if agencies wish to retain this valuable segment of the workforce.

References

2005 outsourcing bills. (2005). *National Conference of State Legislatures.* Retrieved January 20, 2006, from http://www.ncsl.org/programs/employ/outsourcing05.htm

Ahuja, M. K. (2002). Women in the information technology profession: A literature review, synthesis, and research agenda. *European Journal of Information Systems, 11*(1), 20-34.

Albrecht, S. (2005). Leadership climate in the public sector: Feelings matter too. *International Journal of Public Administration, 28*(5/6), 397-416.

Albrecht, S., & Travaglione, A. (2003). Trust in public-sector senior management. *International Journal of Human Resource Management, 14*(1), 76-92.

Balfour, D. L., & Wechsler, B. (1996). Organizational commitment: Antecedents and outcomes in public organizations. *Public Productivity and Management Review, 9*(3), 256-277.

Bluedorn, A. C. (1982). A unified model of turnover from organizations. *Human Relations, 35*(2), 135-154.

Bozeman, B., & Bretschneider, S. (1986). Public management information systems: Theory and prescription. *Public Administration Review, 46*, 475-487.

Bretschneider, S., & Wittmer, D. (1993). Organizational adoption of microcomputer technology: The role of sector. *Information Systems Research, 4*(1), 88-108.

Cammann, C., Fichman, M., Jenkins, G. D., & Klesh, J. (1978). *The Michigan organizational assessment package.* Ann Arbor, MI: University of Michigan, Survey Research Center.

Catlette, B., & Hadden, R. (2000). *Contented cows give better milk.* Germantown: Saltillo Press.

Chan, M. (1989). Intergroup conflict and conflict management in the R&D divisions of four aerospace companies. *IEEE Transactions on Engineering Management, 36*(2), 95-104.

Chen, Y.-C., & Perry, J. L. (2003). *IT outsourcing: A primer for public managers.* Arlington, VA: IBM Endowment for the Business of Government.

Chi, K. S., Arnold, K. A., & Perkins, H. M. (2003). Privatization in state government: Trends and issues. *Spectrum: The Journal of State Government, 76*(4), 12-22.

Couger, J. D., Oppermann, E. E., & Amoroso, D. L. (1994). Changes in motivation of IS managers: Comparison over a decade. *Information Resources Management Journal, 7*(2), 5-13.

Couger, J. D., & Zawacki, R. A. (1978). What motivates DP professionals? *Datamation, 24*(9), 116-123.

Crewson, P. E. (1997). Public-service motivation: Building empirical evidence of incidence and effect. *Journal of Public Administration Research and Theory, 7*(4), 499-518.

Department of Technology & Information. (2004). *2005-2007 strategic plan.* Retrieved March 1, 2006, from *http://dti.delaware.gov/pdfs/strategicplan/06DTIPlan_KeyStrategies_Dec04.pdf*

Durst, S. L. (1999). Assessing the effect of family friendly programs on public organizations. *Review of Public Personnel Administration, 19*(3), 19-33.

Edelman, M. A., Reynolds, D., & Holle, C. M. (1997). *A comparative cost analysis of state government outsourcing: An Iowa case study of drivers' license issuance in rural counties.* Retrieved January 20, 2006, from http:/www.econ.iastate.edu/research/webpapers/NDN0049.pdf

Eisenberger, R., Armeli, S., Rexwinkel, B., Lynch, P. D., & Rhoades, L. (2001). Reciprocation of perceived organizational support. *Journal of Applied Psychology, 86*(1), 42-51.

Eisenberger, R., Huntington, R., Hutchison, S., & Sowa, D. (1986). Perceived organizational support. *Journal of Applied Psychology, 75*, 51-59.

Farrell, D., & Stamm, C. L. (1988). Meta-analysis of the correlates of employee absence. *Human Relations, 41*(3), 211-227.

Fish, K. E., & Seydel, J. (2006). Where IT outsourcing is and where it is going: A study across functions and department sizes. *The Journal of Computer Information Systems, 46*(3), 96-103.

Frank, S. A., & Lewis, G. B. (2004). Government employees: Working hard or hardly working? *American Review of Public Administration, 34*(1), 36-51.

From Vision to Action. (2005). *2006 Michigan IT strategic plan: Appendix G.* Retrieved March 1, 2006, from http://www.michigan.gov/documents/AppendixG

Garson, G. D. (2003). *Public information technology: Policy and management issues.* Hershey, PA: Idea Group Publishing.

Government Accountability Office. (2006). *Offshoring in six human services programs.* Retrieved March 30, 2006, from http//:www.gao.gov/cgi-bin/getrpt?GAO-06-342

Government performance project: Grading the states. (2005). Retrieved June 14, 2005, from http://www.results.gpponline.org

Greaves, J., & Sorenson, R. C. (1999). Barriers to transformation in a higher education organization: Observations and implications for OD professionals. *Public Administration Quarterly, 23*(1), 104-129.

Greenhaus, J. H., Collins, K. M., Singh, R., & Parasuraman, S. (1997). Work and family influence on departure from public accounting. *Journal of Vocational Behavior, 50*(2), 249-270.

Griffeth, R. W., & Hom, P. W. (2001). *Retaining valued employees.* Thousand Oaks, CA: Sage Publications.

Griffeth, R. W., Hom, P. W., & Gaertner, S. (2000). A meta-analysis of antecedents and correlates of employee turnover: Update, moderator tests, and research implications for the next millennium. *Journal of Management, 26*(3), 463-488.

Guttman, D. (2003). Contracting United States government work: Organizational and constitutional models. *Public Organization Review, 3*(3), 281-299.

Hackman, J. R., & Oldham, G. R. (1976). Motivation through the design of work: Test of a theory. *Organizational Behavior and Human Performance, 16*(2), 250-279.

Hackman, J. R., & Oldham, G. R. (1980). *Work redesign.* Reading, MA: Addison-Wesley.

Harris, K. J., Kacmar, K. M., & Witt, L. A. (2005). An examination of the curvilinear relationship between leader-member exchange and intent to turnover. *Journal of Organizational Behavior, 26*(4), 363-378.

Hatcher. T. (1999). How multiple interventions influenced employee turnover: A case study. *Human Resource Development Quarterly, 10*(4), 365-383.

Hutchison, S., Sowa, D., Eisenberger, R., & Huntington, R. (1986). Perceived organizational support. *Journal of Applied Psychology, 71*(3), 500-507.

Igbaria, M., & Greenhaus, J. H. (1992). Determinants of MIS employees' turnover intentions: A structural equation model. *Communications of the ACM, 35*(2), 34-50.

Igbaria, M., Parasuraman, S., & Greenhaus, J. H. (1997). Status report on women and men in the IT workplace. *Information Systems Management, 14*(3), 44-53.

Irving, P. G., & Coleman, D. F. (2003). The moderating effect of different forms of commitment on role ambiguity-job tension relations. *Canadian Journal of Administrative Sciences, 20*(2), 97-106.

Jamal, M. (1990). Relationship of job stress and type-a behavior to employees' job satisfaction, organizational commitment, psychosomatic health problems, and turnover intentions. *Human Relations, 43*(8), 727-738.

Kim, S. (2005). Factors affecting state government IT employee turnover intentions. *American Review of Public Administration, 35*(2), 137-156.

Lan, G. Z., Riley, L., & Cayer, N. J. (2005). How can local government become an employer of choice for technical professionals? *Review of Public Personnel Administration, 25*(3), 225-242.

Lee, P. C. B. (2004). Social support and leaving intention among computer professionals. *Information and Management, 41*(3), 323-334.

Light, P. (1999). *The true size of government.* Washington, DC: Brookings.

Light, P. (2003). *An update on the Bush administration's competitive sourcing initiative.* Retrieved January 20, 2006, from http://www.senate.gov/~gov_affairs/072403light.pdf

Liou, K. T., & Nyhan, R. C. (1994). Dimensions of organizational commitment in the public sector: An empirical assessment. *Public Administration Quarterly, 18*(1), 99-118.

Locke, E. A., Latham, G. P., & Erez, M. (1988). The determinants of goal commitment. *Academy of Management Review, 13*(1), 23-39.

Maranto, R., & Skelley, B. D. (2003). Anticipating change in the higher civil service: Affective commitment, organizational ideology, and political ideology. *Public Administration Quarterly, 27*(3/4), 336-367.

Mathieu, J. E., & Zajac, D. M. (1990). A review and meta-analysis of the antecedents, correlates and consequences of organizational commitment. *Psychological Bulletin, 108*(2), 171-194.

Meyer, J. P., Stanley, D. J., Herscovitch, L., & Topolnytsky, L. (2002). Affective, continuance, and normative commitment to the organization: A meta-analysis of antecedents, correlates, and consequences. *Journal of Vocational Behavior, 61*(1), 20-52.

Michaels, C. E., & Spector, P. E. (1982). Causes of employee turnover: A test of the Mobley, Griffeth, Hand and Meglino model. *Journal of Applied Psychology, 67*(1), 53-59.

Missouri IT environmental scan summary: February. (2005). Retrieved February 22, 2006, from http://dis.wa.gov/news/publications/05-07strategicplan.pdf

Mobley, W. H., Griffeth, R. W., Hand, H. H., & Meglino, B. M. (1979). Review and conceptual analysis of the employee turnover process. *Psychological Bulletin, 86*(3), 493-522.

Mohan, L., Holstein, W. K., & Adams, R. B. (1990). EIS: It can work in the public sector. *MIS Quarterly, 14*(4), 435-448.

Moore, J. E. (2000). One road to turnover: An examination of exhaustion in technology professionals. *MIS Quarterly, 24*(1), 141-168.

Mosher, F. C. (1982). *Democracy and the public service* (2nd ed.). New York: Oxford University Press.

Mowday, R. T., Steers, R. M., & Porter, L. W. (1979). The measurement of organizational commitment. *Journal of Vocational Behavior, 14*(2), 224-247.

Nam, K., Rajagopalan, S., Rao, H. R., & Chaudhury, A. (1996). A two-level investigation of information systems outsourcing. *Communications of the ACM, 39*(7), 36-44.

Newcombe, T. (2002, April 16). Filling the workforce gap. *Government Technology.* Retrieved April 18, 2006, from http://www.govtech.net/magazine/story.php?id=8017&issue=4:2002

Nyhan, R. C. (1999). Increasing affective organizational commitment in public organizations. *Review of Public Personnel Administration, 19*(3), 58-70.

Osterman, P. (1995). Work/family programs and employment relationships. *Administrative Science Quarterly, 40*(4), 681-700.

Overman, S. (2003). Federal, state governments fishing for business process outsourcing bounties. *HR Magazine, 48*(9), 32.

Passori, A. (2000). *Enterprisewide information security best practices.* New York: Meta Group Inc.

Perry, J. L., & Porter, L. W. (1982). Factors affecting the context for motivation in public organizations. *Academy of Management Review, 7*(1), 89-98.

Perry, R. W. (2004). The relationship of affective organizational commitment with supervisory trust. *Review of Public Personnel Administration, 24*(2), 133-149.

Pinkovitz, W. H., Moskal, J., & Green, G. (2001). *How much does your employee turnover cost?* Retrieved May 18, 2006, from http://www.uwex.edu/ces/cced/publicat/turn.html

Porter, L. W., & Steers, R. M. (1973). Organizational, work, and personal factors in employee turnover and absenteeism. *Psychological Bulletin, 80*(2), 151-171.

Rainey, H. G. (1982). Reward preferences among public and private managers: In search of the service ethic. *American Review of Public Administration, 16*(4), 288-302.

Reid, M. F., Riemenschneider, C. K., Allen, M. W., & Armstrong, D. J. (2006). Affective commitment in the public sector: The case of IT employees. In *Proceedings of the 2006 ACM SIGMIS CPR Conference* (pp. 321-332).

Romzek, B. S. (1990). Employee investment and commitment: The ties that bind. *Public Administration Review, 50*(3), 374-382.

Sakar, D. (2005). State and local officials look to outsourcing: Retiring workforce driving new push to contract out services. *Federal Computer Week, 19*(10), 38.

Selden, S. C., & Moynihan, D. (2000). A model of voluntary turnover in state government. *Review of Public Personnel Administration, 20*(2), 63-74.

Sinensky, A., & Wasch, R. S. (1992). Understanding outsourcing: A strategy for insurance companies. *Journal of Systems Management, 43*(1), 32-36.

Somech, A. (2003). Relationships of participative leadership with relational demography variables: A multilevel perspective. *Journal of Organizational Behavior, 24*(8), 1003-1018.

Staats, E. B. (1988). Public service and the public interest. *Public Administration Review, 48*(2), 601-605.

Steers, R. M. (1977). Antecedents and outcomes of organizational commitment. *Administrative Science Quarterly, 22*(1), 46-56.

Steers, R. M., & Mowday, R. (1981). Employee turnover and post-decision accommodation processes. *Research in Organizational Behavior, 3*, 235-283.

Steinhaus, C. S., & Perry, J. L. (1996). Organizational commitment: Does sector matter? *Public Productivity and Management Review, 19*(3), 278-288.

Tett, R. P., & Meyer, J. P. (1993). Job satisfaction, organizational commitment, turnover intention, and turnover: A path analysis based on meta-analytic findings. *Personnel Psychology, 46*(2), 259-293.

Thatcher, J. B., Stepina, L. P., & Boyle, R. J. (2002-2003). Turnover of information technology workers: Examining empirically the influence of attitudes, job characteristics, and external markets. *Journal of Management Information Systems, 19*(3), 231-261.

Thompson, C. A., Jahn, E. W., Kopelman, R. E., & Prottas, D. J. (2004). Perceived organizational family support: A longitudinal and multilevel analysis. *Journal of Managerial Issues, 16*(4), 545-565.

Thompson, J. R., & Mastracci, S. H. (2005). Nonstandard work arrangements in the public sector. *Review of Public Personnel Administration, 25*(4), 299-324.

U.S. Merit Systems Protection Board. (2002). *Assessing federal job-seekers in a delegated examining environment.* Retrieved June 14, 2006, from http://www.mspb.gov/studies/rpt_02-20-02_job-seekers/assessmentmethods.pdf

Virginia Information Technologies Agency. (2006a). *Virginia Information Technologies Agency: About the partnership.* Retrieved June 10, 2006, from http://www.vita.virginia.gov/itpartnership/aboutPartnership.cfm

Virginia Information Technologies Agency. (2006b). *Virginia Information Technologies Agency July 2006 to June 2008 strategic plan.* Retrieved June 10, 2006, from http://www.vita.virginia.gov/docs/pubs/strategicPlan/index.cfm

Washington State Department of Information Services. (2005). *Washington State Department of Information Services: 2005-2007 strategic plan.* Retrieved February 22, 2006, from http://www.oa.mo.gov/itsd/cio/archive/2005EnvironmentalScanSummary.pdf

White, J. D., & Korosec, R. L. (2005). Issues in contracting and outsourcing information technology. In G. D. Garson (Ed.), *Handbook of public information systems* (pp. 407-425). Boca Raton, FL: Taylor and Francis.

Wright, B. E. (2001). Public-sector work motivation: A review of the current literature and a revised conceptual model. *Journal of Public Administration Research and Theory, 11*(4), 559-586.

Wynekoop, J., & Walz, D. (1998). Revisiting the perennial question: Are IS people really different? *The DATA BASE for Advances in Information Systems*, *29*(2), 62-72.

Wynekoop, J. L., & Walz, D. B. (2000). Investigating traits of top performing software developers. *Information Technology & People, 13*(3), 186-195.

Zeffane, R. (1994). Patterns of organizational commitment and perceived management style: A comparison of public and private sector employees. *Human Relations, 47*(8), 977-1011.

Chapter XIII

Computer Tools for Public-Sector Management

Carl Grafton, Auburn University Montgomery, USA

Anne Permaloff, Auburn University Montgomery, USA

Abstract

Almost any public-sector task employing a computer can be accomplished more efficiently with a variety of tools rather than any single one. Basic tools include word processing, spreadsheet, statistics, and database-management programs. Beyond these, Web authoring software, presentation software, graphics, project-planning and -management software, decision analysis, and geographic information systems can be helpful depending upon the job at hand.

Introduction

The use of computer technology in government taps into three sometimes incompatible concepts: government responsiveness to the public, bureaucracy, and technocracy. The tensions between the first two have long been a staple of textbooks and scholarly work in public administration and organization theory (Blau & Meyer, 1971; Borgmann, 1988; Gullick, 1996; Rosenbloom & Kravchuk, 2002). At first, when all computers were mainframes, the technocratic perspective (rule by experts) appeared to bolster Weberian bureaucracies (Ellul, 1964; Freeman, 1974). Even today, computers are often used by bureaucrats to perform routine tasks efficiently or analysts to rationalize policy, and most of this chapter is taken up by descriptions of some of the tools available to them. However, today's computers are employed in far more ways and by many more members of all parts of government than they were a few years ago. The bureaucracy is less centralized just by virtue of the widespread access of government personnel to information and their ability to process that information.

Changes wrought by computers may go beyond bureaucratic decentralization. Eugene J. Akers (2006) speculates that government organized along Weberian bureaucratic lines is increasingly out of step with public expectations of a transparent and responsive service-oriented government. Similarly, Carl Grafton and Anne Permaloff (2005) depict what they call Jeffersonian budgeting: understandable government budgets available on the Internet with which the news media and the public can hold public officials accountable. In addition, Christa Slaton and Jeremy Arthur (2004) describe ways to facilitate public participation in government administration using computer technology.

This chapter concerns computer applications and information technology in government other than financial accounting software, which deserves a chapter of its own. Topics covered include Web publishing, spreadsheets, statistics packages, database management, presentation software, project-planning and -management software, decision analysis, graphics for illustrations, and geographic information systems. Since most readers are likely to have substantial word-processing experience, it would be unproductive to devote much space to this topic.

A Variety of Tools

To make the most of their time and talents, computer users in the public sector or virtually any other setting should have access to more than one tool for nearly any task that extends much beyond typing a short memo. Access to a variety of tools is usually more productive than having the latest version of a single one.

Word-Processing and Web Authoring Software

Word-processing programs are designed primarily for generating print and graphic images on paper; Web authoring programs do the same thing for the Internet. Web pages are generated using HTML (hypertext markup language) sometimes in conjunction with supplemental tools such as Java, a programming language.

When a browser such as Microsoft Explorer reads a file containing HTML code (called tags) and text, it displays the file on the computer monitor according to formatting information in the tags (e.g., whether text is centered or in bold face or whether a separate file containing a graphic image is to be merged with the text). The marketplace offers a variety of text editors primarily designed to generate text with HTML tags (see Kent, 2000, for a list). Ordinary word-processing software can do so as well, but specialized HTML editors contain more features that ease the process of Web-page creation or maintenance.

Most government agencies have adopted particular HTML editors that employees are expected to use. Government agencies often appoint individuals responsible for Web-page work partly to preserve consistency in appearance and also out of concerns for security. The majority of employees will submit text generated with an ordinary word processor along with graphic images to accompany the text to these specialists who will then convert these files into Web pages.

The authors' experience in using government Web sites suggests that they are of three basic types: marketing, informational, and interactive. Marketing sites are designed for such purposes as to attract students to universities and visitors to cities. Marketing sites usually contain photographs scanned from prints or downloaded from a digital camera together with attractively arranged text and various decorative graphics. A certain amount of taste is required to make the marketing part of a Web site attractive, and the assistance of someone with a background in art might be helpful. Peter Kent (2000) lists Web sites that collect badly designed Web sites including one called *Web Pages That Suck: Learn Good Design by Looking at Bad Design*. While the bad Web-site collections dramatize how Web designs can go horribly wrong, fledgling designers will also want to visit sites of organizations similar to theirs to see what works.

Most Web pages presented by government agencies are meant to be entirely informational. Beyond basic neatness, the only significant design considerations are clarity and consistency. For example, the Web sites of most states and cities describe tax structures. These sites typically provide an overview and links that lead to more detailed information such as how taxes can be calculated and paid. These pages are usually quite informative, but sometimes a variety of fonts and formats are employed. For example, in one state, some pages contained buttons while others used simple lists, giving the whole enterprise a somewhat unprofessional appearance. Sometimes information is simply difficult to find, and built-in Web-site search engines require overly precise wording.

Interactive Web sites are the most difficult to build (Kent, 2000). Interactive sites gather information from users. They may take orders for goods or services, maintain an inventory of goods and services available, and accept credit-card information, among other tasks. An example is an online university class-registration system. Interactive sites require specialized programming skills beyond the reach of most public administrators. It is critical that potential users test all Web sites, especially interactive ones. Information and procedures that seem clear to programmers may be sources of confusion and irritation to those unfamiliar with a Web site's logic.

Richard Heeks (2006) observes that failed or flawed Web sites and other IT applications may be found at all levels of government. The central cause of failure is that the planning, construction, implementation, and maintenance of IT systems are important to many actors including agencies with differing responsibilities and agendas, elected members of the executive and legislative branches, and competing vendors, but participation is often limited to engineers, programmers, and high-level management (Coursey & Killingsworth, 2005). Most government employees lack IT backgrounds and tend to surrender Web-site and IT-system development and maintenance to technical personnel. If a system is to serve the needs of all stakeholders including government agency clients, participation in system design must be widespread.

Spreadsheets

No desktop computer tool is more widely used for numerical data storage and analysis than the spreadsheet family of programs. Desktop computers are ubiquitous in homes and workplaces, and spreadsheet programs such as Excel reside on many hard disks as part of a suite of programs; however, a large percentage of public managers know little about spreadsheet operations beyond basic budgeting uses.

Programs such as Excel present the user with a series of worksheets grouped as a book of related information. Each worksheet is a grid of columns and rows. The rectangular intersection of a column and row is a cell. Three major kinds of information may be entered into a cell: text, numbers, and formulas that perform calculations on the numbers or on other formulas. Excel, by far the most widely used spreadsheet program, is also able to utilize Visual Basic, the program language common to the products that make up the Microsoft Office Suite. In addition to performing calculations based on user-inputted formulas, spreadsheet programs can produce graphics such as pie charts, bar charts, and time-series graphs, and they can be used to query a database stored in the spreadsheet. They can also perform statistical analysis with a variety of built-in tools and add-ins (defined below). In addition, the spreadsheet program's basic worksheet is one of the formats used for inputting data into a variety of other programs. It is also used as a format for downloading information (e.g., budget information and election results) from public Web sites.

Benefit-Cost Analysis

Benefit-cost analysis compares the benefits and costs of various choices and selects the choice that yields the highest net benefit (Ammons, 2002; Boardman, Greenberg, Vining, & Weimer, 2001; Stokey & Zeckhauser, 1978). In practice, most benefit-cost problems involve capital expenditures where benefits and costs are received and spent over a period of years.

The first step in performing a benefit-cost analysis is to list sources of benefits and costs associated with a project (Panayotou, n.d.). In this initial phase, the focus is on categories, not dollar amounts. For example, a benefit-cost analysis of a proposed monorail in Seattle listed the following benefits to monorail riders: travel-time savings, parking savings, automobile cost savings, and reliability. Those who continued to drive would enjoy the benefit of lessened traffic congestion (DJM Consulting, 2002; Slack, n.d.).

The second step is to evaluate the relationship between costs and benefits. The analyst is studying the causal relationship (or lack thereof) between expenditures and the benefits they yield (Panayotou, n.d.; Schmid, 1989). For example, the Seattle monorail benefit-cost analysis included an estimate of ridership partly based on comparisons with rail systems in other cities (People for Modern Transit [PMT], n.d.). The three largest benefits (travel-time savings, parking savings, automobile cost savings), which constituted more than 86% of claimed benefits, were heavily dependent on a ridership estimate that could only have been an approximation, a point ignored in a PowerPoint briefing presented by a consulting firm employed by advocates of the monorail. A benefit-cost analysis that fails to provide a range of estimates of critical parameters should be regarded with deep suspicion.

The next step is to establish direct effects of a project: the dollar benefits to project users and costs to build the project. When the market is functioning, project costs are simply off-the-shelf costs (e.g., concrete for highway construction; Levin, 1983; Nas, 1996). If market failure exists (e.g., leading-edge technology provided by a small number of competitors), cost valuation becomes more difficult (Panayotou, n.d.).

In some cases, project benefits are comparable to products or services sold in the private sector. An example is public housing. The value of a public-housing apartment will be close to that of a similar apartment offered by the private sector. If the government decides to offer units at a discount, the benefit to the renter will be the difference between the private-sector value of the apartment and the rent actually being charged. However, project benefits are often not directly comparable to private-sector market values. Thus, for example, a monorail may offer a faster and safer route. The problem is how to value time saved and deaths averted. In the case of the Seattle monorail, travel time was valued at $10.10 per hour, an amount that was half the mean wage rate in that area in 2002 (Slack, n.d.). The value of a life saved is more problematic, and the benefit-cost analysis literature offers many ways of doing so, most of which produce widely varying results (Fuguitt & Wilcox, 1999; Nas, 1996). Advocates of the Seattle monorail claimed $6.3 million in savings from reduced motor-vehicle accidents, but apparently these figures were confined to property damage and not loss of life.

The indirect effects of a project impact individuals or groups who are not users of project output (Schmid, 1989). With the monorail, those who can avoid driving automobiles through Seattle's crowded streets and parking them at high cost are direct beneficiaries. Those whose property values are increased or decreased or who experience other economic impacts because of the monorail are experiencing indirect benefits or costs (Marsden Jacob Associates, 2005; Tindall, 2005). Note that a single individual can enjoy both direct and indirect benefits or pay both direct and indirect costs. The distinction between direct and indirect effects is useful for organizing a benefit-cost study, but whether a given effect is placed in one or the other category should not have an impact on the results.

The two most common numerical indicators used in benefit-cost analysis are net benefits and the benefit-cost ratio. Net benefits are the difference between benefits and costs calculated on a present-value basis. The benefit-cost ratio is benefits divided by costs also on a present-value basis. The present value is the present-day equivalent of a stream of costs or benefits over time. The Seattle monorail net-benefit figure claimed by DJM Consulting in its 2002 briefing was $67 million. However, the benefit-cost ratio was only 1.04. In other words, the project would yield only $1.04 for every $1.00 spent. The 1.04 ratio represents a very thin margin. With slight cost overruns and tiny mistakes in ridership estimates, a 1.04 ratio could easily have flipped to less than the 1.00 breakeven point, and net benefits would have been negative tens of millions of dollars instead of positive. Again, the failure of this briefing to present ranges of estimates should have raised red flags. In fact, Seattle voters ultimately defeated the project.

The basic perspective of benefit-cost analysis is that of society as a whole (Fuguitt & Wilcox, 1999). So, for example, in a project's planning and construction phase, planners and engineers are benefiting in the form of fees and salaries, but from society's perspective, their benefits are costs. Society does not begin to benefit from a project until it is complete.

Large projects often affect geographical, racial, and other groups differently (Boardman et al., 2001; Nas, 1996; Schmid, 1989). Indeed, benefit-cost analysis' common society-wide perspective not only ignores such differential effects, but it may be biased against the disadvantaged. For example, other things being equal, benefit-cost analyses of alternative routes for a highway will favor a route plowing through economically depressed neighborhoods because property acquisition costs will be relatively low. For that reason, the federal government discourages the use of benefit-cost analysis by states in making routing decisions (Clinton, 1994; Federal Highway Administration, 1998).

The formula for calculating present value for any benefit or cost stream is:

$$PV = S_0/(1+r)^0 + S_1/(1+r)^1 + S_2/(1+r)^2 + ... + S_n/(1+r)^n,$$

where S_n represents any sum of money (benefit or cost) in year n, and r is the discount rate. The value of n may be the physical life of a project or its technologically useful life.

Performed on a hand calculator, the benefit-cost analysis of one or two streams of figures using the above formula is of no difficulty, but an actual benefit-cost analysis performed to aid decision making (as opposed to providing decorative support for decisions already made) typically involves sensitivity or "what if" analysis applied to combinations of benefits, costs, and discount rates. The basic question being posed by any kind of sensitivity analysis is how the result will be affected by change in inputs—in this case, benefits, costs, and the discount rate. If the estimates for each of these are all performed using a hand calculator, one is facing a considerable amount of work. With a spreadsheet such a prospect represents no problem.

Figure 1 shows a typical benefit-cost situation. In the first 2 years, benefits are zero because the project in question (e.g., a building) is under design and construction. The design and construction costs are relatively high compared to annual benefits. Because high design and construction costs are incurred in the first few years, they are discounted relatively little by the present value formula. The smaller benefit figures are more and more heavily discounted as they extend further into the future. Thus, the higher the discount rate, which governs the severity of the discounting, the worse the project will look as calculated either by net benefits (benefit minus cost) or the benefit-cost ratio (benefit divided by cost). Obviously, results also depend on cost and benefit estimates.

The table at the top of Figure 1 contains six sets of pessimistic, most likely, and optimistic benefits and costs. We are assuming a range of discount rates from 3 to 5% increasing in one half of 1% increments. The table below the top table contains all possible combinations

Figure 1. Benefit-cost analysis using a spreadsheet

	A	B	C	D	E	F	G	H	I	J
1					**Benefit-Cost Estimates**					
2					(millions of dollars)					
3										
4										
5						**Year**				
6			0	1	2	3	4	5	6	
7	**Pessimistic benefits**		0	0	3	3	3	3	3	
8	**Most likely benefits**		0	0	3.5	3.5	3.5	3.5	3.5	
9	**Optimistic benefits**		0	0	3.8	3.8	3.8	3.8	3.8	
10	**Pessimistic costs**		1.5	14	0.2	0.2	0.2	0.2	0.2	
11	**Most likely costs**		1	12	0.1	0.1	0.1	0.1	0.1	
12	**Optimistic costs**		0.9	10	0.08	0.08	0.08	0.08	0.08	
13										
14										
15			=NPV(C19,$D$7:$I$7)+$C$7							
16						=NPV(C19,$D$10:$I$10)+$C$10				
17					**Present Value**					
18	**Benefits**	**Costs**	**Rate**	**Benefits**	**Costs**	**Net**				
19	Pessimistic	Pessimistic	0.03	$13.34	$15.98	($2.64)		=D19-E19		
20	Most likely	Pessimistic	0.03	$15.56	$15.98	($0.42)				
21	Optimistic	Pessimistic	0.03	$16.90	$15.98	$0.91				
22	Pessimistic	Most likely	0.03	$13.34	$13.10	$0.24				
23	Most likely	Most likely	0.03	$15.56	$13.10	$2.47				
24	Optimistic	Most likely	0.03	$16.90	$13.10	$3.80				
25	Pessimistic	Optimistic	0.03	$13.34	$10.96	$2.37				
26	Most likely	Optimistic	0.03	$15.56	$10.96	$4.60				
27	Optimistic	Optimistic	0.03	$16.90	$10.96	$5.93				

Figure 2. Pivot table output

		Present Value Net Benefits				
		Rate				
Benefits	Costs	0.03	0.035	0.04	0.045	0.05
Most likely	Most likely	2.47	2.24	2.02	1.80	1.59
	Optimistic	4.60	4.36	4.12	3.90	3.68
	Pessimistic	-0.42	-0.63	-0.84	-1.03	-1.23
Optimistic	Most likely	3.80	3.55	3.30	3.06	2.83
	Optimistic	5.93	5.67	5.41	5.16	4.91
	Pessimistic	0.91	0.68	0.45	0.23	0.01
Pessimistic	Most likely	0.24	0.06	-0.12	-0.30	-0.47
	Optimistic	2.37	2.18	1.98	1.80	1.62
	Pessimistic	-2.64	-2.81	-2.98	-3.13	-3.29

of benefits, costs, and discount rates; it is used to build a pivot table (see below) to display the results in a readable fashion. The formulas in the present value columns for benefits and costs all resemble the ones in cells D19 and E19. The formulas in Column F all resemble the one in cell F19.

The syntax of the present value (NPV) function is =NPV(rate,range), where rate is the discount rate and range is a range of cells. Column C contains the discount rates. As the Excel's help utility notes, the NPV function begins discounting with the first cell in the range specified in the formula. Thus, if the block D10:I10 was replaced with C10:I10, the NPV function would begin the discounting 1 year too early. The cell address of the cost for year 0 is subtracted separately (cell C10). There is at least one way to enter the benefit-cost data that would make keying in the formulas more efficient than what we show in Figure 1, but our presentation demonstrates the basic principles more clearly.

The pivot table in Figure 2 summarizes all the combinations of benefits, costs, and discount rates within the specified ranges. Note that along each row, the present value figures decline with increasing discount rates, as they must. This distinctive pattern does not guarantee that the functions in Figure 2 were entered correctly, but it suggests that they were. This table would give decision makers a sense that optimistic or most likely assumptions would produce positive results even at the highest calculated discount rate, but that some inclusion of pessimistic benefit or cost assumptions can generate negative results.

Pivot Tables

We saw that Figure 1 displays part of a worksheet containing four dimensions: benefits characterized as pessimistic, most likely, and optimistic; costs with the same labels; discount rates; and net benefits. There is no way to display all of these data on a single two-dimensional worksheet. It would be even worse if there were an additional dimension such as geographical area. The pivot table in Figure 2 allows us to see benefit and cost categories in rows and rates in columns. Had it been necessary to include geographical categories, they could have been added as well. Potential reader confusion limits the number of variables that can be included in a single pivot table, but it is an efficient way to lay out multidimensional data sets. We lack the space to explain the steps required to build a pivot table, but many reference works (e.g., Berk & Carey, 2004) and Excel help screens explain it clearly. The pivot-table procedure allows us to include and exclude whatever data we want.

The data filter function offers another way to suppress or include data. The data filter function is invoked in Excel by clicking in succession on Data, Filter, and Autofilter. This choice automatically generates menus at the top of a worksheet—one menu for each column. For example, in Figure 1, the menu for the benefits column (beginning in cell A18 and going downward) contains the three benefits categories (pessimistic, most likely, and optimistic). By selecting, for example, pessimistic benefits and in the next column pessimistic costs, the worksheet would seem to eliminate all other benefits and costs, giving the decision maker a quick look at the worst-case scenarios at various discount rates for this project.

Spreadsheets and Statistics

Spreadsheets possess statistical computing capabilities adequate for many public-administration tasks. Excel can perform basic descriptive statistics, curve smoothing, analysis of variance, multiple regression, random number generation, sampling, *t*-test, and others. These capabilities are part of the Analysis ToolPak, an add-in that comes with all versions of Excel that are likely still to be in use. An add-in is a computer program that when installed merges with a base program (Excel in this case) making the add-in an integral part of the base program. Excel's Analysis ToolPak can be supplemented by another add-in called StatPlus, available with *Data Analysis with Microsoft Excel* (Berk & Carey, 2004). Various editions of this reasonably priced text contain versions of StatPlus designed to work with specific versions of Excel and Windows. This book is perhaps the most clearly written and informative introduction to spreadsheet usage and introduction to statistics on the market. Relatively advanced statistical analysis cannot be done readily or at all with Excel, but for

Table 1. Example of data arrangement to be read by a statistics package

	A	B	C	D
1	YEAR	MINERS	FATALTIES	FATLPERMINER
2	1935	56,5316	1,242	0.002197
3	1936	58,4494	1,342	0.002296
4	.	.	.	.
5	.	.	.	.

many administrative jobs, it may be adequate. Advanced work is more conveniently performed using statistics packages to be discussed later.

Spreadsheets are better tools than statistics packages for data entry, data editing, data storage, and printing. A statistics package is a computer program optimized for statistical analysis. Generally speaking, the best statistics packages (e.g., SYSTAT, SPSS, and NCSS) are more suitable for statistical analysis than are spreadsheets, but spreadsheets are superior for printing raw data as well as performing a wide variety of nonstatistical quantitative chores. If spreadsheet data entry follows simple rules specified by all leading statistics package manufacturers, the data may be exported to the statistics package; leading statistics packages can read spreadsheet files almost as easily as they can their native format.

Data to be read by a statistics package should be arranged so that the top row is devoted to variable names and each subsequent row is an observation. For example, if the analyst is examining U.S. coal-mine fatalities over time, data would be arranged as seen in Table 1.

Elsewhere on the worksheet, the definitions of the variables can be entered along with data sources and other information. It is especially convenient to enter such information in the form of comments in cells containing variable names. Some statistics packages allow such comments to be added to a data set, but many do not.

Note that the FATLPERMINER variable in Column D was calculated with a formula that reads =C2/B2. This formula (entered in D2) was copied downward to the last row of data. Someone new to spreadsheet work might think that the copied versions of this formula would all generate the result 0.002197, but spreadsheet formulas should not be read literally in terms of specific cell locations, but in terms of the relationship among cells. So, for example, the formula in D2 is correctly interpreted to mean to take the number in the Cell One to the left of D2, divide it by the number in Cell Two to the left of D2, and display the result in D2. This is called á relative cell reference formula. When this formula is copied, it maintains this relationship wherever it is put unless the user writes an absolute cell reference formula.

If it is determined that a variable (column) is no longer needed, it can be easily eliminated in a spreadsheet. If a transposition of variables (columns) can aid readability, this is much more easily accomplished with a spreadsheet than a statistics package.

Data sets can easily number 50, 100, 200, or more variables. To avoid mistakes while inputting data, it is important that a key variable, such as YEAR in this example, remain in view. With many statistics-package data-entry screens, when more than approximately five variables have been added, the first column (variable) scrolls to the left out of sight. The invisibility of the YEAR variable could easily result in the user entering data for 1999 when it should be in the 2000 row. With a spreadsheet, the YEAR column may be frozen so that it is always in view. Similarly, with some statistics packages, once data have been entered for roughly the first 20 years, variable names scroll away upward. In a spreadsheet, variable names can also be frozen in place. One can also view two entirely different parts of the worksheet with a split window, a feature we have not seen in any statistics-package data-input screen.

Many statistics packages lack even elementary print formatting capabilities producing crude-looking or confusing raw-data printouts. The statistics package output may be saved in a format readable by a word-processing program (or a spreadsheet) where the cleanup of layout and fonts may be accomplished. A spreadsheet allows complete control over what appears on which printout page. It is far superior to a statistics package in terms of the appearance of the printout.

Curve Smoothing and Forecasting

A time-series data set is "a sequence of observations taken at evenly spaced time intervals" (Berk & Carey, 2004, p. 408). Although it is characterized by serious theoretical weaknesses, the single moving average is one of the most common tools of applied time-series analysis, especially among financial securities analysts. The single moving average is easily calculated and plotted using a spreadsheet.

The time-series data in Figure 3 represent fatalities per coal miner in the years 1935 to 2005. The data are relatively noisy. To highlight trends obscured by year-to-year fluctuations and to perform rudimentary forecasting, it can be smoothed by calculating the mean of data over a number of time periods, in this case 10 years (see Berk & Carey, 2004, pp. 424-426). The first moving average data point (cell F11) is the mean of the first 10 years of raw data; the second moving average data point omits the first year of raw data and includes the 11th, and so forth.

The distance between the moving average curve and the raw-data curve is sensitive to changes in trend. Because each point on the moving average curve represents the past, each point on the moving average will be above raw data when the trend is down as it is in most of the years shown in Figure 3. Note the sharp spike in 1968. This data point might have reflected a data-entry mistake, but in fact it was due to an increase in fatalities (222 miners in 1967 and 311 in 1968) combined with a decrease in miners (139,312 in 1967 and 134,467 in 1968; Mine Safety and Health Administration, n.d.). The moving average can easily be calculated by entering the formula and copying it or by using the Excel trend line dialog box.

In financial forecasting (e.g., the Dow Jones Industrial Average), moving averages of 30, 60, or even 90 days are common. In such calculations, each day is treated as being equally important; that is, data from a day 30, 60, or 90 days ago is accorded as much weight as yesterday's data. It seems likely that yesterday's data are somewhat more important than data from a day before that, which in turn is more important than data from 3 days ago, and so forth. A family of curve-smoothing techniques known as exponential smoothing takes this likely truth into account by systematically weighting more recent time periods more heavily

Figure 3. Curve smoothing with spreadsheets

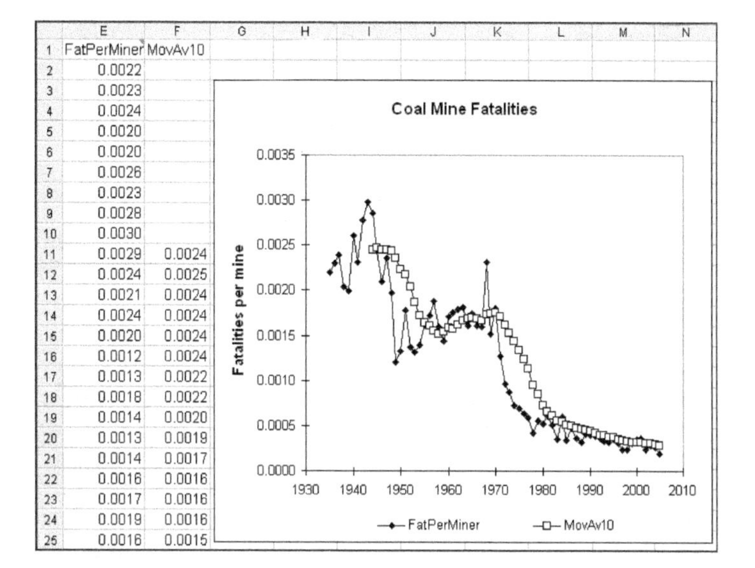

than earlier ones (Fosback, 1987). Excel also contains a number of exponential smoothing techniques (see Berk & Carey, 2004, pp. 427-429).

Many time-series data sets contain several components. As we have already seen, one such component may be a trend, a relatively long-term tendency to move upward or downward. Another time-series component may be seasonality: In a given month or quarter, the data may generally be average, below average, or above average for the year. City sales-tax revenues are often seasonal with, for example, revenues for November and December being above average for the year because of holiday business. Within a quarter or month, there may also be weekly patterns or other cycles. Finally, time-series data also contain random variation. A family of techniques called complex time-series analysis breaks data of this sort into its component parts (trend, seasons, and cycles) and reassembles those parts into the future generating a forecast. Such forecasts reproduce the long-term trend of the data set and the ups and downs of seasons and cycles to produce impressive-looking graphs. Of course, the accuracy of such forecasts depends on whether past trends, seasonal patterns, and cycles are maintained. Excel and StatPlus can almost automatically perform complex time-series analysis using trend and season but not cycle. The statistics package NCSS contains one of the most sophisticated time-series analysis modules that we have seen among general-purpose statistics packages (see as follows).

Sampling

Spreadsheets contain functions that can be used to generate random numbers convenient for sampling. In Excel, the function RAND() generates a random number greater than or equal to 0 and less than 1. For example, entering =RAND() into three cells yielded 0.917547614, 0.364495084, and 0.209362913. The numbers are evenly distributed. In other words, there is an equal likelihood that any digit will appear in any of the decimal places.

The Excel help utility advises the following: "To generate a random real number between a and b, use: RAND()*(b-a)+a." The generation of integers (whole numbers) instead of decimals can be accomplished by combining the above technique with the INT function, which rounds down to the nearest integer. For example, random numbers between 1 and 500 are generated by the formula INT(RAND()*500)+1.

The RANDBETWEEN function is an even easier way to generate evenly distributed random integers. The syntax is =RANDBETWEEN(low,high), where low is the smallest integer that will be returned and high is the largest. Worksheets can be designed to generate simple random samples and systematic samples with or without stratification. With the spreadsheet rows representing cases or subjects, each row may be numbered quickly, and those numbers may be used with a formula to generate a simple random sample. The stratified systematic sample is first developed by sorting the data on one or more variables (for example, sex). Using a formula to generate a skip interval (population size divided by sample size), it is easy to generate another formula for sampling the list once simple random sampling within the first interval generates a random start number. If the data were sorted by sex, the resulting sample will have numbers of males and females proportionate to their numbers on the list.

Spreadsheets contain a calendar with dates represented by integers that can be used to generate random dates between desired limits. The number 1 represents January 1, 1900, something that the reader can confirm by keying 1 into any cell and formatting it as a date (Format, Cell, Number, Date). The generation of random dates is accomplished by determining what integers represent the desired dates and setting up the correct INT(RAND()) or RANDBETWEEN formula.

Linear Programming

Linear programming is a technique for minimizing or maximizing a variable in the face of constraints. We might, for example, be attempting to minimize cost in a government cartography shop by making the most efficient use of equipment and personnel or to maximize the use of city garbage trucks by minimizing travel time between where they start in the morning, pickup locations, and dump locations (both examples are actual applications). Spreadsheets can perform such calculations. Explaining how data for a particular situation are arranged for linear programming solution in a spreadsheet requires far more space than is available here, but many texts on advanced spreadsheet usage provide guidance (Albright, Winston, & Zappe, 2006; Gips, 2002; Weida, Richardson, & Vazsonyi, 2001).

Database Applications

Everything regarding spreadsheet usage up to this point has concerned databases, but there are specific data-storage and -manipulation operations known collectively as database management. The traditional database structure is identical to the column (variable) and row (observation) arrangement discussed above except that in database management, columns are known as fields and rows as records. Inventories, mailing lists, and room assignments in buildings are examples of the kind of information stored and processed as part of database-management operations.

Spreadsheets can be used for this purpose if the number of records does not exceed the maximum number of rows in the spreadsheet. This amount varies depending on spreadsheet manufacturer and version, and it also may be limited by the amount of memory (RAM) in the computer. The practical maximum number of records is also governed by how the user plans to manipulate data; many operations can take additional space. With the spreadsheet software found on most computers of average memory capacity, a database larger than 5,000 records is probably better processed using a database-management program such as Microsoft Access or the easier to use and more reasonably priced Alpha-Five. Database-management programs are better suited to manipulating large data sets than spreadsheets. Database-management programs are also better suited for creating customized data-entry screens.

Statistics Packages

As noted previously, spreadsheets can perform statistical calculations and database-management programs can as well, but if those operations are many in number, involve relatively sophisticated techniques, and/or involve large data sets, statistics packages such as SYSTAT, SPSS, or NCSS are far superior to spreadsheets and database-management programs. For some calculations, statistics packages are the only tool that can be used by the nonprogrammer.

In public-administration applications, a statistics package might be used for elementary operations such as descriptive statistics. Even for this purpose, a statistics package is more convenient than a spreadsheet if work is to be done on more than a few dozen variables. Some policy analysis work requires more advanced techniques.

Regression Analysis

Regression analysis can be employed to explain the relationships between and among variables, and it can be used alone or in concert with other techniques for forecasting. Regression analysis is one of the most commonly described techniques in the statistics literature, and there is no point in our duplicating the many fine treatments of this subject (Berk, 1998; Berk & Carey, 2004; Garson, 2002; Schroeder, Sjoquist, & Stephan, 1986; Wilkinson, Blank, & Gruber, 1996). An example of one application of regression analysis concerns the relationship between the change in per-capita income and the change in university enrollment for the states. This example is part of a university lobbyist's claim that a state's economic well being was directly proportional to the number of people enrolled in universities. Using the language of statistics, a state's economic health was the dependent variable being acted on by the independent variable, higher education enrollment. Regression analysis allows the user to determine how much, if any, variation in a state's economic health is explained by variation in higher education enrollment.

Regression analysis is used in the fields of human-resource administration and justice and public safety to analyze whether discrimination based on race, sex, or other illegal factors is statistically likely. In this instance, regression analysis with several independent variables (multiple regression) is often the tool of choice (Berk & Carey, 2004). Regression analysis also can be used for forecasting time-series data, with time serving as the independent variable (horizontal axis). The resulting regression line or the regression equation can be used to forecast the dependent variable, which can be anything from tax revenues to jail inmate populations.

Interrupted Time-Series Analysis

Interrupted time-series analysis, an application of regression analysis, is an especially helpful analytical tool for policy analysis (Welch & Comer, 1988). For example, the authors (Permaloff & Grafton, 1995) used it to test the claim that George Wallace's first term as governor had a positive effect on education spending in Alabama, and Bloom (1999) estimated educational program impacts on student achievement.

With interrupted time-series analysis, data must be available for a number of time periods before and after an event such as the election of a new governor or the enactment of a coal-mine safety law. This extremely useful technique measures whether the event has made a statistically significant difference and the nature of that difference. There are two basic kinds of statistically significant differences that an event can produce: an abrupt increase or reduction and a long-term increase or reduction. In addition, an event may produce no abrupt change and/or no long-term change. These alternatives yield nine possible combinations: no immediate change but a long-term increase, no immediate change but a long-term reduction, and so forth.

As an example, we can apply this technique to the data in Figure 3. Figure 3 shows a sharp spike in fatalities per coal miner that occurred in 1968 followed by a dramatic reduction after passage of the Coal Mine Safety Act of 1969 (CMSA). The question is whether the CMSA improved coal-mine safety. The analysis begins by adding two variables to the data set. One variable (CV) is a counter variable that begins at zero in 1935 and remains zero for each successive year until 1970 when it becomes one and remains one for the remaining years of the data set. A multiple regression analysis including CV will result in an indication of statistical significance connected with CV if there is an immediate or abrupt shift in fatalities per miner. The second variable is a post-policy intervention counter variable (PPICV). Like CV, it begins at zero in 1935 and remains zero for each successive year until 1970 when it becomes one and, unlike CV, it then increases by one in each successive year. There will be an indication of statistical significance for PPICV if continuing change occurs after CMSA.

Figure 4 shows SYSTAT output for the multiple regression analysis interrupted time series. The P(2 Tail) column shows that the CV variable is statistically significant beyond the 0.001 level, indicating that the drop after 1968 is statistically significant. On the other hand, the PPICV variable is not statistically significant showing that the long-term reduction in fatalities after 1968 is not statistically significant. However, this is only a first cut at the analysis. Many questions remain. One concerns the 1968 spike. It alone could be the origin of statistical significance for CV. We eliminated the impact of the 1968 spike by replacing it with the mean of fatalities per miner for 1967 and 1969. The resulting analysis

Figure 4. Interrupted time series data output from SYSTAT

```
Dep Var: FATPERMINER   N: 71   Multiple R: 0.929   Squared multiple R: 0.863

Adjusted squared multiple R: 0.857   Standard error of estimate: 0.000

Effect         Coefficient    Std Error     Std Coef Tolerance      t   P(2 Tail)

CONSTANT             0.053        0.010         0.0         .      5.203    0.000
YEAR                -0.000        0.000        -0.654     0.120   -5.013    0.000
CV                  -0.001        0.000        -0.343     0.250   -3.797    0.000
PPICV                0.000        0.000         0.042     0.187    0.405    0.686

                    Analysis of Variance

Source             Sum-of-Squares   df  Mean-Square    F-ratio        P

Regression                0.000      3        0.000    141.127      0.000
Residual                  0.000     67        0.000
-------------------------------------------------------------------------------

*** WARNING ***
Case           36 is an outlier        (Studentized Residual =        3.280)

Durbin-Watson D Statistic     0.814
First Order Autocorrelation   0.590
```

still showed CV to be statistically significant but this time at the 0.001 level. Nevertheless, the hypothesis that CMSA resulted in a one-time drop in fatalities is supported. (PPICV remains not statistically significant.)

The lack of statistical significance for PPICV may be in part due to the 1935 starting point that we selected. Soon after 1935, a sharp drop in fatalities began that continued until roughly 1950. The period of 1950 to 1970 was marked by a gradual increase in fatalities. After 1970, fatalities resumed their downward path. What would happen if we began the analysis with 1950 instead of 1935? The result is that both CV and PPICV were statistically significant beyond the 0.001 level suggesting that CMSA had both a short term and long term effect.

Questions regarding how to treat data spikes (or drops) and at what year the data series should be said to begin are very common in interrupted time-series analysis. Answers are matters of judgment. The 1968 spike was real and certainly had to be included in a regression run, but it was an unusual figure that invited another run with the spike deleted. Similarly, the original data set from which we took these data began in 1900. We assume that it would not be appropriate to run the analysis that far back; presumably, coal-mine technology improved considerably since 1900, making data of that era meaningless. But at what time point should the analysis begin? If the authors possessed greater knowledge of mining technology, we could have selected a starting point that marked what might be termed contemporary technology. The only certainty is that this data set required multiple analyses and required that they be reported.

It should be noted that we omitted another factor that should be covered in a serious analysis of coal-mine safety: Surface mining is inherently safer than deep mining, and surface mining has increased over the years. A serious analysis would treat them separately. Again, this is the kind of judgment that must be made in analysis of this sort.

Curve Smoothing and Forecasting

SYSTAT, NCSS, SPSS, and other statistics packages contain a wide variety of curve-smoothing and forecasting techniques, some of which have already been discussed in connection with spreadsheets. Generally speaking, this kind of analysis is more conveniently done in a statistics package than a spreadsheet, and statistics packages offer a greater variety of analysis techniques.

Complex Time-Series Analysis

The topic of complex time-series analysis was introduced earlier in the section on spreadsheets. The NCSS statistics package contains one of the easiest to use and most sophisticated complex time-series analysis modules on the market as well as possibly the clearest documentation available (Armstrong, Collopy, & Yokum, 2005; Jarrett, 1987; NCSS, 2004).

Database Management

Database-management software such as Microsoft Access or Alpha-Five from Alpha Software Corporation is commonly used for personnel records, inventory management, or whenever large numbers of records will be stored. Database-management software is also ideal when data are to be entered by a computer novice. Data-input screens can be designed that include instructions for the user to help ensure that the correct types of information are keyed into particular fields. To some degree, each field can also be designed so that incorrect types of information cannot be entered. A spreadsheet can be modified in a similar fashion, but the expertise required to accomplish this is somewhat greater than is needed with a database-management program, especially one intended for ease of use such as Alpha-Five.

With database-management software, large numbers of records can be sorted and queried and particular records and particular fields can be printed or exported to other software (e.g., statistics, spreadsheet) for further processing. Database-management programs also are ideal for the generation of large numbers of mailing labels and form letters (Schmalz, 2006).

Access, Alpha-Five, and many other database-management programs are relational; they can link and draw information from two or more differently structured databases (as long as the databases have a common field) as if those databases were one (Grafton & Permaloff, 1993). This powerful feature is not shared with the other categories of software represented here.

On a recent project for a state government agency, we received two data files in Excel for each of 19 counties. The files had no common column or field. We used Excel's LEFT and RIGHT functions to extract portions of one column in one data file to create a variable or field common to one found in the second data file. We then created one merged data file for each county using Alpha-Five. The Alpha-Five file is in dBASE format and readable by Excel and the statistics packages we use most often. We employed Excel pivot tables and sorting commands to generate descriptive data required for the project by the federal government. The databases were exported into SYSTAT for multiple regression and other analysis, including cross-tabulation with control variables and associated statistics. The extraction option in SYSTAT allowed us to create specialized subsets of the data files quickly and easily. Portions of the SYSTAT analysis output were saved in text format and read into WordPerfect for final report writing.

Large organizations sometimes buy specialized relational database-management programs set up for their specific applications. Such programs may cost thousands, if not hundreds of thousands, of dollars to buy, install, and maintain. Often a training component is included in the installation fee. Database-management programs are frequently used to store personnel records. Another function is the processing of government regulatory agency records. For example, a state regulatory board may have multiple responsibilities: licensing applicants to a profession, disciplinary functions including the receipt of complaints and their investigations, monitoring of continuing-education providers, and oversight of educational training programs. Each area would be served by different staff personnel with differing record-keeping needs. As long as each area has at least one common field such as Social Security number as an identifier linking the databases, one set of records could be queried for information needed by another area of the organization. For example, the licensing section may need to know the current status of disciplinary actions to determine if a renewal should be granted. With relationally linked databases, licensing personnel can query the separate components of the database as if they were a single database. At the same time, access to specific fields can be granted or blocked on a need-to-know basis.

We have observed three pitfalls with database management in government. If standard database software is adapted by a government agency employee for use by the agency (all such software is designed to be tailored to fit specific needs), it is important that at least one other employee be taught how the customization was implemented and backups be made of the complete system, and, of course, all data. If the employee who performs the customization leaves the agency and the computer on which it is running crashes, an agency can be left without access to its records or worse yet the records may disappear. In fact, this has happened. Events of this sort are less likely to occur if the program is sold and maintained by a commercial developer assuming that the developer remains in business.

Another database-management pitfall is the mutation of the database over a period of years with no records kept of changes made. The original database structure may be clearly described by paper or online documentation, but incremental modifications over a period of years with no accompanying editing of the documentation can produce a database difficult for newcomers to use.

A third pitfall is a database that does not reject grossly incorrect data entries. No software can detect slight numerical mistakes in a field intended to hold numbers such as a 1 instead of a 2, for example, but any modern database-management system ought to reject such mistakes as two-digit numbers entered where there should be three or letters entered where numbers are needed. Also, a database-management program should force input of data into a field containing mandatory information before allowing the user to complete follow-on fields. Database-management systems that permit gross mistakes will produce a database full of inaccuracies, especially when hundreds of people are keying in data.

Data Mining

Data mining is a collection of techniques primarily intended to handle extremely large secondary data sets. A secondary data set is by definition not collected with analysis as the primary objective, so the way this transactional information is gathered and structured may not be ideal for traditional statistical analysis. The Transportation Security Administration (TSA) provides an example of a large secondary data set. This agency uses credit reports and more than 100 commercial and government databases to identify potential terrorists among airline passengers (Rotenberg & Hoofnagle, 2003). Other examples of governmental use of data mining come from law enforcement, higher education, and tax administration (SPSS, n.d.).

Large data sets require enormous computer capacity and they sometimes are in the form of multiple related files (the TSA example above fits this characteristic). These factors mean that traditional statistical techniques cannot be applied without complicated sampling (Hand, 1998). Another problem with large secondary data sets is that they almost certainly contain contaminated data requiring mechanized (because they are so large) data cleaning.

The secondary nature of the data sets also means that they may be characterized by selection bias (some geographical regions may be overrepresented) and nonstationarity (the population may be changing in subtle ways). These problems can also arise in smaller data sets, but they are more difficult to detect and correct in large data sets (Hand 1998).

Another complication with data mining is that it is often exploratory in nature (Statsoft, 2003). Analysts are often not testing relationships derived from theory as they do (or should) in academic research. Instead, they are frequently searching for unknown clues that lead to answers to vaguely defined questions (Hand, 1998). The sheer size of data-mining data sets combined with the frequent lack of theory also makes analysts vulnerable to discovering spurious relationships.

Data mining requires data-mining software, large computers, and specialists. Our sense is that an army of particularly hungry dealers and consultants awaits any government agency venturing into this field.

Project Planning and Management

Project planning and management programs are designed to assist in planning and managing complex projects in terms of time schedules as well as budgets, personnel, and other resources. This category of software can be used to schedule and estimate personnel requirements and budgets for construction jobs, landscape projects, conference planning, scheduling activities in a complex lawsuit, or any other enterprise with a clearly defined beginning and end. Over the years, the widely used Primavera Project Planner (P3) has received some of the most positive reviews, but Microsoft Project is easier to use. A Google search on "project planning software" will yield many other products, some of which offer free trial downloads.

Project planning requires that a project be broken down into building blocks called activities or tasks. An activity or task is the responsibility of one person, involves one kind of work, can be scheduled, and uses the same resources from beginning to end.

Gantt charts are one of two major graphical devices used by this family of programs. A Gantt chart is a bar graph with a horizontal bar representing each task in a project. Each bar's length is proportional to the time the task requires. The beginning of a bar is located at the start date of a task, and the end of the last bar is located at the date the last task should be complete.

The PERT (Project Evaluation Review Technique) chart is the other major graphical tool used by project-planning and -management programs. It is a better tool for highlighting relationships among tasks, but the Gantt chart is more space efficient, allowing more tasks to be displayed on screen or on a single sheet of paper. Programs of this sort can also display bar charts and tables showing personnel plans and budgets.

Information is entered in these programs in forms. The user keys in task names, duration, personnel, and other resources used to complete the task. The software calculates start and finish dates, personnel plans, budgets, and so forth.

Frequently, the first draft of a project plan results in an end date that is too late. These programs automatically highlight the longest path (the critical path) in a plan where a number of tasks are being executed simultaneously. In concentrating changes on the critical path, an administrator can reduce the completion time of the overall project by focusing attention where it will do the most good. Project planning and management programs are also used to monitor progress once a project is under way. Project-wide implications of early or late task completions are quickly observed.

Decision Analysis

Decision analysis is a graphical form of applied probability theory useful for analyzing situations in which uncertainty plays a part. The most common situation to which this technique is applicable has a decision maker trying to choose between two or more policies, each of which is associated with a variety of benefits and costs and probabilities that those benefits and costs will occur. In decision-analysis terminology, a benefit is a positive payoff and a cost is a negative payoff.

As an example, imagine that a basketball team is behind by two points, and it has time for only one shot. Assume that there is a 50% chance of hitting a two-point shot that would tie the game and a 33% chance of hitting a three-point shot that would produce a victory. Also, assume that the team has a 50% chance of winning in overtime. Which shot should the coach call (Clements, 2006)?

Figure 5 shows a decision tree that represents all possible futures in this situation. This decision tree was entered and calculated in an Excel add-in called PrecisionTree, manufactured by Palisade Corporation. It is available either from the manufacturer or in a trial version in an excellent and reasonably priced (for what it contains) textbook called *Data Analysis and Decision Making with Microsoft Excel* (Albright et al., 2006). This volume also comes with a version of the powerful and easy-to-use simulation tool @RISK.

The decision tree begins on the left with the rectangle (a choice node) representing the coach's choice. The choice to take the two-point shot leads to a chance node (a circle) offering a 50% chance of hitting the shot and a 50% chance of missing. We use 0 as the value of a game loss and 1 as the value of a victory. A miss at the two-point shot produces a payoff

Figure 5. A decision tree

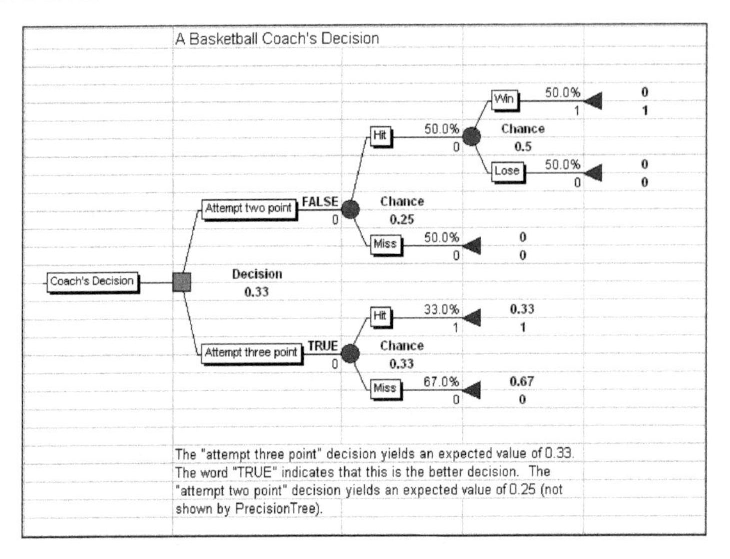

of zero. If the two-point shot is made, there follows another chance node offering a 50:50 chance of a win and a loss. If the coach's basic two-point choice could be made repeatedly as an experiment or simulation, the game would be won only 25% of the time. Instead of running this choice as an experiment or simulation, PrecisionTree can calculate the probabilities and payoffs. In this case, the coach should go for the three-point shot, which offers a 33% chance of victory.

Data such as the names of the branches, probabilities, and pay-offs are entered in the PrecisionTree program one form at a time. PrecisionTree draws the tree and calculates the result. The program also allows the user to conduct sensitivity analyses (with some difficulty at first) to determine the effect on the result of differing pay-off values (which would not arise in this case) and probabilities. Although this example is extremely simple, decision trees can be quite complex.

Graphics for Illustrations

Illustrations often play an important part in reports, presentations, budgets, and other documents. Specialized software can be of great assistance, augmenting the graphics features built into the software discussed previously (Permaloff & Grafton, 1996).

Flow Charts and Organization Charts

A family of programs such as RFFlow and Visio is used to create flow charts, organization charts, and similar diagrams featuring boxes (of all shapes) and lines (solid, dotted, with or without arrowheads). Diagrams are created by the user selecting box shapes from menus, keying in labels (the boxes expand to fit labels), and connecting the boxes with lines also selected from menus. Once placed on the screen, boxes and lines can be rearranged as aesthetics and logic dictate.

The choice of programs depends on features needed, budget, and time available for learning. For example, comparing RFFlow and Visio, the former is the least expensive ($49 vs. $199) and by far easier to learn, but it is limited to creating flow charts and organization charts. Visio can render a greater variety of shapes, and it can be tied to powerful Visual Basic programs. Visio, like all Microsoft products, subjects the user to an intrusive online product-registration process lacking in RFFlow. As with most other products discussed in this chapter, a Web search yields many other flow-chart and organization-chart generation programs, many of which can be downloaded.

We do not recommend that most public administrators attempt to employ a computer-aided design (CAD) program for occasional diagram work unless the individual is already trained in its use. CAD programs can do far more than specialized flow-chart programs, but they exact an even higher price than Visio in terms of learning time, and they require many steps and much more time to complete tasks (see Permaloff & Grafton, 1996.)

Scanned Graphics

Documents can sometimes be improved with the inclusion of scanned material such as photographs. Scanners range in price from $50 to approximately $650, but even the least expensive are sufficient for many applications. Flatbed scanners are probably the easiest to use. The flatbed scanner resembles a small photocopy machine. The original image is placed in the scanner that generates a digitized version of the image that can be saved on disk and inserted into application programs including word processing, spreadsheet, database management, and project planning. In some cases, the image can be dragged directly from the scanning program into an application program without the need to save on disk.

Presentation Programs

The use of computers and computer projection hardware or large monitors for presentations is gaining in popularity as the cost of projection equipment declines. Although 35mm slides, the old mainstay of presentations, are virtually gone, computer presentations are built on what might be termed virtual slides. Transparencies, another kind of slide, are still used, but they appear to be regarded as passé despite their continued practicality, readability on a large screen, and low cost.

A computer slide may contain any combination of text and graphics, including sound and motion. A single slide may contain elements from several sources. Menu choices or built-in templates determine the background on which text and/or graphics will appear. That background can be any color or texture. Similarly, each slide's basic layout (whether there is a centered heading, bulleted text, a graphic, columns, etc.) is chosen from menus or templates.

It is easy to include so many clever graphics and sound effects in a computer presentation that the audience is distracted from content or even irritated by excesses. Indeed, a former chairman of the Joint Chiefs of Staff became so annoyed with elaborate PowerPoint presentations he reportedly attempted to ban the program's use in the military.

Another drawback of slide presentations is audience boredom, a major reason some presenters employ what they hope are entertaining effects. One clever and subtle solution to the boredom problem is the use of two projectors and two screens. Image 1 appears on Screen A and then Image 2 on Screen B (with Image 1 still seen on Screen A). Image 3 then replaces Image 1 on Screen A, and so forth. This idea's effectiveness must literally be seen to be believed. It reduces boredom and eliminates the need for gimmicky dissolves from slide to slide.

Geographic Information Systems

Geographic (or geographical) information system (GIS) programs assemble, store, and display data that are specific to geographical areas such as countries, states, counties, or zip codes. Data might include information regarding population, economics, numbers of users of government agency services, and crime rates. Data are typically displayed in the form of maps in which geographical areas are colored or patterned to represent various levels of whatever is being measured (Garson & Vann, 2001; Orford, Harris, & Dorling, 1999; Vann & Garson, 2001; Vann & Garson, 2003).

Some GIS programs such as MapInfo and BusinessMAP are free standing while others are attached to software such as Excel or Visio. GIS packages vary widely in terms of file formats that they can read, data-manipulation features, display capabilities, and price, so considerable research is required to make an informed purchasing decision. A list of resources including GIS programs, boundary files, and data files is available at the Web site of the Department of Geography, University of Edinburgh (http://www.geo.ed.ac.uk/home/agidict/welcome.html).

Long-Term Storage

Editable data files (e.g., word processing or worksheets) and scanned images of printed pages can be stored digitally for archival purposes. But on what medium should archived documents be stored? One convenient but unsafe possibility is CD-R (compact disk-recordable) or CD-RW (compact disk-rewritable) disks. Unfortunately, these media may not be suitable for long-term storage. CD-R and CD-RW disks are not the same as CD-ROMs (compact disk–read only memory), which are commonly used for commercial software and music. CD-ROMs are manufactured by molding polycarbonate with pits that constitute digital information. The pits are then coated with a reflecting layer of metal. That layer is then coated with protective plastic. Even these relatively tough disks can be rendered unreadable by scratches, but with careful handling, they can probably last for decades.

Recordable or rewritable disks probably have a much shorter life span because they are physically more delicate. CD-Rs use transparent dye, not pits, as their data-storage method. When the disks are "burned," thc computer's laser darkens the dye. CD-RWs use an alloy that changes state when the recording process occurs (Ancestry.com, 2000). CD manufacturers claim that recordable and rewritable disks can last for decades, but Kurt Gerecke, an IBM storage specialist, estimates that inexpensive and high-quality disks only last approximately 2 to 5 years, respectively (Blau, 2006). Gerecke also cautions against relying on a hard drive for archival purposes. Here the problem is not surface degradation, but the longevity of the disk bearings. Gerecke recommends magnetic tapes for long-term storage. Tape life spans range from 30 to 100 years depending on their quality (Blau; Practical PC, n.d.).

Storage media formats represent another and different kind of problem for archiving. Consider how difficult it would be to locate a disk drive that can read a 5¼-inch floppy disk much less the larger format that preceded it. The safest materials for long-term storage may still be paper and microfilm or microfiche.

Conclusion

Most tasks in government employing a computer can be accomplished more efficiently with a variety of tools. Happily, most of those tools are under $500. Of the software discussed here, only customized database-management programs, data-mining software (which often require highly trained staff or hired consultants), and high-end geographic information systems go past the $1,000 mark.

References

Akers, E. J. (2006). *A study of the adoption of digital government technology as public policy innovation in the American states.* Unpublished doctoral dissertation, Auburn University, Auburn, AL.

Albright, S. C., Winston, W. L., & Zappe, C. J. (2006). *Data analysis and decision making with Microsoft Excel* (3rd ed.). Stamford, CT: Duxbury, Thomson Learning.

Alpha-Five [Computer software]. (n.d.). Burlington, MA: Alpha Software Corporation. Retrieved from http://www.alphasoftware.com

Ammons, D. N. (2002). *Tools for decision making.* Washington, DC: CQ Press.

Ancestry.com. (2000). *The life span of compact disks.* Retrieved March 26, 2006, from http://www.ancestry.com/learn/library/article.aspx?article=2131

Armstrong, J.S., Collopy, F., & Yokum, J.T. (2005). Decomposition by causal forces: A produce for forecasting complex time series. *International Journal of Forecasting, 12* (pp. 25-36).

Berk, K. N. (1998). *Introductory statistics with SYSTAT.* Upper Saddle River, NJ: Prentice-Hall.

Berk, K. N., & Carey, P. (2004). *Data analysis with Microsoft Excel: Updated for Windows XP.* Belmont, CA: Brooks/Cole–Thomson Learning.

Blau, J. (2006, January 10). Storage expert warns of short life span for burned CDs: And don't count on hard disk drives for long-term storage, either. *Computerworld.* Retrieved March 20, 2006, from http://computerworld.com/hardwaretopics/ storage/story/ 0,10801,107607,00.html

Blau, P. M., & Meyer, M. (1971). *Bureaucracy in modern society.* New York: Random House.

Boardman, A. E., Greenberg, D. H., Vining, A. R., & Weimer, D. L. (2001). *Cost-benefit analysis: Concepts and practice.* Upper Saddle River, NJ: Prentice-Hall.

Borgmann, A. (1988). Technology and democracy. In M. E. Kraft & N. J. Vig (Eds.), *Technology and politics* (pp. 54-74). Durham, NC: Duke University Press.

BusinessMAP 4 [Computer software]. (n.d.). Dallas, TX: CMC International. Retrieved from http://www.businessmap4.com

Clementine [Computer software]. (n.d.). *Chicago: SPSS.* Retrieved March 26, 2006, from http://www.spss.com/downloads/Papers.cfm?ProductID=Clementine&DLT...

Clements, J. (2006, March 15). Net gains: How watching basketball can improve your approach to investing. *Wall Street Journal*, p. D1.

Clinton, W. J. (1994). *Executive Order 12898: Federal actions to address environmental justice in minority populations and low-income populations.* Retrieved March 19, 2006, from http://www.fhwa.dot.gov/legsregs/directives/orders/6640_23.htm

Coursey, D. H., & Killingsworth, J. (2005). Managing e-government in Florida: Further lessons from transition and maturity. In G. D. Garson (Ed.), *Handbook of public information systems* (2nd ed., pp. 331-343). Boca Raton, FL: CRC Press, Taylor & Francis Group.

DJM Consulting. (2002). *Benefit-cost analysis of Seattle monorail proposal: Elevated Transportation Company Board PowerPoint briefing.* Retrieved March 20, 2006, from http://archives.elevated.org/documents/Cost_Benefit_Analysis_DJM.pdf

Ellul, J. (1964). *The technological society.* New York: Vintage.

Federal Highway Administration. (1998). *FHWA actions to address environmental justice in minority populations and low-income populations* (Directive 6640.23). Retrieved March 19, 2006, from http://www.fhwa.dot.gov/legsregs/directives/orders/ 6640_23.htm

Fosback, N. (1987). *Stock market logic.* Fort Lauderdale, FL: Institute for Econometric Research.

Freeman, D. M. (1974). *Technology and society.* Chicago: Markham.

Fuguitt, D., & Wilcox, S. J. (1999). *Cost-benefit analysis for public sector decision makers.* Westport, CT: Quorum Books, Greenwood Publishing Group, Inc.

Garson, G. D., & Vann, I. B. (2001). Resources for computerized crime mapping. *Social Science Computer Review, 19*, 357-361.

Gips, J. (2002). *Mastering Excel: A problem-solving approach* (2nd ed.). New York: John Wiley.

Grafton, C., & Permaloff, A. (1993). Statistical analysis and data graphics. In G. D. Garson & S. S. Nagel (Eds.), *Advances in social science and computers* (Vol. 3, pp. 267-284). Greenwich, CT: JAI Press.

Grafton, C., & Permaloff, A. (2005). Analysis and communication in public budgeting. In G. D. Garson (Ed.), *Handbook of public information systems* (2nd ed., pp. 463-488). Boca Raton, FL: CRC Press, Taylor & Francis Group.

Gullick, L. (1996). Notes on the theory of organization. In J. M. Shafritz & J. S. Ott (Eds.), *Classics of organization theory* (pp. 86-95). Fort Worth, TX: Harcourt Brace College Publishers.

Hand, D. J. (1998). Data mining: Statistics and more? *American Statistician, 52*, 112-118. Retrieved March 26, 2006, from http://www.amstat.org/publications/ tax/hand.pdf

Heeks, R. (2006). *Implementing and managing eGovernment.* London: Sage.

Jarrett, J. (1987). *Business forecasting methods.* Oxford, England: Basil Blackwell.

Levin, H. (1983). *Cost-effectiveness: A primer.* London: Sage.

Marsden Jacob Associates. (2005). *Frameworks for economic impact analysis and benefit-cost analysis.* Retrieved March 20, 2006.

MapInfo [Computer software]. (n.d.). Troy, NY: MapInfo Corporation. Retrieved from http://www.mapinfo.com

Mine Safety and Health Administration. (n.d.). *Coal fatalities from 1900 through 2005.* Retrieved March 23, 2006, from http://www.msha.gov/stats/centurystats/ coalstats.htm

Nas, T. F. (1996). *Cost-benefit analysis.* London: Sage.

NCSS [Computer software]. (n.d.). Kaysville, UT: Number Cruncher Statistical Systems. Retrieved from http://www.ncss.com

Orford, S., Harris, R., & Dorling, D. (1999). Geography information visualization in the social sciences: A state of the art review. *Social Science Computer Review, 17*, 289-304.

Panayotou, T. (n.d.). *Basic concepts and common valuation errors in cost-benefit analysis.* Cambridge, MA: Harvard University, Institute for International Development. Retrieved August 14, 2001, from http://www.eepsea.org/publications/specialp2/ACF2DB.html

Permaloff, A., & Grafton, C. (1995). *Political power in Alabama: The more things change.* Athens, GA: University of Georgia Press.

Permaloff, A., & Grafton, C. (1996). Computer tools for crafting clear and attractive diagrams in social science. *Social Science Computer Review, 14*, 293-304.

People for Modern Transit. (n.d.). *Breakeven on operations: A myth.* Retrieved April 24, 2006, from http://www.peopleformoderntransit.org/pmtadmin/home.nsf/e93bad-dc3ba8bf0a88256d42000f7a73/58502

Practical PC. (n.d.). *CD-R media lifespan.* Retrieved March 26, 2006, from http://www.practicalpc.co.uk/computing/storage/cdrlifespan.htm

Primavera Project Planner (P3) [Computer software]. (n.d.). Bala Cynwyd, PA: Primavera Systems, Inc. Retrieved from http://www.primavera.com

RFFlow [Computer software]. (n.d.). Loveland, CO: RFF Electronics. Retrieved from http://www.rff.com

Rosenbloom, D. H., & Kravchuk, R. S. (2002). *Public administration.* Boston: McGraw Hill.

Rotenberg, M., & Hoofnagle, J. (2003). *Memo to representatives Adam Putnam and William Clay, House Government Reform Subcommittee on Technology, Information Policy, Intergovernmental Relations, and the Census, March 25, 2003.* Retrieved March 10, 2006, from http://www.epic.org/ privacy/profiling/datamining3.25.03.html

Schmalz, M. (2005). *Integrating Excel and Access.* Cambridge, UK: O'Reilly Media.

Schmid, A. A. (1989). *Benefit-cost analysis.* Boulder, CO: Westview.

Schroeder, L. D., Sjoquist, D. L., & Stephan, P. E. (1986). *Understanding regression analysis: An introductory guide.* Beverly Hills, CA: Sage.

Slack, B. (n.d.). *Cost/benefit analysis in practice.* Retrieved March 20, 2006, from http://people.hofstra.edu/geotrans/eng/ch9en/appl9en/ch9a3en.html

Slaton, C. D., & Arthur, J. L. (2004). Public administration for a democratic society: Instilling public trust through greater collaboration with citizens. In M. Malkia, A. Anttiroiko, & R. Savolainen (Eds.), *eTransformation in governance* (pp. 110-130). Hershey, PA: Idea Group.

SPSS. (n.d.). Chicago: Author. Retrieved from http://www.spss.com

Statsoft. (2003). *Data mining techniques.* Retrieved March 27, 2006, from http://www.statsoft.com/textbook/stdatmin.html

Stokey, E., & Zeckhauser, R. (1978). *A primer for policy analysis.* New York: W. W. Norton & Company.

SYSTAT [Computer software]. (n.d.). Richmond, CA: Systat Software, Inc. Retrieved from http://www.systat.com

Tindall, C. (2005). *The monorail episode redux.* Retrieved March 20, 2006, from http://uchicagolaw.typepad.com/faculty/2005/11/the_monorail_ep.html

Vann, I. B., & Garson, G. D. (2001). Crime mapping and its extension to social science analysis. *Social Science Computer Review, 19,* 417-479.

Vann, I. B., & Garson, G. D. (2003). *Crime mapping: New tools for law enforcement.* New York: Peter Lang.

Weida, N. C., Richardson, R., & Vazsonyi, A. (2001). *Operations analysis using Microsoft Excel.* Pacifica Grove, CA: Duxbury.

Welch, S., & Comer, J. (1988). *Quantitative methods for public administration: Techniques and applications.* Pacific Grove, CA: Brooks/Cole.

Wilkinson, L., Blank, G., & Gruber, C. (1996). *Desktop data analysis with SYSTAT.* Upper Saddle River, NJ: Prentice-Hall.

Chapter XIV

Computers and Social Survey Research for Public Administration

Michael L. Vasu, North Carolina State University, USA

Ellen Storey Vasu, North Carolina State University, USA

Ali O. Ozturk, North Carolina State University, USA

Abstract

The integration of social survey methods into public-administration research and practice is the focus of this chapter. Coverage applies to other social science disciplines as well. This chapter reviews the use of computers in computer-assisted survey research (CASR), computer-assisted interviewing, computer-assisted telephone interviewing (CATI), computer-assisted personal interviewing (CAPI), and survey research methods. The chapter takes the perspective of total survey error.

Introduction

Survey research has been a pivotal methodology for academic social science research since World War II. Today, survey research is integral to research and practice in public administration (Folz, 1996; Fowler, 2001). Simply stated, surveys are a form of interviewing. Surveys are individual interviews typically targeted at a single respondent or unit of analysis. The enterprise essentially involves the art of asking questions (Payne, 1951). The questions that are asked constitute variables in the language of research. The purpose of asking these questions is to establish relationships between and among independent and dependent variables, and, typically, to test a series of hypotheses derived from some body of theory. The question and answer process, integral to surveys, is also a form of measurement and is subject to errors of measurement. This article takes the total error perspective. Some of the errors in surveys are more or less minimized by information technology, while others are inherent in the nature of the survey process.

According to the American Association for Public Opinion Research (AAPOR), a scientific sample survey is different from a nonscientific one, or a survey not based on probability theory. They define a scientific survey to help the media, and, through them, the public to distinguish between surveys (AAPOR, 2001; Weisberg, 2005). AAPOR lists four principal characteristics of a scientific sample survey or poll as follows:

- **Coverage:** A scientific survey samples members of the defined population in a way such that each member has a known nonzero probability of selection.
- **Sampling:** A scientific survey collects data from a sufficient number of sampled units in the population to allow conclusion to be drawn about the prevalence of the characteristic in the entire study population with desired precision at stated level of confidence (e.g., 90 or 95 %).
- **Nonresponse:** A scientific survey uses reasonable tested methods to reduce and account for unit and item nonresponse error (difference between characteristics of respondents and nonrespondents) by employing appropriate procedures for increasing unit and item response rates and making appropriate statistical adjustments.
- **Measurement:** A scientific survey uses reasonable tested methods to reduce and account for errors of measurement that may rise from question wording, the order of questions and categories, the behavior of interviewers and of respondents, data entry, and the mode of administration of the survey.

All surveys rely on the answer to questions. The types of questions asked in the field of public administration are extensive. They may concern community aesthetics, growth management issues, budget priorities, dimensions of program effectiveness, and feedback from citizens, constituents, or customers. In public administration, the researcher typically asks questions to capture data and use the data in some policy analytic way. The traditional approach to capturing the data from the question and answer interviewing process previously described was carried out with a paper and pencil, and hence was called Paper and Pencil Interviewing (PAPI; Dufour, Kaushal, & Michaud, 1997). The advent of computers, of course, promised advantages over the paper and pencil approach that included decreased cost, increased convenience, and quality. It also promised ways to reduce some of the errors of measurement inherent in the survey research process. Many of these promises have been realized. Other sources of error are inherent in the total survey process and are over and above the scope of technology; they will always be part of the survey process (Weisberg, 2005). For example, no amount of technology can provide useful information from questions that are not valid or reliable.

This chapter will focus on ways in which computers can enhance the survey research. This chapter will look at this topic through the lens of public administration research and practice. The chapter will discuss issues broadly as they relate to survey research. For more extensive software reviews (typically discussed down to the keystroke level), the reader is advised to consult, for example, *Social Science Computer Review* (http://www.sagepub.com).

This chapter specifically reviews the use of computers in computer-assisted survey research (CASR), computer-assisted telephone interviewing (CATI), computer-assisted personal interviewing (CAPI), and survey research on the Internet. The chapter also discusses these issues within the context of *total survey error* and how to minimize it. Because issues in the field are constantly evolving, the reader is advised to consult Web sites that provided current and updated survey information (Garson, 2006).

The Survey Research Process

As stated previously, survey research has been central to social science research since World War II. Currently, technological changes are giving new strength to an old workhorse. In order to understand the real and potential impact of computers on the survey process, it is necessary to understand the survey process. Subsequently, we will outline the steps or stages of the survey process and identify potential sources of measurement error inherent in survey research.

All methodological techniques seek to minimize the total error of measurement in an attempt to gain some understanding about the relationship between and among variables. All fail in some important respects. Even in experimental research, relationships between variables need to be replicated many times prior to being accepted. In ex post facto research, where both the independent and dependent variables have already occurred in time (survey research is one type of ex post facto research), the measurement problems are significantly greater (Bradburn & Sudman, 1991; Lyberg, 1989, Weisberg, 2005). Typically, in surveys we seek to reduce total survey error in search of meaningful relationships. This total survey error is all the error that can seep into a study that affects the survey's accuracy, that is, the ability to mirror something in the real world.

Interestingly, Weisberg (2005) indicates that the term error is usually thought of as a synonym for mistake, but in the survey context, it refers to the difference between an obtained value and the true value, usually the true value for the larger population of interest. This total survey error is the result of sampling error, nonresponse problems, interviewer bias, biased questions, nonsampling error, and other general errors of measurement. Moreover, even well-constructed (valid and reliable) survey questions offer very little depth or ability to probe or drill down when compared, for example, to a focus group (Folz, 1996; Lyberg et al., 1997; Rea & Parker, 1997; Sudman & Bradburn, 1982: Weisberg). A list of the sources of error in the total survey error approach follows:

1. **Sampling error:** This error occurs when a sample of the population rather than the entire population is surveyed. There can be systematic bias when nonprobability sampling is employed, while probability sampling has the advantage of permitting mathematical computation of the sampling error. Moreover, what is typically reported as error in scientific surveys is sampling error (e.g., plus or minus 3%), which is actually only one source of error, and not necessarily the largest found in many surveys (Converse & Traugott, 1986). However, scientific survey research does provide the breadth that is the result of being predicated on some form of sampling design that allows the researcher to generalize to population parameters based on sample statistics.
2. **Coverage error:** When the list from which the sample is taken (known also as the sampling frame) does not correspond to the population of interest, coverage error occurs.
3. **Unit nonresponse error:** This type of error occurs when the designated respondent does not participate in the survey, thereby limiting how representative the actual respondents are of the population of interest.
4. **Item nonresponse error:** The error occurs when the respondent participates but skips some questions. For instance, "Don't Know" (DK) responses are an important item nonresponse problem, making it necessary for researchers to decide how to deal with missing item data in the data analysis.
5. **Measurement error:** A measurement error occurs when the measure obtained is not an accurate measure of what was to be measured. Measurement errors can be observed in two categories: measurement error due to the respondent (depends on how accurate research questionnaire is designed and whether the respondent gives an accurate answer to survey questions) and measurement error due to the interviewer (occurs when effects associated with the interviewer lead to inaccurate measures).
6. **Postsurvey error:** It has been discussed whether it is really a part of the survey process or not, but some describe that this type of error occurs in processing and analyzing survey data.
7. **Mode effects:** As a survey administration issue, the selection of survey mode (such as face-to-face interviewing, phone surveys, mailed questionnaires, etc.) could have some effect on the obtained valid, reliable, and useful results.
8. **Comparability effects:** The survey administration issue involves the difference between survey results obtained by different survey organizations, in different nations, or at different points in time.

In a well-conducted survey, a researcher wants to minimize the above error in his or her work, but the degree to which error can be reduced is limited by practical constraints. These constraints are cost, time, and ethics (Weisberg, 2005). Time is also an important consideration in survey research since survey projects are often time sensitive. If the information or data are supposed to be collected within a certain time, surveys should be conducted properly to achieve research goals. For instance, preelection surveys must end before the election actually occurs. In addition to time limitations, some attitudes and facts will change over time, so it is necessary to measure them in a timely fashion (Weisberg). As stated previously, many of these sources of error are not specifically addressed by advances in technology and are more in the domain of appropriate survey management.

The steps or stages in the survey research process provide a framework for understanding the degree to which the computer can be integrated into and enhance the survey process, specifically, how computing can cut costs, improve quality, and minimize some errors of measurement. These steps or stages of the survey process are presented in various forms in survey textbooks (e.g., Babbie, 1990; Czaja & Blair, 2005; Folz, 1996; Rea & Parker, 1997; Rossi, Wright, & Anderson, 1983). In the next section, we provide our own summary of these steps.

Conceptualization, Planning, and Management

The purpose and methods of the survey research design are defined and established at this stage. The variables are conceptualized, operational definitions developed, scale and indexes created, literature reviewed, theoretical constructs established, hypotheses developed, and statistical procedures selected. During this stage, the purpose(s) of the research (exploration, description, explanation) is defined. The type(s) of information required is determined (attitudinal, behavioral, demographic, etc.). Who the surveyed will be and by what method (Web, telephone, mail, in person) is decided. Moreover, a variety of logistical issues are addressed, for example, what personnel are required (field staff, technical staff, senior professional staff) and what financial resources are required.

Arguably the most important stage in the survey research process, it is at this stage that various management decisions need to be made. All surveys require good management. The larger the survey, the more complex the management system is. Failures in survey management affect total survey error. Information technology has a role at this stage in the research, even if it is for no other reason than making pervasive use of computing in modern management. The management complexity involved in dealing with people and data in a large survey project is very well summarized in the following quote from the *Handbook of Survey Research*:

Not only people have to be managed. There are also tasks of managing schedules, deliverables, final products, and of course, budgets. But much of what is interesting about the management of schedules, products, and budgets concerns people management. For example, because surveys are labor intensive, schedules generally can be accelerated only by adding more staff labor. Budgets, as we have already observed, are based mostly on project payrolls. To monitor the budget is largely to attend to how much labor is going into different tasks. Seldom will cost overruns be the result of faulty estimates of non-labor costs; they frequently will result from underestimates of the staff effort required to administer a complex questionnaire or trace respondents or code open-ended answers. Poor management of staff labor will directly affect schedules and budgets more than anywhere else in the survey process. Of course, deliverables and products, whether scholarly monographs policy evaluations, or simply clean data, depend upon assigning the right people to the correct task with appropriate resources and creating the incentives that motivate performance. (Rossi et al., 1983, p. 134)

Sampling

The basis of scientific survey research is probability theory. The basis of probability theory is random selection. Random selection is typically defined as a process where every element (typically a person in a survey) has an equal or known probability of inclusion in the sample. The ultimate goal of sampling is to secure a representative sample, one that includes every important variable in the same proportion that would be present in the total population. A specific goal of a representative sample is to produce a sample in which no group is systematically under or overrepresented. Since researchers seldom have exact information about population parameters, they use the logic of probability theory to make statistical inferences about larger populations predicated on a single sample—the one they have selected for a given research project (Czaja & Blair, 2005; Deming, 1950; Kish, 1995; Lyberg et al., 1997). The sample statistics (for example, the mean age of 1,140 members of a sample selected from the roster of the American Institute of Planners), in effect, becomes an estimate of a population parameter (the age of all members of the American Institute of Planners).

The difference between a sample statistic and a population parameter is known as sampling error. The degree of precision, essentially a statistical statement about how close our estimate (sample mean age) is to the real population parameter, can be determined in advance using probability theory. Usually expressed as plus or minus 3, 4, or 5%, precision is a measure of how much error the researcher is willing to tolerate. In addition to tolerated error, the researcher is also interested in a statement of confidence that a given statistical estimate really lies within the plus or minus error range specified. This is stated as the degree to which we are confident that a sample estimate is within a certain number of standard errors of the actual population parameter. Confidence levels are usually expressed in terms of 95 or 99 samples of 100. Sampling designs (simple random, systematic, stratified, cluster, and multistage) specify methods for selecting cases from a sampling frame in a manner that insures that the random selection required for probability theory is achieved. Finally, some sampling designs require disproportionate sampling and weighting of cases (Kish, 1995).

In planning a survey, determining the desired sample size is a crucial decision involving cost and time. In survey research, sampling error is inevitable since the results for a sample will not correspond exactly to those for the full population (Weisberg, 2005). It is desirable to decrease sampling error, but that can be expensive since interviewing more people to reduce sampling error directly increases costs as well as time for completion of the survey. In addition to time and cost, Weisberg explains two ethical constraints involved in sampling in his book, *The Total Survey Error Approach*, as follows:

One (ethical constrain) is the importance of including in a survey report not only a description of the sampling but also a reminder that nonsampling errors must be considered when the validity of any conclusions based on survey evidence is assessed. The other (ethical constrain) is the need to remember that the sampling method is more important than the sample size. In the era of Internet surveys, it is easy to fall in to the same trap as the Literary Digest surveys of the 1930s: imagining that a large number of responses can substitute for a probability sample in yielding representative results. The results of an Internet survey are not necessarily valid even if millions of people participated in the survey. Regrettable, this is a lesson that seems to have to be learned again with every innovation in data collection modes, whether for mail surveys, call-in polls, or Internet surveys. (p. 257)

Designing the Survey Instrument

The essential objective in designing questions for survey research is to elicit responses from respondents that are valid and reliable. In other words, questions are supposed to measure what you want them to measure, and produce results that are consistent over time. Failure to produce valid and reliable questions constitutes another potential source of error in surveys. In effect, in seeking valid and reliable questions, we seek to avoid the inherent bias evident in the answer to the question, "Have you stopped beating your wife?" This question illustrates one common feature of a biased question in that it "channels" the answer to a question in a specific direction; in this case, the respondent incriminates himself with either a yes or a no answer. In addition to this type of outright bias, there are other potential errors in the survey process that are more subtle. These errors confound the ability of the researcher to interpret the response to a question by potentially changing response patterns because of the form of the question rather than its context. This bias may be the result of "instructions, question wording, question order, response choices, or the format of the instrument" (Folz, 1996, p. 87; also, Czaja & Blair, 2005). We will briefly review the major categories of such errors.

Survey research is not free from measurement error since question wording is never perfect. Weisberg (2005) believes that the criteria for good survey questions are reliability, validity, and usefulness. He believes that sometimes a researcher realizes that a question has not been worded perfectly, but it is still useful, such as when it is decided that it is better to maintain old question wording so that time trends can be analyzed, even if some alterations could lead to a better question.

Outside of sampling error, question-order effects have been identified as the most frequently cited explanation for errors in surveys (Schuman & Presser, 1981). Question order is a general term used to describe the extent to which answers to a given question are influenced by the questions that precede them in a survey. The question-order effect is typically explained as the contextual effect that can be produced by asking a series of questions that produces a frame of reference for the respondent (Czaja & Blair, 2005). An example is asking a series of general questions about the commitment to free speech and then asking whether a known communist should be allowed to speak in the public schools. Question-order effects are real and have been documented empirically to affect response outcomes (Schuman & Presser). An analogous problem is that of response order or the primacy or recency effect: the tendency of the respondent to choose the first or last alternative in a response set (Czaja & Blair, 2005; Lyberg et al., 1997; Schuman & Presser; Schwarz & Hippler, 1991; Sudman & Bradburn, 1982). Weisberg (2005) warns survey researchers particularly about Internet surveys regarding formatting problems. He points out that questionnaires and response alternatives may look different depending on the respondent's browser and screen resolution, making the survey less standardized than the researcher may realize. Additionally, survey researchers should pay attention to question structures and answer formats for Internet surveys so interviewees can reach clear instructions on all questions and see all answers on the screen at the same time.

The two major types of questions typically employed in surveys are forced choice and open ended. The former provides a list of mutually exclusive and exhaustive categories to the respondent. The latter, as implied, offers the respondent the opportunity for an extended narrative. Close-ended or forced-choice questions are subject to a response-set bias, an example of which is a long series of agree or disagree questions for which the respondent initially tries earnestly to answer the questions, but because of boredom or fatigue, he or she speeds through the rest of the questions by marking all agree or disagree.[1] Social desirability is another response bias that affects questions that have strong social norms for compliance associated with them, for example, answers to questions about voting. Finally, there are the very important concerns relating to "non-attitudes"; one empirical study found the following:

Our analysis of questions about two issues unknown to the American public leads to several important conclusions. First, a substantial minority of the public—in the neighborhood of 30%—will provide an opinion on a proposed law that they know nothing about if the question is asked without the DK option. This figure is certainly lower than the "majority" some times bruited about, but it is obviously large enough to trouble those assessing attitudes or beliefs concerning public issues. (Schuman & Presser, 1981, p. 158)

The fact that respondents will express attitudes about issues of which they have no knowledge has lead many to utilize screen or filter questions that determine if the respondent has any knowledge about the subject before questions about his or her attitude are solicited.

Pretesting the Instrument

One practical way to ensure that many potential factors that can ultimately impact *total survey error* are minimized prior to the start of a survey is to pretest the questionnaire on a subsample of the population of interest. This allows the researcher to determine if there are any general problems in the instructions, wording of questions, the order of questions in terms of context effect, response-set bias, or the general design or format of the questionnaire. It is tempting to assume that these factors can be discovered without a dry run. However, experience shows this is not so. Rarely does a pretest not lead to some modification that ultimately reduces total survey error.

Weisberg (2005) also points out that length poses another important constraint in surveys since researchers invariably want to ask more questions than is feasible to ask in an interview. Survey costs depend on the length of the interview in addition to the number of respondents. That is why pretests are useful for determining the approximate length of the interview. If the pretests run long, the survey organization may inform the researcher that a researcher should cut the number of questions or the number of respondents in order to stay within the budget. In addition, it is equally important that researchers understand that respondents are more likely to engage in satisficing behavior during a long questionnaire, so a shorter

Table 1.

Field	Variable Name	Variable Description	Value	Value Label
(1)	Gender	Respondent's Sex	0 = M 0 = F	1 = Male Respondent 2 = Female Respondent

questionnaire will get higher quality data. However, survey researchers should make sure that the survey length will fit within available cost and time constraints considering theoretical and empirical aspects of their study.

Data Entry Creating the Codebook

Data are the information gathered during the survey. This data acquisition is perhaps the point at which informational technology can be most useful. The case or the unit of analysis for most surveys is typically the individual respondent. The questions that constitute the survey questionnaire are variables (Czaja & Blair, 2005; Folz, 1996). Creating the codebook is the process of assigning a unique variable name, variable description, value, and value label to each item, and identifying DK or missing variables. These variables are also measured at a given level of measurement (nominal, ordinal, interval, or ratio). These variables have names and descriptions, and the values are typically data code(s). The process also typically involves stipulating the number of fields the variable will occupy. An example is the variable gender (see Table 1).

Coding decisions also potentially effect total survey error and should be implemented before the survey is implemented (Czaja & Blair, 2005; Folz, 1996). Both open- and closed-ended (forced-choice) questions need a coding scheme that will provide a framework for statistical and conceptual analysis.

Data Analysis and Report Writing

The final stage in the survey process is the analysis of data and the writing of reports. Typically, the analysis process begins with getting a univariate breakdown of the data from the survey. These frequencies include the variables, the description of the variables, the values of the variable, and their labels. DK and missing values are identified. This is called the machine codebook and typically breaks down the variables by valid and cumulative percentages. Bivariate analysis compares an independent variable with a dependent variable. Multivariate analysis involves more than three variables and the concept of statistical control.

In this final stage of survey research, estimation error could occur when the wrong statistical techniques are applied to the data. For instance, a number of researchers believes that the use of linear regression analysis on ordinal or nominal data is inappropriate, although some researchers prefer linear regression when it seems to be giving similar results as more appropriate techniques. In addition, another type of estimation error occurs when a researcher uses techniques that assume normally distributed variables when the variables are not normally distributed (Weisberg, 2005). Obviously, statistical software (SPSS or SAS) is a very important part of this step in the process. SPSS and SAS can give researchers control over measures of the form of the distribution(s).

The last part of the postsurvey process is the reporting of the survey, which includes a final summary of the project and analyzing the data. A survey researcher should be aware of reporting error, which occurs when the results of a survey are reported incorrectly or incompletely. Incomplete reports may imply that survey findings are stronger than they actually are, such as when survey results are reported as factual without admitting that the response rate was low (Weisberg, 2005). Finally, in the physical format and graphic format (including tables, bar charts, histograms, etc.) of the final report are important elements of the survey process that are also greatly enhanced by statistical software. In fact, the quality of the presentation and content of a final report are greatly enhanced by information technology.

Computer-Assisted Survey Research

In outlining the stages in the survey research process in the previous section, we have attempted to provide a framework for understanding how integrating the computer can enhance the survey process, specifically, how computing potentially cuts costs, improves quality, and minimizes some errors of measurement. By far, the most significant single influence on the practice of survey research to occur in since the 1990s is the development of the powerful personal computers (PCs) at an affordable cost. Today most researchers and practitioners who have access to a CPU (central processing unit) with sufficient random access memory (RAM) and a large-enough disk drive can perform almost any analysis on a desktop without being tied to a large mainframe computer. Moreover, as we shall see subsequently, while the collection of primary or original data for most large surveys still typically requires more infrastructure than a single PC, the analysis of primary or secondary data (data previously collected and, for example, available via the Internet) has been revolutionized by the current power of desktop PCs and available software (Clark & Maynard, 1998; Czaja & Blair, 2005; Folz, 1996). Finally, using today's PCs, a thoroughly professional final report, complete with first-rate graphics and tabular materials, is possible using readily available software like Microsoft Office and SPSS or SAS for Windows.

Computer-assisted survey research is a general term for the integration of the computer into different levels of the survey process. CASR was first established at larger university survey research programs (Skronski, 1990). Computer-assisted surveys have numerous advantages. In addition to automating the large amount of drudge work associated with the manual implementation of surveys, CASR techniques can help to reduce total survey error. First of all, CASR helps eliminate inter-interviewer bias since the computer provides exactly the same instrument in exactly the same way each time. If desired, however, the computer can also randomly administer two ostensibly similar but differently worded instruments to test instrument validity by the split-halves method. This has both basic research and practical applications. On the one hand, it allows researchers to test empirically hypotheses about, for example, the inclusion or omission of a DK or the inclusion of the middle position between two polar anchor points. As a practical matter, this feature can give the survey researcher certain control over question-order and response-set bias. Computer-assisted interviews allow automatic branching to further questions contingent upon the respondent's earlier answers, helping make interviews not only more efficient, but also more conversational in tone. Given that the computer does not suffer from fatigue, and can search through a complex questionnaire very fast, this can reduce errors in complex branching structures common to some survey designs.

While most research shows no difference in responses received for computerized vs. pencil-and-paper instruments, some research has noted extreme responses to some items with computer surveys than with paper-and-pencil surveys. In addition, electronic data collection exhibits more stability across levels of methodological variables (Helgeson & Ursic, 1989). Kiesler and Sproull (1986), in an experimental comparison of electronic surveys compared to traditional pen-and-paper mailed surveys, found that the computer method led to more honest and detailed results. Researchers have found that closed-end responses in the electronic survey were less socially desirable and tended to be more extreme than were responses in the paper survey, possibly suggesting greater respondent honesty. Open-ended responses that could be edited by respondents were relatively long and disclosing. Moreover, the recent development of computer software that creates category systems extracted from textual materials holds enormous promise for using open-ended questions. The traditional way to analyze open-ended responses in survey research has been content analysis. Content analysis seeks categories of meaning from, for example, open-ended verbatim interviews of respondents. This process is complex, subjective, and can add to total survey error by producing errors related to inter-rater reliability. New software, based on formal linguistic processes for analyzing category systems, holds great promise for formalizing the analysis of textual responses to open-ended questions by requiring adherence to category development principles (which have been developed empirically over time based on grammar and semantics research) and offering an empirical foundation for what has traditionally been a qualitative process. This will clearly reduce total survey error by increasing validity and reliability in the development of response sets for closed-ended questions and in the analysis of open-ended questions (Litkowski, 1997).

It is also relevant to note that computerized versions of standard psychological tests have been in use for several years and are now widely accepted as equivalent to pencil-and-paper instruments (e.g., Nurius, 1990). Researchers have compared computer-administered surveys with other methods of survey administration to determine if differences exist in respondents' assessment of survey mode or responses and have found very few differences. Most research concludes response rates were very similar by mode. Fully computerized surveys also eliminate coding errors that often plague manual survey research (Czaja & Blair, 2005; Folz, 1996). The computerization of survey research functions to preserve anonymity, thereby improving the veracity of results. It appears less threatening and more neutral to the subject. CASR provides more timely feedback to participants and research sponsors alike, helps prevent interviewer bias and participant intimidation (e.g., when participants are the sponsor's employees or students), and reduces overall costs of the information-gathering process. Finally, it is possible with computerized surveys to measure response-delay intervals, which have been shown to be correlated with the faking of responses. CASR software may thus record response latency times, compare them with norms, and branch to probe questions when latency times are outside the normalized range (George & Skinner, 1990).

In addition, CASR is currently evolving computer-based management systems that assist in the survey research project. Statistics Canada, for example, has developed a computer-based case-management system (CMS) that performs three main functions. It routes cases for analysis during the survey process from interviewer to the head office. It tracks the status of a survey at a given point in time including describing the status of interviews. CMS gives options to interviewers to make appointments and records specialized notes. CASR has evolved to such a point that online codebook browsing and conversation survey analysis is now possible. The University of California at Berkeley has developed a system to encourage undergraduate research. Using their codebook browser with its point-and-click features, the student can click on a particular variable such as "Catholic," and the question text window will jump to the first question with that text. The student can then select a second variable and do a cross-tabulation. This technology is presently restricted to rectangular data formats (He & Gey, 1996).

Finally, CASR has the further advantage of moving the data-editing process to the interview itself, which will diminish cost and timing constraints for survey researchers. Moreover, the computer can be programmed to catch implausible combinations of answers and have the interviewer correct one of the entries or ask the respondent to resolve the apparent inconsistency (Weisberg, 2005). In short, even though computerization adds new complexities to survey researchers' task, it also ensures the completeness and consistency of the data that are conveniently collected.

Computer-Assisted Personal Interviewing

Computer-assisted personal interviewing (CAPI), is a term for methods in which the interviewer brings a portable computer on which CAPI software has been loaded to the interview site. Typically, the interviewer reads questions from it and enters responses directly. CAPI allows interviewers to conduct face-to-face interviews using a portable computer (generally notebook PCs, handheld PCs [HPCs], and palmtop PCs). After interviews, the interviewers send the data to a central computer, either by data communication or by sending a data disk using regular mail. CAPI can also include a computer-assisted self-interview (CASI) session where the interviewer hands over the computer to the respondent for a short period and remains available for further instructions in the field (United Nations Economic and Social Commission for Asia and Pacific [UNESCAP], 2004). Pioneered in Europe by Statistics Sweden (Lyberg, 1989) and the Netherlands Central Bureau of Statistics, CAPI has been used in this country for the Nationwide Food Consumption Survey (Rothschild & Wilson, 1988) and the National Health Interview Survey (Baum & Rowe, 1989). Statistics Canada has recently perfected a CAPI system that involves an initial in-person interview. Over 1,000 interviews were done with portable computers to conduct the Labor Force Survey at the household level. This CAPI interview is then followed by five telephone interviews (Dufour et al., 1997). Research shows that CAPI increases data quality, although it appears that CAPI is initially more expensive than traditional paper-and-pencil interviewing (Baker, 1992). In terms of survey research methodology, the role of interviewer is highly critical in enhancing the survey data quality. Some argue that inexperienced interviewers may direct much of their attention to keeping the computer running and getting the answers correctly. If using the computer weakens the relation between interviewer and respondent, the interview will not be conducted adequately, and thus the data quality may suffer (UNESCAP).

Virtually every major American survey organization has or is developing a CAPI system. CAPI is currently a well-established computer application with a number of specific advantages that impact data quality and aid in the reduction of total survey error (Baker, 1992; Baker, Bradburn, & Johnson, 1995; UNESCAP, 2004). CAPI is used for the following:

- **To reduce the time needed to collect and process survey data:** Computers can help us perform all the steps to collect and process data faster. The capacity to integrate several of these steps, for example, editing, coding, data entry, cleaning, and in some cases receipt and low-level sample management, into a single process reduces the elapsed time between survey design and analysis.
- **To determine optimum question formulation:** CAPI offers new question-formulation possibilities to provide each respondent with a unique sequence of questions by randomizing the order of questions in a scale. This eliminates the systematic question-sequence effect. In addition, response categories can be randomized, which prevents question-format effects.
- **To exert greater control over the survey process and therefore improve the quality of the information collected:** Errors, both by interviewers and respondents, can be detected more quickly and resolved, often with the help of the respondent.
- **To reduce survey costs:** The efficiency created by integrating formerly discrete tasks into the single step eliminates the keying, editing, and much of the costly postprocessing that is required for paper-and-pencil surveys.
- **To implement more complex questionnaire designs than are possible with paper and pencil:** Computers can deal with much more complex skip patterns and use previously collected information more effectively than can human beings working only with paper and pencil.
- **To provide more assurance of confidentiality:** Interviewees tend to respond to sensitive questions (such as drinking habits; sexual, political, and religious preferences; etc.) more than with other interview methods. It is especially true when a portable PC is used to enter responses vs. writing them down on a form with the respondent's identifying information.

Multimedia applications are making their appearance in the area of computerized survey research including CAPI. These multimedia applications have both practical and measurement implications for survey research. Research Triangle Institute, NC, for example, has experimented with adding audio to the interviewing process. The process works as follows. The respondent listens to a digitally recorded version of questions and answers with choices through headphones. The respondent subsequently records his or her answer directly in to the computer, essentially making the process a self-interview. This technology allows maximum privacy. It has a number of applications in areas where the answers required are highly sensitive, for example, sexual behavior (Cooley, Turner, O'Reilly, Allen, Hammill, & Paddock, 1996). It also has a number of advantages that potentially reduce total survey error. First, it does not require the respondent to be literate, only that he or she can hear. It allows multilingual interviewing without requiring multilingual interviewers. It allows the type of controlled branching previously mentioned, automatic range checking, and the automated production of data files for analysis. Finally, this system (hardware and software) always produces a standardized questionnaire in all languages in which it is administered (Cooley et al.). This use of audio is not only innovative but has a number of potential uses that will proliferate as the technology decreases in cost and increases in quality.

Computer-Assisted Telephone Interviewing

Computer-assisted telephone interviewing is a hardware and software technology that has become sufficiently inexpensive and convenient and will no doubt grow substantially in the next decade. Though CATI has been available since the early 1970s (A. Fink, 2005), it now provides a convenient and cost-effective method of obtaining interview data. Once used on minicomputers like the VAX, CATI today is implemented on microcomputers, with support provided by database and statistical tools for related analyses (Crispell, 1989). CATI is an interactive computer system that leads interviewers to ask questions by controlling branching to or skipping among questions. It validates the data as being keyed into the computer system immediately by the interviewer. The interviewers are more personalized, and probing, questioning, and the use of historic data are standardized (UNESCAP, 2004).

The advantages of CATI are several after the initial investment in computing hardware and software and telephone lines (much of which may be used for other purposes and therefore may not represent additional costs). CATI allows projects to be completed faster and at a lower cost, particularly in terms of labor costs. It also can deal with complex questionnaires with branching formats. Since it permits only the entry of valid codes, it reduces total survey error. It also lets the researcher immediately track the respondents' profile (Anderson & Magnan, 1995; Folz, 1996; Weisberg, 2005). In addition, if respondents are properly familiarized with the method, CATI will be perceived as more anonymous in nature and hence is better suited to obtaining candid responses. This is even more true of another form of computer-assisted survey, in which employees enter information in voting-booth style using a computerized survey that never asks for identifying information. CATI varies substantially with the specific software used. We may differentiate broadly between two forms, however. Partial CATI still requires human telephone interviewers, but the interviewers are prompted on the computer screen with the questions to ask and they enter data directly into the survey database, eliminating an extra coding step. Full CATI eliminates human interviewers altogether; this technology is sometimes called TDE (touch-tone dial entry) or ATI (automated telephone interviewing). The computer poses the questions to the respondent using either voice emulation or recordings, and the respondent answers by pressing buttons on a touch-tone phone; the software translates these tones and the corresponding responses are entered directly in the survey database (Werking, Tupek, & Clayton, 1988). TDE has been used by the Bureau of Labor Statistics to collect some of its current employment statistics data, for example. Other advantages cited for CATI are simplification of interviewer training and closer interviewer supervision in partial CATI designs, more rapid changing of survey instruments, easier use of multiple forms, more rapid availability of data, as well as improved cost control and record keeping.

More sophisticated uses of CATI also allow branching within a questionnaire contingent on employee responses, re-presentation of past responses to assist in recall, online calculations using prior responses to change the content of items currently being posed, and the availability of standardized help screens and prompts where the explication of items is desired. Other features of CATI software are sample generation from telephone databases, call scheduling (e.g., handling time zones, making callback appointments), accepting incoming callbacks, management reports (e.g., on interviewer productivity or refusal rates), quota control in stratified designs, menu-driven questionnaire-authoring systems, pretest administration and analysis, question order randomization, text insertion (e.g., the respondent's name), dynamic generation of choices (respondent-supplied information from prior items appears as choices in subsequent items), revision editing (users can back up), and the handling of respondent skips and qualifiers. For a discussion of advance CATI techniques, see Groves (1988). For a discussion of optimal call scheduling with CATI, see Greenberg and Stokes (1989), and Weeks, Kulka, and Pierson (1987). Finally, the Questionnaire Programming Language (QPL), public-domain software developed by the U.S. General Accounting Office, is accessible to public administrators and is extremely useful in conducting CATI surveys (Anderson & Magnan, 1995).

In addition to all the ongoing advantages, CATI systems can be used in conjunction with random digit dialing (RDD) for sampling purposes. RDD is far superior to relying on available lists such as outdated phone directories that, obviously, do not contain unlisted numbers. Recall that probability sampling requires that all sampling elements have an equal or known probability of inclusion. The assumption along with that of adequate response rate are assumptions that underlie many commonly employed statistical tests. RDD is one way to reduce sampling error and increase precision because it increases the probability that every household with a phone (regardless of if it has a listed phone number) can be sampled (Best, Krueger, Hubbard, & Smith, 2001; Couper, 2000a; Gunn, 2002). RDD typically begins by establishing all sampling prefixes that are isomorphic with a given geographical area (for example, 380, 467, 319, etc. for Cary, North Carolina) and then uses some randomized process to select the other four digits. As a practical matter, a number of commercial sampling services exist that will draw RDD samples for specified localities. These companies typically use an algorithm that produces sample lists of phone numbers in proportion to the number of lines in each of the prefixes in the telephone companies' jurisdiction.

CATI is enhanced by the appearance of massive databases of phone numbers, such as the CD-ROM products *Disc America*, with 100 million residential and business names, addresses, and phone numbers, and *Disc America: Business*, a CD-ROM with 10 million U.S. business names, addresses, and phone numbers, with SIC codes and Boolean searching. A similar resource is the *Associations CD*, a CD-ROM version of the 18-volume *Encyclopedia of Associations*, including 90,000 associations worldwide, 22,600 U.S. associations, 8,200 international associations, 60,000 U.S. subnational organizations, and 11,000 association periodicals. It comes with software for producing mailing labels for survey research or marketing purposes, and has a built-in autodialer for computer-assisted telephone interviewing. In addition to CD-ROM databases of telephone numbers, similar data are available online. For instance, researchers in nonprofits may be interested to know that the *Encyclopedia of Associations* is also online on DIALOG, a leading information vendor (File 114). For survey research purposes, this file can be used to identify groups offering computerized membership lists.

The Internet

The use of surveys on the Internet is exploding. As of this writing, a simple Google search rendered 87,000 hits on Internet surveys. Few highly innovative academic survey research applications exist that use the Internet. However, as Dillman and Bowker (2001) observe, programmers and information-technology professionals, not survey research specialists, have done much of this development. As a consequence, not enough attention as been given to four major sources of error in surveys previously discussed. These errors continue to exist and cast doubt on the reliability and validity of many Web surveys. Recall our previous discussion of a probability sample being one in which every member has an equal and known probability of inclusion. Consequently, surveys on the Web that seek valid and reliable answers must adhere to solid principles of survey design. Many of the 87,000 hits on the Google search, clearly, would not meet these criteria.

As Dillman and Bowker (2001) note, the advent of Web surveying makes possible the delivery of the survey instrument to hundreds of thousands of people and processes the results for very little cost. Not surprisingly, they observe, this process is accelerating; however, this process does not, cannot, and will not replace carefully drawn samples. Collecting survey data on the Internet is currently evolving, and, we believe, will develop at breakneck speeds in the next 5 years. There are millions and millions of users on the Internet in the United States and worldwide, and, while this sampling frame of Internet users is biased toward well-educated, upper class males, this is changing daily. Data collection on the Internet is fast, inexpensive, capable of utilizing the multimedia capabilities of computing, and, allows for the immediate processing of data (Gunn, 2002; Zhang, 1991). Many of the technical problems of survey research on the Internet are being streamlined as we speak. The Internet provides the researcher with access to a worldwide group of respondents, a sampling frame that is growing daily. However, the problem remains: How does a respondent get into your sample and to whom can you generalize?

It is still critical that the researcher be able to defend the fact that the findings of the sample can be generalized. The form the Internet survey takes varies widely. This is the source of many problems in using the Internet for surveys. Until the disconnect between how a survey is broadcast by the Web designer and how it is being read by the end user is resolved, the problem of using the Web for surveys will persist. This problem is more or less tied to the form of the survey (Gunn, 2002). An inexpensive form of Internet survey is the e-mail survey, in which the questionnaire is simply embedded in an e-mail message. However, some of the most innovative uses of Internet surveys involve multimedia, in particular, survey participants who are on broadband networks can be sent streamed video (Weisberg, 2005).

In terms of coverage error, it is clear that not everyone has access to the Internet, and compared to telephone or area probability samples, Web surveys offer relatively poor coverage of the general household population (Fricker, Galesic, Tourangeau, & Yan, 2005). In fact, the majority of countries in the world does (Dillman & Bowker, 2001). However, the world is, as Thomas L. Friedman says, flat. Fiber optic connections are growing daily as is bandwidth that can act to medicate many of the problems. More and more access to the Internet is an almost certain scenario. Nonetheless, the current state of affairs has implications for both sampling error and coverage area. It is clear some sampling frames do have adequate access to the Internet and, therefore, can be candidates for valid and reliable Web surveys. Examples include those with traditional professions, college students, and people with higher levels of education. As a general bit of advice to researchers, when considering a Web-based survey, repeatedly ask the question, to whom am I attempting to generalize? Sorting through the answer(s) to this question will in many cases determine the appropriateness of using the Web. As a general guideline, avoid all procedures that allow for self-selection. This is a difficult, but not impossible task. In fact, the Munich Public Health Service in Germany reports its experience in collecting data by direct mailing on the Internet using features of the World Wide Web (Swoboda, Muhlberger, Weitkunat, & Schneeweib, 1997). The Munich Public Health Service has experimented with transferring traditional survey research methodology onto the Internet. A parsing program was employed to scan all newsgroup messages for e-mail addresses and to store those messages on a file. In this way, the researchers were able to overcome the fact that there is no complete e-mail directory of users on the Internet. All returned questionnaires were stored in a single concatenated file. Data were then converted to an SAS file for analysis. As we said, the Internet provides the researcher with access to a worldwide group of respondents, a sampling frame that is growing daily (Dillman & Bowker; Swoboda et al.). Clearly, we are only seeing the beginning of direct mail using the Internet. However, current research shows that Web surveys tend to have lower response rates than traditional mail methods (Crawford, Couper, & Lamias, 2001). Kwak and Radler (2000), for instance, found that their survey on comparable samples of college students had a response rate of 42% for mail and 27% for Web surveys. Similarly, Guterbock, Meekins, Weaver, and Fries (2000) determined a response rate of 48% for mail and 37% for Web surveys.

Turning to the issue of nonresponse, researchers deal with nonresponse in a scientific survey by using reasonable tested methods to reduce and account for unit and item nonresponse error. Dillman (2000) argues that research procedures minimizing nonresponse rates on the Web can be developed in a generalized sense. However, the Web offers some unique problems to users trying to finish the Web questionnaire. While we list these issues, the reader is advised to read Dillman's article in its entirety prior to doing a Web survey. In a Web survey, researchers should consider people's relative computer literacy; for example, do all users know what to do with a top-down menu? Do they understand radio buttons? A Web survey should allow the user to see all response choices without scrolling up and down, and these choices should be offered to the respondent so that all answer choices seem appropriate. The respondent should know how much longer the questionnaire is and have a clear and defined set of instructions (Dillman & Bowker, 2001). Dillman and Bowker provide a very useful matrix that covers all the sources of error and the design principles used to address them in the general area of measurement. The most important caution they offer, in our view, is that Web surveys should present questions in form similar to paper-and-pencil formats that have now become standard. They also indicate the need to restrain the use of color to ensure navigational flow and measurement properties. They also offer other design tips beyond the scope of this article.

A second major impact on survey research of the Internet is in the analysis of secondary data. Secondary data are those that were collected for purposes other than for which they are currently being analyzed. This secondary analysis generally has as its purpose in presenting new findings (findings different from those contained in the original final report). Given the costs of collecting primary data, the analysis of secondary data can be very important to public-administration researchers and practitioners. The ability of public administrators who have a PC and a connection to the Internet to analyze large amounts of data collected according to the structures of a well-executed sampling design and using questions that have been constructed and pretested for the types of errors discussed in this chapter is an exciting development. Moreover, access to these data sources are moving toward the point-and-click technology found in the Windows environment (Clark & Maynard, 1998). Public-opinion archives worldwide are in the process of making access to end users more user friendly, using tools like Java programming. Using the common gateway interface (CGI) and Web browsers, the user can request pages and basic statistical analysis of archived survey data. Having data online makes it possible to pursue multiple databases from one's own PC. In addition, a variety of search engines are available to assist the researcher in locating the data they need by entering key words and subject headings related to the research endeavor. Many data archives such as the Roper Center currently providing data sets to users can be seen over the Internet using FTP (file transfer protocol). Among the major data producers and data archives are the Gallup Organization (http://www.gallup.com), the National Opinion Research Center (NORC, http://www.norc.uchicago.edu/), and the Interuniversity Consortium for Political and Social Research (ICPSR), University of Michigan (http://www.icpsr.umich.edu/).

Conclusion

Survey research is a methodological technique in public administration. Like all methodological techniques, it seeks to minimize the total error of measurement in an attempt to gain some understanding about the relationship between and among variables. All methodological techniques fail in some important respects. Even in experimental research, relationships between variables need to be replicated many times prior to being accepted. In ex post facto research, where both the independent and dependent variable(s) have already occurred in time (survey research is one type of ex post facto research), the measurement problems are significantly greater. Typically, in surveys we seek to reduce total survey error in search of meaningful relationships. This total survey error is all the error that can seep into a study that affects that survey's accuracy, that is, the ability to mirror something in the real world. As we have indicated, computers and information technology can help to reduce some elements of this error. We have also indicated that good sampling and survey design techniques can only address some other errors.

Finally, information technology is important at all steps in the survey research process from management to report writing. In this sense, information technology cannot be overlooked as an important element in the tool kit of the modern researcher.

References

Adobe Page Maker [Computer software]. (2001). Adobe Systems.

American Association for Public Opinion Research (AAPOR). (2001). Annual membership meeting. *Public Opinion Quarterly, 65*, 7-78.

Anderson, R., & Magnan, S. (1995). The questionnaire programming language (QPL): An overview with examples of call management. *Social Science Computer Review, 13*(3), 291-303.

Arksey, H., & Knight, P. T. (1999). *Interviewing for social scientists: An introductory resource with examples.* Thousand Oaks, CA: Sage Publications.

Babbie, E. (1990). *Survey research methods* (2nd ed.). Belmont, CA: Wadsworth.

Baker, R. (1990). Applications of new computer technology in survey research: An overview. In *Proceedings of the Conference on Advanced Social Sciences Computing*, Williamsburg, VA.

Baker, R. (1992). New technology in survey research: Computer assisted personal interviewing (CAPI). *Social Science Computer Review, 10*(2), 145-157.

Baker, R., Bradburn, N. M., & Johnson, R. A. (1995). Computer-assisted personal interviewing: An experimental evaluation of data quality and cost. *Journal of Official Statistics, 11*(4), 413-431.

Baum, M., & Rowe, B. (1989). Uses of CATI at NNCHS. Proceedings of the National Field Technologies Conference, St. Petersburg, FL.

Behling, O., & Law, K. S. (2000). *Translating questionnaires and other research instruments: Problems and solutions* (Quantitative applications in the social sciences, No. 133). Thousand Oaks, CA: Sage Publications.

Berot, J. C., & McClure, C. R. (1996). Electronic surveys: Methodological implications for using the World Wide Web to collect survey data. In S. Hardin (Ed.), *Global complexity: Information, chaos, and control* (pp.173-192). Medford, MA: Information Today.

Best, S. J., Krueger, B., Hubbard, C., & Smith, A. (2001). An assessment of the generalizability of Internet surveys. *Social Science Computer Review, 19*, 131-145.

Bourque, L. B., & Fielder, E. P. (2002a). *How to conduct self-administered and mail surveys* (Vol. 3). Thousand Oaks, CA: Sage Publications.

Bourque, L. B., & Fielder, E. P. (2002b). *How to conduct telephone surveys* (Vol. 4). Thousand Oaks, CA: Sage Publications.

Bourque, L. B., & Clark, V. A. (1992). *Processing data: The survey example* (Quantitative applications in the social sciences, No. 85). Thousand Oaks, CA: Sage Publications.

Bradburn, N., & Sudman, S. (1991). The current status of questionnaire design. In P. Biemer, R. Groves, L. Lyberg, N. Mathiowetz, & S. Sudman (Eds.), *Measurement errors in surveys* (pp. 29-40). New York: John Wiley.

Clark, R., & Maynard, M. (1998). Using online technology for secondary analysis of survey research data. *Social Science Computer Review, 16*(1), 58-71.

Code of professional ethics and practices. (1986). Ann Arbor, MI: American Association for Public Opinion Research.

Comer, D. (1997). *The Internet book.* Upper Saddle River, NJ: Prentice Hall Inc.

Converse, J. M., & Presser, S. (1986). *Survey questions: Handcrafting the standardized questionnaire* (Quantitative applications in the social sciences, No. 63). Beverly Hills, CA: Sage.

Converse, P., & Traugott, M. (1986). Assessing the accuracy of polls and surveys. *Science, 234*, 1094-1098.

Cook, T., & Campbell, D. (1979). *Quasi-experimentation: Design and analysis issue for filed studies.* Boston: Houghton Mifflin.

Cooley, P., Turner, C., O'Reilly, J., Allen, D., Hammill, D., & Paddock, R. (1996). Audio-CASI: Hardware and software considerations in adding sound to a computer assisted interviewing system. *Social Science Computer Review, 14*(2), 197-204.

Couper, M. P. (2000a). Usability evaluation of computer-assisted survey instruments. *Social Science Computer Review, 18*(4), 384-396.

Couper, M. P. (2000b). Web surveys: A review of issues and approaches. *Public Opinion Quarterly, 64*, 464-494.

Couper, M. P., Baker, R. P., Bethlehem, J., Clark, C., Martin, J., Nicholls, W., et al. (1998). *Computer assisted survey information collection.* New York: John Wiley & Sons.

Crawford, S., Couper, M. P., & Lamias, M. J. (2001). Web surveys: Perceptions and burden. *Social Science Computer Review, 19*(2), 146-162.

Crews, M., & Feinberg, M. (2002). Perceptions of university students regarding the digital divide. *Social Science Computer Review, 20*(2), 116-123.

Crispell, D. (1989). People talk computers listen (Using computer assisted interviewing). *American Demographics, 11*(8), 1.

Czaja, R., & Blair, J. (2005). *Designing surveys: A guide to decisions and procedures* (2nd ed.). Thousand Oaks, CA: Sage Publications.

Deming, W. (1950). *Some theory of sampling.* New York: John Wiley & Sons.

Dillman, D. (1978). *Mail and telephone surveys: The total design method.* New York: John Wiley.

Dillman, D. (2000). *Mail and Internet surveys: The tailored design method.* New York: John Wiley & Sons.

Dillman, D. A. (1999). *Mail and Internet surveys: The tailored method* (2nd ed.). New York: John Wiley and Sons.

Dillman, D. A., & Bowker, D. K. (2001). The Web questionnaire challenge to survey methodologists. In U.-D. Reips & M. Bosnjak (Eds.), *Dimensions of Internet science.* Lengerich, Germany: Pabst Science Publishers. Retrieved March, 22, 2006, from http://www.sesrc.wsu.edu/dillman/zuma_paper_dillman_bowker.pdf

Dufour, J., Kaushal, R., & Michaud, S. (1997). Computer-assisted interviewing in a decentralized environment: The case of household surveys at Statistics Canada. *Survey Methodology, 23*(2), 147-156.

Ehrlich, H. (1969). Attitudes, behavior, and the intervening variables. *American Sociologist, 4*(1), 29-34.

Eiler, J., Nelson, W., Jensen, C., & Johnson, S. (1989). Automated data collection using bar code. *Behavior Research Methods, Instruments, and Computers, 21*(1), 53-58.

Fink, A. (2002a). *How to ask survey questions.* Thousand Oaks, CA: Sage Publications.

Fink, A. (2002b). *How to design survey studies* (Vol. 6). Thousand Oaks, CA: Sage Publications.

Fink, A. (2002c). *The survey handbook* (Vol. 1). Thousand Oaks, CA: Sage Publications.

Fink, A. (2005). *How to conduct surveys* (3rd ed.). Thousand Oaks, CA: Sage Publications.

Fink, A., & Kosecoff, J. (2005). *How to conduct surveys: A step-by-step guide* (3rd ed.). Thousand Oaks, CA: Sage Publications.

Fink, J. (1983). CATT's first decade: The Chilton experience. *Sociological Methods and Research, 12*(2), 153-168.

Folz, D. (1996). *Survey research for public administration.* Thousand Oaks, CA: SAGE Publications, Inc.

Fowler, F. J., Jr. (2001). *Survey research methods* (3rd ed.). Thousand Oaks, CA: Sage Publications.

Fox, J. A., & Tracy, P. E. (1986). *Randomized response: A method for sensitive surveys* (Quantitative applications in the social sciences, No. 58). Thousand Oaks, CA: Sage Publications.

Fricker, S., Galesic, M., Tourangeau, R., & Yan, T. (2005). An experimental comparison of Web and telephone surveys. *Public Opinion Quarterly, 69*(3), 370-392.

Garson, D. (2006). *Survey research.* Retrieved March 2006 from http://www2.chass.ncsu.edu/garson/pa765/survey.htm

Gelman, A., King, G., & Liu, C. (1998). Not asked and not answered: Multiple imputation for multiple surveys. *Journal of the American Statistical Association.*

George, M., & Skinner, H. (1990). Using response latency to detect inaccurate responses in a computerized lifestyle assessment. *Computers in Human Behavior, 6*(2), 167-175.

Gilliland, J., & Kinchen, S. (1987). Microcomputers for survey data-entry and analysis. *Population Index, 53*(3), 374.

Greenberg, B., & Stokes, S. (1989). *Developing an optimal call scheduling strategy for a telephone survey.* St. Petersburg, FL: National Field Technologies Conference.

Groves, R. (Ed.). (1988). *Telephone survey methodology.* New York: John Wiley and Sons

Groves, R., & Dillman, D., Eltinge J. L., & Little, R. J. A. (n.d.). *Survey nonresponse.* New York: John Wiley and Sons.

Groves, R., & Mathiowetz, N. (1984). Computer assisted telephone interviewing: Effects on interviewers and respondents. *Public Opinion Quarterly, 48*(1B), 356-369.

Gunn, H. (2002). Web-based surveys: Changing the survey process. *First Monday, 7*(2). Retrieved March 15, 2006, from http://www.firstmonday.dk/issues/issue7_12/gunn/

Guterbock, T. M., Meekins, B. J., Weaver, A. C., & Fries, J. C. (2000). *Web versus paper: A mode experiment in a survey of university computing.* Paper presented at the annual meeting of the American Association for Public Opinion Research, Portland, OR.

He, J., & Gey, F. (1996). Online codebook browsing and conversational survey analysis. *Social Science Computer Review, 14*(2), 181-186.

Helgeson, J., & Ursic, M. (1989). The decision process equivalency of electronic versus pencil-and-paper data collection methods. *Social Science Computer Review, 7*(3), 296-310.

Holden, R., & Hickman, D. (1987). Computerized versus standard administration of the Jenkins-Activity-survey (form-T). *Journal of Human Stress, 13*(4), 175-179.

Ingels, J. (1989). *Microcomputers and field management at NORD.* Paper presented at the National Field Technological Conference, St. Petersburg, FL.

Kiesler, S., & Sproull, L. (1986). Response effects in the electronics survey. *Public Opinion Quarterly, 50*(3), 402-413.

Kimmel, A. (1988). *Ethics and values in applied social research.* Newbury Park, CA: Sage.

Kish, L. (1995). *Survey sampling: Wiley classics library.* New York: John Wiley & Sons.

Krueger, R. (1998). *Analyzing and reporting focus group results.* Thousand Oaks, CA: SAGE Publications.

Kwak, N., & Radler, B. T. (2000). *Using the Web for public opinion research: A comparative analysis between data collected via mail and the Web.* Paper presented at the Annual Meeting of the American Association for Public Opinion Research, Portland, OR.

Lee, E. S., & Forthofer, R. N. (2005). *Analyzing complex survey data* (Quantitative applications in the social sciences, Vol. 71, 2nd ed.). Thousand Oaks, CA: Sage Publications.

Litkowski, K. (1997). Category development based on semantic principals. *Social Science Computer Review, 15*(4), 394-409.

Long, L., & Vasu, M. (1988). *Public issues in 1988.* New York: Independent Insurance Agents of America.

Lucas, P., Mullen, P., Luna, C., & McInory, D. (1977). Psychiatrist computer interrogators of patients with alcohol-related illness: A comparison. *British Journal of Psychiatry, 131*, 160-167.

Lyberg, L. (1989). Topic 18.2. In *Proceedings of the 45th Session, International Statistics Institute* (Book 3).

Lyberg, L., Briemer, P., Collins, M., deLeeuw, E., Dippo, C., Schwarz, N., et al. (1997). *Survey measurement and process quality.* New York: John Willey & Sons.

Miller, D., & Salkin, N. J. (2002). *Handbook of research design and social measurement.* Thousand Oaks, CA: SAGE Publications.

Miller, T. I., & Kobayashi, M. (2000). *Citizen surveys: How to do them, how to use them, what they mean.* Washington, DC: ICMA.

Miller, T. I., Kobayashi, M., Caldwell, E., Thurston S., and Collect B. (2002). Citizen surveys on the Web. *Social Science Computer Review, 20*(2), 124-135.

Morgan, D. (1998). *The focus group guidebook.* Thousand Oaks, CA: SAGE Publications.

Nationwide Food Consumption Survey. (1987). A landmark personal interview survey using laptop computers. In *Proceedings of the Fourth Annual Research Conference* (pp. 347-356).

Nesbary, D. (1999). *Survey research and the World Wide Web.* Boston: Allyn and Bacon.

Nicholls, W. (1988). Computer-assisted telephone interviewing: A general introduction. In R. Groves (Eds.), *Telephone survey methodology* (pp. 377-385). New York: Wiley and Sons.

Nurius, P. (1990). A review of automated assessment. *Computers in Human Services, 6*(4), 265-281.

O'Sullivan, E., & Rassel, G. (1995). *Research methods for public administrators.* New York: Longman.

Oishi, S. M. (2002). *How to conduct in-person interviews for surveys* (Vol. 5). Thousand Oaks, CA: Sage Publications.

Payne, S. L. (1951). *The art of asking questions.* Prince University Press.

Peterson, R. A. (2000). *Constructing effective questionnaires.* Thousand Oaks, CA: Sage Publications.

Rea, L., & Parker, R. (1992). *Designing and conducting survey research.* San Francisco: Jossey-Bass Inc.

Rea, L. A., & Parker, R. A. (1997). *Designing and conducting survey research: A comprehensive guide* (Jossey-Bass Public Administration Series). San Francisco: Jossey-Bass.

Rossi, P., Wright, J., & Anderson, A. (1983). *Handbook of survey research.* San Diego, CA: Harcourt, Brace, Jovanovich.

Rothschild, B., & Wilson, L. (1988). National food consumption survey 1987. In *Proceedings of the 4th Annual Research Conference* (pp. 347-356).

Rubin, H. J., & Rubin, I. S. (2004). *Qualitative interviewing: The art of hearing data.* Thousand Oaks, CA: Sage Publications.

Salant, P. (with Dillman, D. A.). (1994). *How to conduct your own survey.* New York: John Wiley and Sons.

Schuman, H., & Presser, S. (1981). *Questions and answers in attitude surveys: Experiments on question form, wording, and context.* New York: Academic.

Schwarz, N., & Hippler, H. (1991). Response alternatives: The impact of their choice and presentation order. In P. Bimer, M. Groves, L. Lyberg, N. Mathiowetz, & S. Sudman (Eds.), *Measurement errors in surveys* (pp. 41-56). New York: John Wiley.

Skronski, M. (1990). Computer-assisted survey methods have a new data collection facility. *Berkeley Computing Quarterly, 2*(2), 4-5.

Sudman, S., & Bradburn, N. (1982). *Asking questions: A practical guide to questionnaire design.* San Francisco: Jossey-Bass.

Swoboda, W., Muhlberger, N., Weitkunat, R., & Schneeweib, S. (1997). Internet surveys by direct mail. *Social Science Computer Review, 15*(3), 242-253.

Tartar, D. (1969). Toward prediction of attitude-action discrepancy. *Social Forces, 47*(4), 398-404.

United Nations Economic and Social Commission for Asia and Pacific. (2004). *Guidelines on the application of new technology to population data collection and capture.* Retrieved March 2005 from http://www.unescap.org/Stat/pop-it/pop-guide/pop-guide.asp

U.S. General Accounting Office. (1991). *Using structured interviewing techniques* (PEMD-10.1.5, 7/91).

Vasu, M. (1998). *Cary growth strategies project.* Raleigh, NC: Triangle Growth Strategies Inc.

Vasu, M., Long, L., & Hughes, D. (1990). Continuous audience response technology combined with survey methods in field research: A description and application. In *Advances in social science and computers* (Vol. 13). Thousand Oaks, CA: Sage Publications.

Waksberg, J. (1978). Sampling methods for random digit dialing. *Journal of the American Statistical Association, 73*(361), 40-46.

Weeks, M. F., Kulka, R. A., & Pierson, S. A. (1987). Optimal call scheduling for a telephone survey. *Public Opinion Quarterly, 51*, 540-549.

Weisberg, H. F. (2005). *The total survey error approach.* Chicago: The University of Chicago Press.

Werking, G., Tupek, A., & Clayton, R. (1988). CATI and touchtone self-response applications for establishment surveys. *Journal of Official Statistics, 4*(4), 349-362.

Willis, G. B. (2005). *Cognitive interviewing: A tool for improving questionnaire design.* Thousand Oaks, CA: Sage Publications.

Winstein, R. M. (1982). The mental hospital from the patient's point of view. In W. R. Gove (Ed.), *Deviance and mental illness.* Thousand Oaks, CA: Sage Publications.

Zhang, Y. (1991). Using the Internet for survey research: A case study. *Journal of the American Society for Information Science, 51*(1), 57-68.

Appendix A: List of Software Cited

ABsurv. (n.d.). Parker, CO: AndersonBell.

Associations CD. (n.d.). Detroit, MI: Gale Research Inc., Book Tower.

Bainbridge, W. S. (1989). *Survey research: A computer-assisted introduction* [Computer software and textbook]. Belmont, CA: Wadsworth, Inc.

Ci2 System. (n.d.). Ketchum, ID: Sawtooth Software.

Conjoint Value Analysis System. (n.d.). Ketchum, ID: Sawtooth Software.

Disc America. (n.d.). Warwick, NY: SilverPlatter Directories.

EnquLte Survey System. (n.d.). Durham, NC: National Collegiate Software, Duke University Press.

EX-SAMPLE+. (n.d.). Columbia, MO: The Idea Works, Inc..

INTERV. (n.d.). Amsterdam: Sociometric Research Foundation.

Measurement and Scaling Strategist. (n.d.). Columbia, MO: The Idea Works, Inc.
MicroCase Aggregate Analysis Program. (n.d.). Seattle, WA: Cognitive Development Inc.
On the Campaign Trail. (n.d.). Washington, DC: Campaigns and Elections, Inc.
ParSURVEY-GST. (n.d.). Costa Mesa, CA: Economics Research Inc.
PC Quest. (n.d.). Northridge, CA: Human Systems Dynamics.
PC/Stargraph. (n.d.). Eagan, MN: Questar Data Systems.
PeopleFacts: The Opinion Processor. (n.d.). San Diego, CA: Shamrock Press.
QPL. (n.d.). Springfield, VA: NTIS.
Quick Tally. (n.d.). Beverly Hills, CA: Quick Tally Systems.
Quizwhiz. (n.d.). Akron, OH: Quizwhiz Enterprises, Inc.
Seek-Easy. (n.d.). Durham, NC: National Collegiate Software, Duke University Press.
SPSS/PC+Categories. (n.d.). Chicago: SPSS Inc.
SPSS/PC+Data Entry II. (n.d.). Chicago: SPSS Inc.
Survey, The. (n.d.). Alpine, UT: Cybernetic Solutions Company.
Survey Data Entry. (n.d.). Forest Lake, MN: Craig Roberts Consulting.
Survey I. (n.d.). Durham, NC: National Collegiate Software, Duke University Press.
Survey Master. (n.d.). Seattle, WA: Masterware.
Survey System, The. (n.d.). Salt Lake City, UT: Creative Research Systems, Academic Computing Specialists.
Synthesis. (n.d.). Ann Arbor, MI: Bauer and Associates.
Topics in Research Methods: Survey Sampling. (n.d.). Iowa City, IA: Conduit, the University of Iowa.
Watson/VIS. (n.d.). Natick, MA: Natural Microsystems Corp.
Word Match. (n.d.). Durham, NC: National Collegiate Software, Duke University Press.

Appendix B: Selected Survey Data Sources

The Inter-University Consortium for Political and Social Research (ICPSR) now offers four online services of use to social-science researchers. ICPSR's *Guide-on-Line* lists all holdings in the print *Guide to Resources and Services* and updates in the *ICPSR Bulletin* on over 1,600 databases. ICPSR's *Variables* lists the full text of items in selected surveys in ICPSR holdings, covering 64,000 variables from 160 databases. ICPSR *Rollcalls* contains descriptions of all U.S. Congress roll-call votes in the last 22 years. SMIS (Survey Methodology Information System) is a bibliographic database on survey methodology that has existed since 1972. To use any of these databases, one must establish an account at ICPSR Member Services (although one need not be an ICPSR member) at 313-763-5010.

POLL is the Public Opinion Location Library at the Roper Center. Available since 1986, this online service allows the retrieval of nearly 100,000 survey items, including surveys by

NORC, Gallup, Garris, NBC, CBS/NYT, and others. Data go back to the mid-1960s. Items may be searched for by topic, subject headings, and key words in the question, date, sponsoring organization, and other fields. For information on use of POLL. Contact the Roper Center by mail at P.O. Box 440, Storrs, CT 06268, or by phone at 203-486-4440.

Cambridge Reports, Trends, and Forecasts covers economic, business, consumer, and public-policy public-opinion data. A $247 subscription brings you their magazine plus access to their database of 10,000 items since 1973. Contact Cambridge Reports at 675 Massachusetts Ave., Cambridge, MA 02139, or at 617-661-0110.

The 1989 General Social Survey (GSS) is the 16th in this much-used series from the Roper Center for Public Opinion Research, containing NORC-sponsored items cumulative since 1972 and including some new items in 1989, including coverage of sexual relations and AIDS. The cumulative file also includes 1988 items from the International Social Survey Program, with respondents from the United States, Austria, West Germany, Ireland, Hungary, Netherlands, Italy, and Great Britain. The Complete Data Set is $110 and the SPSSX System File Control Cards file is an additional $185. Both are available on tape only. You must specify tracking (nine-track EBCDIC or nine-track ASCII), density (6,250 or 1,600), and block-size maximum (default is 32,760). You may provide 2400' magnetic tape, or the Roper Center will provide it for $22 (two tapes required if 1,600 density). You must specify if you want the Cumulative or the Trends version of the data set: Both include data from 1972 to 1978, 1980, and 1982 to 1989, but Trends contains data only for variables repeated at least once. Contact the Roper Center at P.O. Box 440, Storrs, CT 06268, or at 203-486-4440.

The Bureau of the Census from time to time has invited proposals from social scientists to participate through joint statistical agreements in research evaluating the behavioral causes of census undercounts and other topics. Information on census operations may be obtained through L. Brownrigg and L. Shinagawa, Center for Survey Methods Research, Bureau of the Census, Room 433, Washington Plaza Building, Washington, DC 20233 (301-763-7976).

Notes

The principal drawbacks of CASR are startup costs for equipment, software, and training. There is also a greater need than with other methods to familiarize agency employees with the nature of the survey process. Costs, however, vary greatly. It is possible to use very inexpensive microcomputers. Also, it is not out of the question to program one's own survey package. Some commercial packages, such as Survey Master ($49.95), are quite inexpensive. Shareware or user-supported software is also an extremely inexpensive option, as with inexpensive classroom packages like Survey I ($25), or EnquLte Survey System by Pierre Corbeil and FranHois Larocque. One is included free with the software accompanying William Sims Bainbridge's (1989) text, *Survey Research: A Computer-Assisted Introduction.* Most commercial packages cost in the $400 range and up (e.g., PeopleFacts: The Opinion Processor, $395; The Survey, $495; The Survey System, $500; Synthesis, $995; and PC/Stargraph, $2,500 to $4,500 depending on version).[2]

There is a large array of software packages for computer assistance of various aspects of survey research. Without counting the large number of statistical packages relevant to survey research, a few specifically survey-oriented packages can be mentioned here as illustrations of different types:

- ParSURVEY-GST illustrates many packages that can automate the mechanics of survey research. It interfaces with a scanner for rapid data input. The software can handle Likert scales (26 scales), reverse scoring, cross-tabulation, subsamples, and reports. The cost is $495. ParSURVEY-TAB is ERI's $99 module for survey tabulation, including means and standard deviations. ParTEST is a $249 module for item banking, tracking, and test generation. ParSCORE is a $995 module for automatic scoring of tests using scanners. A less expensive alternative is PC Quest, software for questionnaire creation, survey administration, and data analysis. Both Likert scales and demographic scales up to eight categories are supported for up to 400 questions in up to 50 categories. Analysis, which supports grouping, includes frequency distribution, quartile coefficients of dispersion, median, and mode. The cost is $200.
- QPL illustrates inexpensive computerized survey packages with CATI functions. Designed for typical interviews of 150 to 250 questions (taking 20 to 40 minutes to complete), it supports branching, range limits, record editing, and other basic CATI functions. It is a public-domain program originated by the Human Resources Division of the General Accounting Office, and is available from the National Technical Information Service. It is now in Version 2.0.
- SPSS/PC+ Categories illustrates many packages devoted to advanced survey techniques arising from marketing research, such as conjoint and correspondence analysis. These are seen by many in marketing as superior, more flexible procedures than loglinear, canonical correlation, and factor analysis approaches to data analysis. For additional information on CASR and CATI techniques in perceptual mapping, conjoint analysis, and computer interviewing.
- ABsurv illustrates packages closely tied to database-management software. It is a companion to the ABstat statistical package, noted for its close integration with dBASE applications. It handles numeric, character, multiple-choice, all-that-apply, and verbatim open-ended responses. Output includes not only tables but also banner and stub reports (multiple tables sharing the same column headers, which may include multiple variables), common in marketing and other fields. Graphic output includes box and whisker plots, bar graphs, histograms, pie charts, and scatter plots. Its limits are up to 256 items on 16,000 cases, 255 characters per item, 4,000 characters per record, and as many records as will fit on one's disk. It requires 640K, MS-DOS; a 720K drive or hard disk is required if used with ABstat. The cost is $295.
- Microcase Aggregate Analysis Program illustrates classroom-oriented survey packages, in this case incorporating the General Social Survey, probably the most-used survey series in social-science research-methods courses. The cost is $1,000 for the package and 15 years of data. The MAAP program is available separately for $395, as are any years (1972-1988) of the GSS for $95. MAAP provides standard statistics

such as crosstabs and regression, but its strong point is effortless geographic mapping of survey data with the ability to highlight selected data points on the map or scatter plot. From the same publisher is the MicroCase 2 statistical package, which now supports CATI (up to 32,000 items on up to 10 million cases) and includes Survey Showcase and the 1983 General Social Survey for $395.

- Survey Data Entry illustrates software that allows on-screen re-creation of printed questionnaires for purposes of data entry. It also performs data screening checks and branches automatically where appropriate. A competing product is SPSS/PC+ Data Entry II. On data entry, see Gilliland and Kinchen (1987).
- EX-SAMPLE is an expert system to determine sample size. It determines the minimum sample size required for dozens of common statistical analyses and compares it to the maximum possible size given resources of time, money, and personnel. In addition, EX-SAMPLE also helps the user decide which types of analysis to perform. The cost is $195.
- Measurement and Scaling Strategist is an expert system to assist in the operationalization of concepts in the item-construction phase of survey research. After answering a series of questions posed by the computer, the package generates recommendations regarding measurement scales and, if requested, the reasons for these recommendations. The cost is $99.95.
- *Topics in Research Methods: Survey Sampling* is more narrowly focused training software, illustrating the relation of sample estimates to error rate, showing that the precision of an estimate varies as a function of sample size in both simple and stratified random samples. The cost is $50.
- *Survey Research: A Computer-Assisted Introduction* is a college-level introductory course on survey research, built around IBM software that allows students to analyze data from a 40-item questionnaire completed by 200 college students and 200 corporation executives, to design and administer one's own surveys (up to 48 items, 200 respondents), simulate local and national polling processes, and explore crosstabs, recoding, scaling, and reliability.
- Word Match is a content-analysis package that can be used to analyze open-ended survey items, such as those generated by Survey I or contained in dBASE memo fields. It outputs case numbers in order of highest frequency of finds from user search lists, with found words and their counts, plus counts for each file and for entire search. It supports multifile searches, search phrases, and root words. It includes the dBASE III interface program. It costs $45. SeekEasy, a user-supported (shareware) program is an alternative (the cost is $10 plus a $30 registration).
- On the Campaign Trail illustrates an entirely different type of computer assistance, namely training in the interpretation of surveys. This package simulates in considerable detail a fictional Senate race, in which teams of students analyze survey data to make detailed decisions about appropriate media buys. Authored by Murray Fishel, David Gopoian, and J. Michael Stacey and published in 1987 by a campaign consulting firm, it is available for $19.95.

Endnotes

[1] We will not discuss a number of additional potential errors in questionnaire design due to space limitations. Some of the more important omissions include avoiding double-barreled questions, avoiding double negatives, biased terminology like "socialized medicine" or "bureaucrats," measuring the middle position, balance and imbalance in question design, issue intensity, centrality, crystallization, specifics of formatting mail questionnaires, and DK filters. All of these issues are richly elaborated in the references provided.

[2] Where not otherwise mentioned, software discussed in this article is for MS-DOS machines such as IBM microcomputers.

Chapter XV

Geographic Information System Applications in the Public Sector

Douglas A. Carr, Oakland University, USA

T. R. Carr, Southern Illinois University Edwardsville, USA

Abstract

Geographic information systems emerged in the 1970s and have become a significant decision-making tool as their capabilities have been enhanced. This chapter discusses various GIS applications and highlights issues that public managers should consider when evaluating implementation of a geographic information system. GIS applications provide benefits at the planning level by producing maps efficiently, and at the management decision-making level through an ability to geographically display important information for policy-level decisions. While GIS analysis can be a powerful tool, there are a number of issues that pubic managers should consider in order to achieve effective implementation and use of geographic information systems.

Geographic Information Systems

The defining feature of geographic information systems, or GIS, is the ability to integrate a variety of information, such as economic, infrastructure, political, or demographic data, with the geographic locations associated with the data of interest. GIS applications have the ability to access large databases and create maps of almost any combination of data. For example, a map may include locations of blighted or substandard housing, attendance zones for elementary schools, land-use patterns, household income levels, and even consumer spending patterns for individuals within households.

The term geographic information system emerged in the 1970s when it was used to describe a variety of techniques employed to create maps as an aid in data analysis. This application was an outgrowth of the development of various tools such as computer-aided mapping (CAM) and computer-aided design (CAD) systems used primarily by cartographers, draftsmen, and engineers to produce detailed and accurate maps. Through the application of CAM and CAD programs, very precise maps could be drawn and updated quickly to reflect changes in infrastructure, political boundaries, and topography. Cartographers found these new techniques to be an especially efficient addition to their craft. As the use and availability of these techniques increased, other disciplines found new applications for the technology. Public agencies and private-sector organizations discovered that these applications provided the foundation for spatial analysis of geographic data stored in large databases. This application of spatial (geographic) analysis allows policy analysts to present economic, demographic, and other data in map form, which enhances their ability to understand and communicate complex relationships (Huxhold, 1991).

GIS involves a combination of computer hardware, specialized software applications, trained personnel, and the creation and maintenance of a geographically coded database. Implementation of a GIS requires a managerial commitment to data-driven decision making as an aid in reducing uncertainty; using maps to display data combined with statistical analysis allows managers to visualize important geographic patterns in the data. The integration of a wide variety of social, economic, physical-resource, landform, and other data creates the potential to improve the quality of public-sector decision making (O'Looney, 2000).

The continuous development of GIS technologies, such as advances in mapping software, database systems, satellite remote sensing, and global positioning system (GPS) devices, creates dynamic parameters for spatial analysis. The result is that GIS continues to evolve and has an ever-expanding ability to incorporate a wide range of analytic techniques that were separate and distinct in the past. This means that the public administrator is faced with financial and managerial challenges associated with an increasingly useful technology that will continue to consume resources as it evolves. The decision to acquire GIS technology is a two-edged sword. It facilitates information flow, but it also requires a significant commitment of resources in future years if the system is to be maintained and used to its potential. For example, the utility of GIS for decision making is directly related to the quality of information contained in the database. Inaccurate or obsolete information in the database will produce flawed maps with potentially adverse effects when used in decision making. This means that geographic information systems involve a significant long-term investment not only in hardware, software, and personnel training, but also in continuous database development, expansion, and maintenance.

The next section of this chapter uses examples to discuss some potential applications of GIS in the public sector. The following section then addresses technical issues that managers need to understand when implementing a geographic information system. The third section discusses a number of important issues public managers should understand when considering implementing a GIS. The concluding section reflects on issues public managers should consider when evaluating and implementing GIS analysis.

GIS Applications

The literature concerning GIS reflects an expanding array of public-sector applications. This is due to the fact that geography (physical, social, and political) has a profound impact on the activities of government agencies. Diverse activities such as bus transportation routes, school attendance zones, election precincts, location of police and fire stations, economic development planning, and infrastructure construction and maintenance have all been performed with the aid of geographic information systems. The following examples illustrate the utility, adaptability, and flexibility of this technology for public managers.

Emergency Services and Disaster Response

GIS provides municipalities with a useful tool to monitor factors affecting public safety and to advise citizens of potential threats in a timely manner. Minimizing dispatch and arrival time is a crucial element for emergency services such as ambulance, fire, and police. GIS has proven to be an important tool for public agencies in improving response times (Mitchell, 1997). Such a system typically includes a database consisting of several different files containing telephone numbers and street addresses, property lot size and location, street and road networks, and locations of emergency service providers such as hospitals and police and fire stations. With a GIS system in place, the emergency dispatcher can identify the exact location needing service, produce a map with the shortest distance for the emergency response team, and provide that information while the team is in route.

In a different application, Pasadena Water and Power (California) utilized its GIS database to provide advance warnings to residents concerning potential power blackouts during a period of energy supply crisis in 2001. Individuals with health risks benefited from the opportunity to make alternative plans and corporations benefited from the opportunity to manage workforce utilization and maintain productivity due to the advance notifications.

GIS also has the potential to improve responses to local and regional disasters. Following the devastation of Hurricane Katrina in 2005, the U.S. Army Corps of Engineers provided the Federal Emergency Management Agency (FEMA) with GIS maps as an aid in assessing damage, staging equipment for rescue and recovery, and implementing cleanup and rebuilding plans following the natural disaster (Castagna, 2005). International responses to humanitarian emergencies in areas as diverse as Kosovo, Africa, and Afghanistan have also been improved through the use of GIS technology (Kaiser, Spiegel, Henderson, & Gerber, 2003).

Economic Development, Land Use, and Infrastructure Planning

Urban sprawl presents a challenge to local governments in planning the delivery of public services and the impact of development on the environment. GIS models provide a tool for local governments to understand the pattern of urban growth and its consequences (Sostek, 2001). GIS technology provides local governments and regional entities greater understanding of the impact of factors such as employment location, housing types, and housing values on commuting patterns of citizens (Horner, 2004).

Figures 1 and 2 are examples of applying GIS analysis to these issues in the St. Louis, Missouri, metropolitan area. Figure 1 displays commute times and major roads. Downtown St. Louis is on the Missouri-Illinois border, which is on the eastern (right) edge of the map. This figure shows an east-west band of shorter commute times through the middle of the St.

Figure 1. Median commute times and major roadways in the St. Louis, Missouri, metropolitan area (Source: Median commute data is from the 2000 Census Summary File 3)

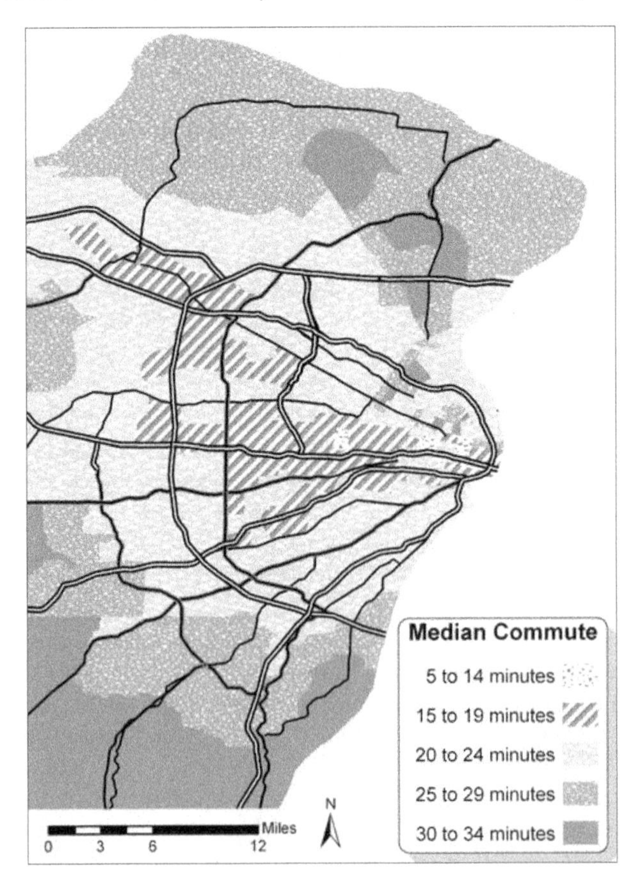

Figure 2. Median house values and major roadways in the St. Louis, Missouri, metropolitan area (Source: Median commute data is from the 2000 Census Summary File 3)

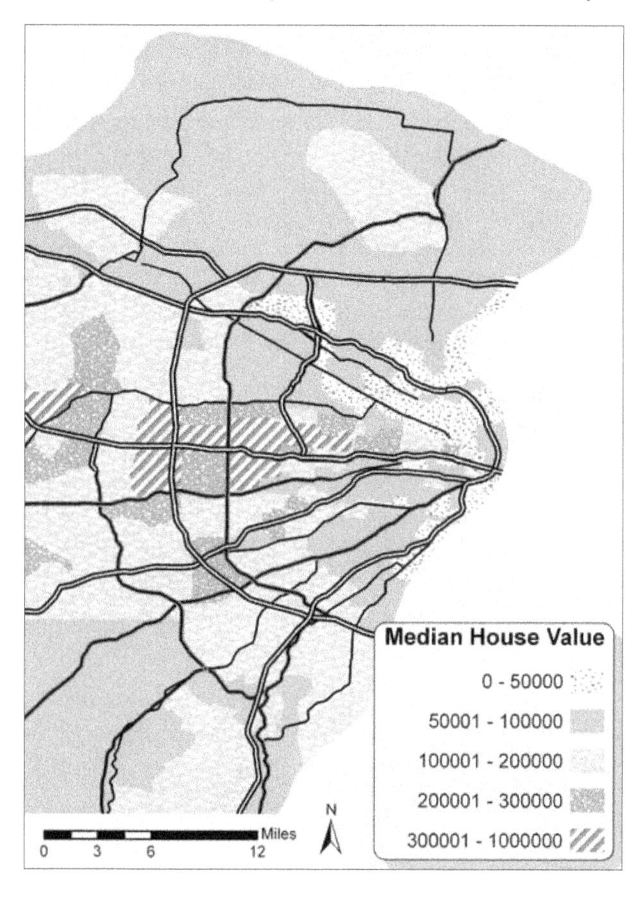

Louis area, with longer commutes for individuals who live north or south of this band. This map shows how the current transportation network encourages westward sprawl by allowing shorter commute times in this east-west band. The results of westward sprawl can also be seen in Figure 2, which shows the same geographic area shaded according to median house values. Desirable areas, indicated by high median house values, are found in an east-west corridor in the middle of the St. Louis area. These observations made from GIS analysis have many applications, including economic development. For example, redevelopment efforts in the areas with low housing values just north of downtown would have to address the factors leading to westward sprawl. Public managers working on economic development could make these important observations through GIS analysis.

The use of GIS provides decision makers with high-quality information for making decisions relating to land use and zoning (Aronoff, 1988). Typically, data files containing such information as geology, topography, property ownership, population density, transportation networks, and commercial activity are linked to produce a variety of maps for analytical purposes. This spatial analysis enhances the ability of decision makers to visualize the potential impact of zoning and other land-use decisions. Another important application of GIS for local governments involves the ability to monitor and track property values and neighborhood characteristics that are indicators of potential social and economic distress, providing an opportunity to develop preventative strategies (Heitgert, 2001; Wade, 2001).

Figure 3 offers an example of applying GIS analysis to local zoning and transportation. This figure shows industrial zoning and major road and rail networks in Boston, Massachusetts. This map could be used to assess how the current transportation network serves the current industrial areas and could help identify regional needs for transportation infrastructure improvements. This map could also help local officials make rezoning decisions. For example, if it was proposed to rezone an area in the city for an industrial land use, this map would help local officials understand the long-term viability of the proposed industrial area by highlighting how it is served by the existing transportation network.

Figure 3. Industrial zoning and major transportation networks in Boston (Source: Zoning data is from the Office of Geographic and Environmental Information [MassGIS], Commonwealth of Massachusetts Executive Office of Environmental Affairs)

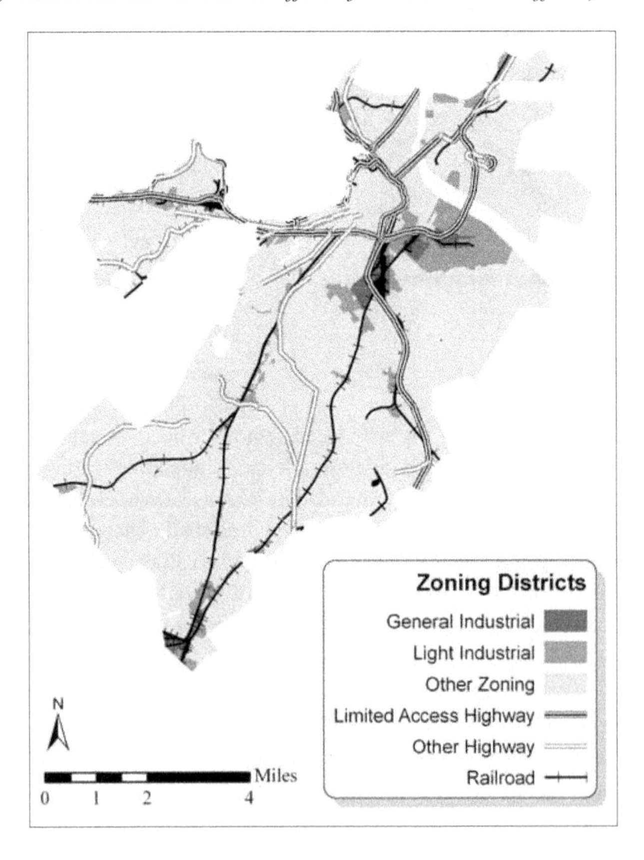

The three maps in this section were created from census data and state data in the public domain. Other readily available data would have allowed us to draw a number of maps indicating such factors as family size, marital status, occupation and employment status, income level, crime statistics, health care facilities, brownfield locations, and voting turnout and behavior. The variety of maps is limited only by the types of databases accessed by the GIS specialist.

GIS has also been used as a tool for increasing the effectiveness and efficiency of property assessment by appraisers (Guttormsen & Guttormsen, 2003) and as a debt management tool for infrastructure planning (Hokanson, 1994). Through an extensive database of current and proposed infrastructure and other capital improvement projects containing items such as cost, projected timelines, types of improvement or construction, location, and funding mechanisms, decision makers can access multiyear maps and assess the impact of infrastructure expansion on debt. With the inclusion of property assessment data, decision makers can evaluate the potential impact of infrastructure projects on tax revenues.

Health Care Planning

Geographic information systems have demonstrated utility for health care planners (Birkin, Clarke, & Wilson, 1996; Lang, 2000). *Medical geography* is a term that reflects this application of GIS. Epidemiology, linkages between poverty and disease, and the impact of access to service on utilization are three major areas in which GIS supports health care planning activities. The utility of GIS as a planning and management tool for local health departments to address epidemiologic issues has been clearly demonstrated. GIS mapping was used to strengthen the ability of local rural health departments in North Carolina to analyze the impact of septic tanks on water quality and disease distribution. These departments have also used GIS analysis to assess the distribution and incidence of diseases such as the West Nile virus and encephalitis, and to develop appropriate strategies for vector control related to other diseases such as rabies (Miranda et al., 2005). The relevance of GIS technology for mitigating any future potential pandemic such as avian flu has been clearly demonstrated by past experiences. Geographic patterns reflecting demand for services, costs associated with service delivery, and allocation of financial and personnel resources are important elements in GIS applications in health care planning (Lubenow, 2001).

Political Campaigns and Reapportionment

Novotny and Jacobs (1997) document the increasing importance of GIS in election campaigns at all levels of government. Campaigns at the national and state level are heavily influenced by maps containing demographic and attitudinal data. Campaign strategy, media advertisements, and resource allocation decisions are all heavily influenced by analysis produced by geographic information systems. The trend continues for political campaigns at all levels of government to increasingly utilize this technology. One consequence for public managers at both the local and state level involves expanded requests for public-domain data that will be used for explicitly partisan political purposes.

The ability of GIS to integrate population data with detailed maps also allows decision makers the ability to examine a variety of plans for revising political boundaries mandated by population changes. GIS technology allows practitioners to create a variety of redistricting plans in an efficient manner for consideration by elected officials. While the technology facilitates the development of numerous potential political boundaries, the ultimate decision will be made in the context of political realities.

National Defense Applications

The continued improvement of GIS technology with remote sensing capabilities has had a significant impact on national defense and military applications (Fandeyev, 2004; Shroder, 2005). The combination of photographic and other remote sensing imagery with geologic databases has resulted in the creation of detailed maps of potential enclaves of suspected Al Qaeda operatives for counterterrorism operations. Public expectations of minimal impact on noncombatant civilians arising from counterterrorism military operations has generated extensive application of GIS technology, significantly improving the accuracy of artillery, missiles, and bombs dropped from aircraft. One result of this is the ability of American and coalition forces to achieve target accuracy in the 1- to 2-meter range. This enhances the capability to minimize or eliminate noncombatant casualties.

Private-Sector Applications

GIS has been used extensively by the private sector to increase operating effectiveness and efficiency. Marketing research applications allow business enterprises to map customer behaviors such as shopping frequency, residence, household income, and driving distances (Pack, 1997; Ross, 1998). One such private-sector company, Buxton, employs extensive data on individual consumer income, spending, driving, and special interests that are labeled collectively as psychographics with a detailed GIS mapping system. The resulting analysis is then marketed to private corporations as an aid in determining retail outlets and to municipal governments as a tool to recruit retailers for their city. This type of information enhances marketing strategies, advertising campaigns, and planning for future growth and expansion (Harder, 1997). Some examples of other private-sector applications include the realignment of sales territories based on demand, employee workload and competition, real-estate acquisition and management, transportation routing, and the use of interactive maps on the World Wide Web for marketing purposes.

Continued GIS Development

Both public- and private-sector applications of GIS continue to expand due to the utility of this management tool (Green, 2000). Advances in computer technology in terms of processing speed and data access, combined with improvements in GIS software, create an environment in which more and more governmental units are able to consider the adoption of geographic information systems as aids in their decision-making processes. When reviewing the literature surrounding GIS, numerous examples of successful applications are documented, while examples of unsuccessful implementation efforts are somewhat rare (Ventura, 1995). However, there is some indication in the literature that GIS has not completely lived up to its advertised potential, and in the words of some critics, "GIS is an unfortunate diversion in our journey towards truth" (Worral & Bond, 1997; Pickles, 1995). Later in the chapter, attention will be given to issues public managers should consider when faced with the tasks associated with developing, maintaining, and expanding a GIS; addressing these issues will help public managers implement an effective geographic information system.

GIS Basics for Public Managers

Public administrators tend to be consumers of GIS analysis rather than actual analysts or experts. In this section we will examine some of the technical issues public managers need to understand when implementing GIS analysis.

Level of Measurement

Just as statistical analysis systems are governed by level of measurement, so are GIS systems. The level of data governs the manner in which variables from a database can be displayed in any map that may be produced. Nominal-level data in a map might include the location of a town, airport, road, river, or political boundary. Nominal data elements can be identified, but comparisons cannot be made indicating that one element is "more than" or "better than" any other element on the map. Ordinal-level data in a map might include size of the town, type of airport (local, national, or international), type of road (interstate highway, main traffic artery, or residential street), size of the river (stream or navigable river), median personal income or assessed property valuation in each county, or the number of hospital beds in a health care facility. With ordinal-level data, comparisons can be made indicating that one element is "more than" or "better than" any other element on the map. Ratio-level data in a map might include indicators of the number of aircraft departures per day or the incidence of criminal activity. With ordinal- and ratio-level data, relatively precise comparisons can be made between various areas on the map. For example, the incidence of crime might be twice as high in one neighborhood as compared to another, or the value of property in one neighborhood might be half that of another neighborhood in the city.

The types of questions that can be answered by GIS analysis are governed by the level of data that have been collected and stored in the supporting databases. Public managers will find that the utility of GIS analysis is increased when careful planning guides the types of data collected for future use in graphic analysis.

Sampling

When GIS analysis is used in the decision-making process, the nature and quality of information in the database has a direct impact on the accuracy of the analysis. Databases used by GIS are spatial in nature. A spatial database records the geographic locations associated with the data. GIS databases may contain spatial information on physical traits, such as landforms, or socioeconomic traits, such as income level or assessed valuation within defined geographic areas. Given the costs associated with obtaining data on every element of the population, some databases tend to rely on data drawn from samples. This means that in the creation of graphic displays, areas that were not included in the sample are assigned estimated values through the use of either interpolation or extrapolation methodologies.

Using appropriate sampling methods in creating a database is critical for producing accurate GIS maps. An unstated but underlying assumption in GIS analysis is that the information in the database accurately represents the characteristics of the area as a whole. If the database contains sample data, accuracy is best achieved when a random sampling technique is used. Reliance on databases that have been constructed based on convenience or some other nonrandom selection technique introduces a potentially serious threat to the validity of any graphic analysis that is produced. As consumers of GIS analysis, public managers can minimize the risk of faulty analysis by understanding the sampling techniques used in the construction of supporting databases.

Map Detail and Construction

Public managers responsible for implementing GIS systems should avoid common pitfalls as identified by Kent and Klosterman (2000). A common mistake occurs when public managers ask GIS professionals to "zoom in" on a particular section of a map, such as a particular neighborhood, to increase detail. Map detail is a function of the database used to create the map. If the database used to create a citywide map was not constructed using neighborhood data, zooming in on a neighborhood will only change the map scale, not increase details represented in the map.

Public managers, as consumers of maps produced by GIS experts, should ask a series of questions about any GIS-produced graphic. Do the maps contain sufficient information relevant to the policy question under study? Impressive maps can be produced that may contain incomplete or irrelevant data. Do the maps contain excessive information that hinders utility as a decision-making tool? GIS has the capability to store and access many layers of information for graphic display. Managers should specify the types of information they desire in advance in order to receive a map displaying appropriate data. In order to avoid clutter, managers should consider requesting multiple maps, each containing different information. Managers should also ask if the GIS maps are missing any necessary elements, such as the map scale, a legend that identifies all map symbols, and an indicator of the sources of data used to produce the map.

Public managers will find GIS analysis to be most effective when they approach it with a critical perspective. By asking questions relating to the purpose and content of the map, this technology will have greater value in the decision-making process.

Issues for Public Managers

The implementation and use of GIS analysis in the decision-making process is influenced by a variety of technical, legal, ethical, and managerial issues (Johnson, 1996; Somers, 1998; Worral, 1994a). Public managers who are considering implementing GIS analysis should consider these issues when making decisions about the use of GIS.

Needs Assessment

An adequate needs assessment process is a time- and resource-consuming activity that can provide the foundation for the implementation of an effective geographic information system. Information concerning system requirements for hardware and software, database requirements, human resources, and provisions for future system expansion are identified during this process. An inadequate needs assessment has the potential to create a number of problems at the implementation stage. If primary emphasis is placed on meeting immediate or short-term needs, inadequate hardware or software may be acquired, severely limiting the system's ability to meet future needs.

GIS systems have an expansive appetite for data, and creating and maintaining GIS databases is often an expensive endeavor. During the needs assessment process, managers evaluate the time and cost associated with developing and maintaining an adequate database. Most governmental units find that their existing databases will require significant cleaning, modification, and a commitment to continuous updating if they are to serve as the foundation for a quality geographic information system. Since data acquisition and maintenance activities consume considerable financial and personnel resources, GIS implementation decisions are heavily influenced by resource allocation decisions made during the needs assessment.

Managers considering GIS implementation should recognize that both the literature concerning GIS and GIS vendors tend to report only successful applications and experiences. Both vendors and public agencies find their interests to be better served by minimizing dissemination of information about mistakes, shortfalls, and disappointments with geographic information systems. This means that an agency considering adopting a GIS should spend significant effort identifying potential missteps concerning GIS implementation.

System Requirements and Design

Development and implementation of an effective GIS requires incorporating perspectives drawn from all levels within an agency. In the design process, the technical, legal, managerial, and conceptual perspectives are utilized to create a system capable of providing the type of analysis necessary to support the decision-making process. A preferred sequence of events would flow as follows: GIS analysis objectives govern selection of software, and then software selection governs selection of both hardware and personnel training. An inherent risk that public managers face is that persuasive representatives from a software vendor will influence the purchase of a less-than-adequate or inappropriate GIS software package that will be unable to deliver the desired type of analysis.

Alternatively, financial or other pressures may draw public managers to the acquisition of a non-GIS program, such as computer-aided drafting or computer-aided mapping program. While these types of programs can provide very accurate graphs and maps, they do not have the ability to support spatial analysis that is integral to a true geographic information system.

System design also incorporates human-resource requirements. Effective GIS requires significant maintenance in terms of ongoing data entry, data cleaning and verification, and data revision. In order to achieve an acceptable return on the investment from a GIS, personnel training and retention costs should be factored into the implementation. Individuals with highly developed GIS skills tend to be very marketable and capable of demanding competitive salary packages. One aspect of the system design involves a strategy to retain skilled employees that have been trained by the agency. Failure to adequately address this personnel issue has the potential to create the need for an ongoing training program for new replacement employees.

Organizational Changes

A geographic information system must be integrated into the operations and functions of the organization if it is to serve its intended purposes. It is at this phase that the implications concerning the impact of the system on agency procedures and functions can be assessed. Effective implementation planning incorporates a top-down view of the design and implementation process (Somers, 1998). The long-term impact of the system on routine functions, resource allocation, and the agency mission should be clearly identified. Because implementation of any system may produce unintended and unanticipated impacts on agency operations and resource consumption patterns, the necessary commitments to accept changes in procedures and to allocate support to the system should be obtained from departments throughout the organization to ensure effective GIS implementation.

Many of the elements associated with successful GIS implementation can be traced to the human factor; employee attitudes and values are very important (Nedovic-Budic & Godschalk, 1996; Ventura, 1995). Bureaucratic resistance to new procedures and techniques can be attributed to a variety of elements. These include fear of change, resistance to requirements to learn potentially difficult skills, concerns about procedures growing out of administrative control and authority, fear that new operations and techniques may expose weakness and inadequacies in current operations, and concern about budgetary impacts of new projects on personnel positions.

Accordingly, effective GIS implementation strategies can be enhanced by providing some combination of the following to employees: an emphasis on personal benefits that can be derived from the change; selecting or incorporating individuals with strong computer skills for the implementation; maximizing employee exposure to GIS operations before, during, and after implementation; and providing reassurances concerning potential impacts on personnel assignments, duties, and employment status.

The implementation of a GIS has the potential to impact the existing external relationships of an agency. Existing intergovernmental relationships may be changed or altered due to changes in the flow of information. These changes may be either vertical in nature (between local, state, and federal bodies) or horizontal in nature (between departments with the same governmental unit). As stated in the previous section, fear of change has the potential to be a significant factor in resisting the implementation of innovative technologies. Effective implementation strategies require public managers to examine potential changes in external and internal relationships.

Communication patterns within departments may be significantly altered with the implementation of a geographic information system. Such change is the product of data being shared with diverse departments within the agency and potentially to external entities as well. As communication patterns change, informal power structures are created, eliminated, or changed. Changes in the flow of information have potentially significant impacts on the institutional environment.

Organizational Politics

Implementation of a geographic information system typically involves building an internal and sometimes external political base (Budic, 1994). Securing the support of upper level management and elected officials is typically a prerequisite in the development of a GIS. Political support ranges form the allocation of budgetary resources necessary for funding to organizational and institutional support associated with changes in the flow of information within and from the agency. Political support is necessary to overcome hesitancy in sharing information between departments and with external consumers, whether they are governmental or are in the private sector.

An emerging issue relates to the power and influence of GIS managers as public agencies increase their usage of and reliance on GIS (Romeo, 2004). GIS is an important decision-making and information communication tool both within organizations and to external groups and organizations. Since communication is an important source of power within organizations, the power and influence of GIS managers within organizations also tends to increase. Internally, GIS managers gain influence on budget and resource allocation decisions to fund their role in the decision-making process. Externally, GIS managers have the potential to build a significant informal power base as they exercise their role of responding to information requests from external clients and other groups.

GIS professional certification is another issue for public managers to consider (Kemp, 2003). As GIS matures as a profession, certification may become a reality. Proponents of certification hold that the knowledge and responsibility of GIS managers (especially in the context of data management and privacy issues) require the creation of professional certification. At the same time, the facts that the profession remains somewhat undefined and that the skills and knowledge base are also being refined and expanded hinder certification. Some universities do provide certificates of completion for a GIS curriculum. Professional certification for GIS managers remains an unresolved issue and any impact on governmental budgets and operations remains to be determined.

Even though GIS represents a potentially valuable advancement in technology, successful implementation requires that public managers exercise considerable political skill.

GIS and Democracy

GIS affords public managers with the ability to access data with greater speed and to engage in a variety of analytical scenarios concerning possible courses of action. This use of technology may result in an unanticipated threat to democracy. Haque (2001) observed that the ability to control technocratic tools by the bureaucracy has the potential to distance societal understanding of government actions. GIS is a source of expertise and has the potential to shift power to public-sector professionals at the expense of elected public officials. One possible strategy to counter this concentration of power over GIS technology has been described as a bottom-up GIS (Talen, 2000). GIS is becoming easier to use and with the advances in technology is less costly. While GIS has traditionally been viewed as a top-down analytical tool used by administrative and technical professionals, it can also be utilized to increase public participation in the planning process. Some communities currently do this by providing a Web site citizens can use to generate their own GIS maps using a simple interface. Under this approach, citizens are afforded the opportunity to use GIS techniques to present their own understanding relative to their own community. Making this technology available to citizens enhances and preserves participatory government, a democratic value.

Access to Public Records

When GIS databases are created by public agencies, they are typically in the public domain and may be accessible by other public or private agencies. For example, public access to GIS output on planned capital spending for infrastructure development can impact the behaviors of real estate developers and fuel speculative land purchases. Effective project planning activities afford the opportunity to investigate the experiences of other agencies with GIS implementation and to establish strategies for dealing with those consequences.

Policy concerning access to GIS data is largely governed by legislation dealing with access to public records in general (Johnson, 1996). Effective GIS systems are supported by extensive databases that are often created from public records that were gathered for purposes originally unrelated to the objectives and needs of the GIS system. For example, property ownership records, property assessment data, property surveys, and zoning data were traditionally collected and maintained for different purposes, and any linkage between records was incidental. With the implementation of a geographic information system, these files may be computerized and linked to economic and personal spending patterns for spatial analysis purposes. The extent to which an agency has an obligation to make these new databases available to citizens and private groups continues to be an issue of concern.

Some public agencies sell GIS data, allowing the opportunity to recover a portion of the costs associated with data collection and database maintenance activities. Because the data obtained from the public agency may be distributed to third parties, selling GIS data presents legal problems if the data contains nonpublic information. In some instances, contracts and licenses have been used in an effort to regulate such activities, but enforcement of any contract involving public information between a government agency and a citizen is somewhat problematic.

Attempts to achieve a degree of cost recovery for GIS development may also conflict with policies promoting open access to public information. There is little uniformity between governmental units in terms of their own policies. Some agencies are committed to open access while others have moved to implement cost recovery policies and procedures (Johnson, 1996). Considerable variation exists between local governments across the United States. One consequence of this variability in policy is that private-sector consumers will continue to seek open access to GIS data rather than engage in cost-sharing partnerships whenever possible.

Differences in dissemination policy increase the difficulty in sharing data across governmental units. A city or county committed to cost recovery is understandably hesitant to share data with another governmental agency committed to open access. During and after GIS implementation, dissemination issues will remain on the policy agenda of the governmental unit.

Public and Private Partnerships

In an effort to reduce the financial burden on public agencies associated with creating and maintaining GIS databases, partnerships with the private sector have been proposed as a viable mechanism for cost sharing (Skurzynski, 1998). Public agencies have access to an extensive array of data but tend to operate in an environment of limited financial resources and other budgetary constraints. The private sector tends to have access to venture and investment capital and has a desire to access the data collected by public agencies.

Proponents of partnerships between public and private agencies argue that the access to data by the public agency and the access to venture capital by the private sector provide an irresistible solution to the financial burden associated with the implementation of GIS. However, such partnerships are not without risk. Negotiating contracts for such a partnership can be a lengthy and expensive process. Parties may possess a range of unrealistic expectations concerning both the process and the ultimate GIS product from the partnership. Elected and other officials may mistakenly view the income from the partnership as new revenues suitable for any number of uses unrelated to GIS. These revenues require protection and should be allocated to the GIS project for both development and maintenance purposes.

Expectations concerning dissemination and ownership of GIS data may not be realistic. Private-sector organizations may confuse the right to possess and use data with data ownership. Due to legal constraints, much of this data, even that contributed by the private partner, may ultimately become considered to be in the public domain and accessible by their competitors.

Privacy Issues

Geographic information systems do raise significant issues relating to the right to privacy (Dobson, 1998). The threat to individual privacy is created by the nature of the databases used by GIS systems. It is not necessary to gather new information about an individual to create a relatively accurate personal profile for any given individual. For example, property records, tax assessment data, credit-card purchase information, and other demographic data can be combined to describe the behavioral patterns of neighborhoods as well as individuals. The issue of privacy becomes increasingly important when partnerships between public agencies and private-sector companies facilitate the combination of extensive consumer-related (private) and citizen-related (public) records.

While GIS may be viewed as a potential threat to privacy, social surveillance unrelated to GIS continues to expand. Video cameras are widely used for security and law-enforcement purposes, and satellite photography allows detailed monitoring of a range of daily activities by nations and by individuals. Even though other threats to privacy may be largely ignored, there remains considerable concern with the impact of GIS systems on this basic right (Jain, 2003). This has given rise to the concerns about the "socialization" of GIS (McLafferty, 2004). This refers to the expanded utilization of GIS to gather, organize, and distribute information that can provide detailed information about the background, attitude, behaviors, and preferences of individuals. One result of this capability is that what was once considered to be confidential information about an individual or a family may now become an integral component of a larger database potentially accessible to a wide range of public and private users. The significant variation in privacy laws between the states has created an environment in which safeguards for protecting individual privacy may be porous or nonexistent. GIS is therefore a technology that potentially impacts broader social and economic relationships within society.

Liability Issues

Public agencies may be required to assume some responsibility for the accuracy and quality of the data provided to secondary users. This issue of liability is increased when an agency vends the data to others even on a cost-recovery basis. Attempts to restrict assess to GIS data to qualified users may, in turn, raise additional liability issues. Given the litigious nature of society today, a high probability exists that public agencies will be held liable for erroneous decisions made by consumers of GIS data if those decisions were partly the product of inaccurate data.

Evaluating and Implementing GIS in the Public Sector

Proponents of GIS implementation argue that the technology affords public managers with the capacity to "reinvent local government" because it enables the spatial analysis of data (Wilson, 1995). While GIS implementation can enhance communication and decision making, it is also a costly endeavor. Ongoing GIS analysis requires qualified personnel, specialized computer software, and ongoing database development and management. Public managers interested in implementing a geographic information system should carefully consider both the benefits of GIS and the cost of its implementation and maintenance.

While it is important to understand both the costs and benefits of GIS, public managers should be aware that measuring benefits to justify the costs associated with GIS applications for local governments can be a difficult and daunting task (Worrall, 1994b). Costs associated with GIS implementation tend to be highest during development and then continue at a reduced level over the life of the system. These costs are relatively easy to assess and document. Benefits, on the other hand, tend to be more difficult to quantify and typically are produced over the life of the system after implementation. Managers should enter into the process with an understanding that hard data justifying implementation may be somewhat elusive. GIS evaluation efforts should be conducted with sensitivity to benefits that can be measured, such as cost savings that are associated with records maintenance or produced by automation of previously labor-intensive mapping procedures, productivity gains associated with the adoption of new procedures, improvements in the development or enforcement of regulations, and savings or income generation through enhanced management of infrastructure resources. Specific procedures can be established at the point of implementation to begin the process of documenting benefits.

The expectations of various participants in a GIS project tend to vary, and these differences in expectations impact the ultimate evaluation of the usefulness of GIS in decision making (Somers, 2001). Users and public managers may have unrealistic expectations concerning costs, the time frame for project completion, and a lack of appreciation for the complexities associated with creating a database to support a GIS project. It is important not to make an uncritical assumption that sufficient time and resources are available to complete a particular project within an arbitrary time frame. GIS managers may also unrealistically expect both elected officials and public managers to have the same goals and to be equally satisfied with the same product. This means that addressing in advance the expectations of those involved is an important step in an effective GIS project. This can be achieved when a GIS project is planned in advance to consider administrative, technical, and political factors.

Public managers will increase the probability of obtaining satisfactory GIS results when they accomplish the following:

- Work with elected officials and GIS professionals to develop a GIS strategy appropriate for the issue or policy area under investigation. The expectations of various actors in the project should be addressed at this point.
- Create a plan with specified milestones to measure progress. The absence of such evaluation will increase the probability that the GIS project will not meet expectations.
- Provide sufficient management support to the GIS team. In the absence of sufficient resources or guidance, the project may fall behind schedule or drift from the original goals that were established when the project was initiated.
- Adopt a flexible attitude toward the project. Expectations may change during the life of the project, and the organizational and political climate may change as well. Public managers should be prepared to adapt to a range of unanticipated changes throughout the life of the project.

With these cautionary notes in mind, there is little doubt that geographic information systems do have significant potential to improve the operations of governmental agencies. The literature and promotional data contain a seemingly endless record of GIS success stories, and this trend is expected to continue.

References

Aronoff, S. (1989). *Geographic information systems: A management perspective.* Ottawa, Ontario, Canada: WDL Publications.

Birkin, M., Clarke, G., Clarke, M., & Wilson, A. (1996). *Intelligent GIS: Location decisions and strategic planning.* New York: John Wiley & Sons, Inc.

Budic, Z. D. (1994). Effectiveness of geographic information systems in local planning. *Journal of the American Planning Association, 60*, 244-254.

Castagna, J. (2005). A picture is worth a thousand lives. *Engineer, 35*(4), 24-25.

Dobson, J. (1998). Is GIS a privacy threat? *GIS World, 11*, 20-21.

Fandeyev, A. G. (2004). Using GIS in support of missile forces and artillery. *Military Thought, 14*(2), 108-110.

Green, R. W. (2000). *GIS in public policy.* Redlands, CA: Environmental Systems Research Institute.

Guttormsen, S., & Guttormsen, H. (2003). GIS, GPS, county uses, concerns, and considerations. *Assessment Journal, 10*(1), 15-23.

Haque, A. (2001). GIS, public service, and the issue of democratic governance. *Public Administration Review, 61*, 259-265.

Harder, C. (1997). *GIS means business.* Redlands, CA: Environmental Systems Research Institute.

Heitgert, J. (2001). Using GIS and demographics to characterize communities at risk: A model for ATSDR. *Journal of Environmental Health, 64*(5), 21-23.

Hokanson, J. B. (1994). Planning and financing infrastructure using GIS technology. *Government Finance Review*, 19-21.

Horner, M. (2004). Spatial dimensions of urban computing. *The Professional Geographer, 56*(2), 160-173.

Huxhold, W. E. (1991). *An introduction to urban geographic systems.* New York: Oxford University Press.

Jain, D. (2003). A discussion of spatial data privacy issues and approaches to building privacy protection in geographic information systems. *Assessment Journal, 10*(1), 5-13.

Johnson, J. P. (1996). Case studies of dissemination policy in local government GIS agencies. *Computers, Environment, and Urban Systems, 19*, 373-389.

Kaiser, R., Spiegel, P., Henderson, A., & Gerber, M. (2003). The application of geographic information systems and Global Positioning Systems in humanitarian emergencies: Lessons learned, programme implications and future research. *Disasters, 27*(2), 127-140.

Kemp, K. (2003). Why GIS professional certification matters to all of us. *Transactions in GIS, 7*(2), 159-163.

Kent, R., & Klosterman, R. (2000). GIS and mapping: Pitfalls for planners. *Journal of the American Planning Association, 66*(2), 189-198.

Lang, L. (2000). *GIS for heath organizations.* Redlands, CA: Environmental Systems Research Institute, Inc.

Lubenow, A. (2001). GIS technology helps pinpoint patients. *Health Management Technology, 22*(1), 54-55.

McLafferty, S. (2004). The socialization of GIS. *Cartographica, 39*(2), 51-53.

Miranda, M., Silva, J., Galeano, A., Brown, J., Campbell, D., Coley, E., et al. (2005). Building geographic information system capacity on local health departments: Lessons from a North Carolina project. *American Journal of Public Health, 95*(2), 2180-2185.

Mitchell, A. (1997). Zeroing in: *Geographic information systems at work in the community.* Redlands, CA: Environmental Systems Research Institute.

Nedovic-Budic, Z., & Godschalk, D. R. (1996). Human factors in adoption of geographic information systems: A local government case study. *Public Administration Review, 56*, 554-567.

Novotny, P., & Jacobs, R. H. (1997). Geographical information systems and the new landscape of political technologies. *Social Science Computer Review, 15*, 264-285.

O'Looney, J. (2000). *Beyond maps GIS and decision-making in local government.* Redlands, CA: Environmental Systems Research Institute, Inc.

Pack, T. (1997). Mapping a path to success. *Database, 20*, 31-35.

Pasadena uses GIS to warn residents of blackouts. (2001). *American City and County, 54*, 15, 54.

Pickles, J. (1995). *Ground truth: The social implications of geographic information systems.* New York: Guildford Press.

Romeo, J. (2004). View at the top. *American City & Country, 119*(11), 36-41.

Ross, J. R. (1998). Geography lessons: Why you should teach customers about GIS. *Reseller Management, 21*, 84-88.

Shroder, J. (2005). Remote sensing and GIS as counterterrorism tools in the Afghanistan war: Reality, plus the results of media hyperbole. *The Professional Geographer, 57*(4), 592-597.

Skurzynski, J. (1998). Public/private partnerships hurdle data costs. *GIS World, 11*, 12-18.

Somers, R. (1998). Building your GIS from the ground up. *American City and County, 113*, 14-41.

Somers, R. (2001). Measuring success, managing expectations. *Public Management, 83*(3), 18-21.

Sostek, A. (2001). Bringing sprawl to life. *Governing, 15*(3), 30-32.

Talen, E. (2000). Bottom-up GIS: A new tool for individual and group expression in participatory planning. *Journal of the American Planning Association, 66*, 279-294.

U.S. Census Bureau. (2000). *Census 2000, summary file 3 (SF 3).*

Ventura, S. J. (1995). The use of geographic systems in local government. *Public Administration Review, 55*, 461-471.

Wade, B. (2001). Citywide GIS will warn of distressed properties. *American City and County, 116*(14), 35.

Wilson, J. P. (1995). Reinventing local government with GIS. *Public Works, 126*, 38-39, 83.

Worrall, L. (1994a). Incorporating GIS into strategic management in local government. *Local Government Policy Making, 21*, 15-24.

Worrall, L. (1994b). Justifying investment in GIS: A local government perspective. *International Journal of Geographic Information Systems, 8*, 545-565.

Worrall, L., & Bond, D. (1997). Geographical information systems, spatial analysis and public policy: The British experience. *International Statistical Review, 65*, 365-379.

Chapter XVI

You Have Mail, but Who is Reading It? Issues of E-Mail in the Public Workplace

Charles L. Prysby, University of North Carolina at Greensboro, USA

Nicole D. Prysby, Attorney at Law, USA

Abstract

The increasing use of electronic mail in the workplace has generated important legal questions for public organizations. The legal questions concerning e-mail in public institutions and agencies fall into two basic categories: (1) issues of employee privacy regarding e-mail messages, and (2) public access to e-mail under applicable freedom-of-information legislation. While the employer has broad legal grounds for reading workplace e-mail, at least if there is some legitimate business reason for doing so, employees frequently feel that such monitoring is an excessive invasion of their privacy, and the result sometimes is organizational conflict over these privacy issues. These privacy concerns have generated demands for greater protection of employee privacy in this area, and some states have responded with legislation that covers e-mail in the workplace. Government organizations also must treat at least some of their e-mail as part of the public record, making it open to public access, but this also can lead to conflict between public administrators, who may feel that much of their e-mail represents thoughts that were not intended for public disclosure, and external groups, such as the press, who feel that all such information belongs in the public domain. State laws vary considerably in terms of how they define the types of e-mail messages that are part of the public record, some being far more inclusive than others. Given the uncertainty and confusion that frequently exist regarding these legal questions, it is essential that public organizations develop and publicize an e-mail policy that both clarifies what privacy expectations employees should have regarding their e-mail and specifies what record-keeping requirements for e-mail should be followed to appropriately retain public records.

Introduction

The increasing reliance on e-mail in the workplace has generated important legal questions, so much so that most experts strongly recommend that organizations adopt explicit policies concerning e-mail use. Public organizations in particular must be concerned about the legal ramifications of e-mail. The legal questions concerning e-mail in public institutions and agencies fall into two basic categories: (1) issues of employee privacy regarding e-mail messages, and (2) public access to e-mail under applicable freedom-of-information legislation. We discuss both of these topics in this chapter, attempting not only to outline current legal thinking in the area, but also to raise questions that public managers and policy makers should consider.

Many of the legal issues surrounding the use of e-mail are direct extensions of principles that apply to other forms of communications. In fact, much of the law that governs e-mail is not legislation that was written explicitly to cover this particular form of communication. Issues of the privacy of employee e-mail messages, for example, are analogous to issues of the privacy of employee phone calls or written correspondence. Similarly, the right of the public to have access to governmental e-mail messages is a direct extension of the right to have access to written documents. Of course, there are questions about exactly how legal principles that were established for older communication technologies should be applied to a new one, but our understanding of this topic is broadened if we appreciate the application of legal principles across communication media.

Privacy Issues

Most employees probably believe that they should have some privacy regarding their e-mail messages at work. These workers undoubtedly believe that it would be highly inappropriate for their supervisors to listen to their personal phone conversations at work or to open and read their personal correspondence, so by extension they may also feel that supervisors should not be reading their e-mail without permission, except in clearly defined legitimate cases. Many employees use their work e-mail system to send personal messages, both internally and externally, or they mix personal and professional items in the same message, in much the same way that both may be mixed together in a phone conversation with a colleague. Employees may believe that they are entitled to privacy in these matters, and the fact that passwords are required to access their computer accounts, and thus their e-mail, may be considered confirmation of this belief (Dixon, 1997; Greengard, 1996).

Regardless of what many employees might believe should be the case, their legal right to privacy is quite limited when it comes to e-mail messages. The possible basis for a right to privacy of e-mail messages from the scrutiny of the employer could come from several sources. First of all, the Fourth Amendment prohibits the government from unreasonable searches and seizures, and this restricts public- (but not private-) sector employers. Second, federal legislation, most notably the Electronic Communications Privacy Act of 1986, provides some protection for communications. Third, many states may have their own constitutional and statutory provisions, which may even go beyond what the U.S. Constitution or federal laws stipulate. Finally, under common law, an individual may assert a tort claim for invasion of privacy. However, the application of the above legal and constitutional principles to workplace e-mail is extremely limited, as we shall see.

The various legal protections of an individual's privacy stem from general societal beliefs that individuals are entitled to privacy. Unwanted intrusion into an individual's personal affairs violates the respect and dignity to which an individual is entitled (Adelman & Kennedy, 1995). The concept of privacy is a complex one that is not easily captured with a simple definition. Doss and Loui (1995) argue that privacy has at least three distinct meanings: confidentiality, anonymity, and solitude. The first two of these are relevant for our topic. Confidentiality refers to the right to keep personal information private. Anonymity refers to the absence of unwanted attention. Individuals frequently send communications (via e-mail, telephone calls, or surface mail) that are for the intended recipient only. When a third party intercepts such a communication without permission, one or both individuals suffer a loss of confidentiality because personal or private information is divulged to others, and both suffer a loss of anonymity as they now are the subject of undesired attention (Doss & Loui). Of course, putting these general philosophical principles into a workable legal framework can be difficult.

Fourth Amendment Protections

The Fourth Amendment protects citizens against unreasonable searches by government officials, a protection that extends to unreasonable searches of public employees by their employers. Since federal, state, and local government employees constitute about one sixth of the labor force, this protection is relevant for a large number of employees. However, the Fourth Amendment provides only limited protection in this situation. First of all, an employee has no valid objection to a search by the employer unless the employee has a reasonable expectation of privacy in the situation, a legal principle that applies not just to e-mail, but to other aspects of the workplace (White, 1997). For example, the employee may or may not have a reasonable expectation of privacy regarding his or her desk drawers or file cabinets. In one case, the U.S. Supreme Court ruled that a public hospital employee did have a reasonable expectation of privacy regarding the desk and file cabinets in his office (*O'Conner v. Ortega*, 1987). The court also said, however, that such expectations would have to be determined on a case-by-case basis, depending on the particular facts of each case (Cole, 1997). The determination of whether a reasonable expectation of privacy exists would depend on a variety of factors, including the context of the search, the uses to which the searched area is put, and the societal expectation of the extent to which the area deserves protection from governmental intrusion (Prysby & Prysby, 1999). For example, several years after the *O'Conner* holding, the Ninth Circuit Court of Appeals found that a civilian employee in the Navy had no reasonable expectation of privacy in the contents of his desk given the high level of security maintained by his employer (*Schowengerdt v. United States*, 1991/1992). In contrast, the Second Circuit Court of Appeals held in 2001 that a New York state employee did have a reasonable expectation of privacy in the content of his workplace computer because the employee worked in a private office with a door and had exclusive use of the desk, filing cabinet, and computer in the office (*Leventhal v. Knapek*, 2001). Similarly, the Fifth Circuit Court of Appeals found that a city fire marshal had a reasonable expectation of privacy in the files on his office computer because he kept his office door locked and installed passwords to limit access to his computer files (*United States v. Slanina*, 2002).

An employee's expectation of privacy may be diminished by an employer's policy stating that it can inspect the employee's computer. For example, the Seventh Circuit Court of Appeals held that when an employer's policy stated that it had a right to inspect employer-owned laptops used by employees, the policy "destroyed any reasonable expectation of privacy" of the employee (*Muick v. Glenayre Electronics*, 2002).

Even if the employee is able to assert a reasonable expectation of privacy, the employer may still have a right to read the employee's e-mail messages, as the Fourth Amendment only prohibits unreasonable searches. The general rule is that a search must be reasonable both at its inception and in its scope. A search is reasonable at its inception if there are reasonable grounds to suspect that it would turn up evidence of work-related misconduct or if it is necessary for a noninvestigatory work-related purpose. A search is reasonable in its scope if the extent of the search is reasonably related to the accomplishment of the objectives of the search (Jenero & Mapes-Riordan, 1992). In order for the search to be reasonable, the employer must also conduct the search in a fashion that is designed to obtain the information without unnecessarily intruding into the individual's privacy (Cole, 1997). The Supreme Court ruled in *O'Conner v. Ortega* (1987) that even though the employee had a reasonable expectation of privacy regarding his file cabinets and desk drawers, the employer nevertheless had a right to search these locations because the search was not unreasonable (Cole; Cozzetto & Pedeliski, 1997). While the *O'Conner* case did not involve e-mail, the principles are clearly extendable. An employer's search of an employee's e-mail files is reasonable if the employer has a necessity or valid business reason to conduct the search (Cole; White, 1997). Such a reason would exist in a variety of circumstances. One situation would be where the employer needs to retrieve information to conduct business, such as if an important message must be obtained from an absent employee's electronic mailbox. Another possibility would be a situation where an employee was suspected of violating organizational policies or otherwise engaging in inappropriate or illegal conduct, and where searching the individual's e-mail might reasonably provide the employer with evidence pertaining to the suspected activity. For example in *United States v. Slanina* (2002), although the employee did have a reasonable expectation of privacy in his computer files, the court analyzed the search using *O'Conner* as a guideline and determined that the search was justified and reasonable, based on the facts of the case.

Federal Legislative Protection

The Electronic Communications Privacy Act (ECPA), enacted by Congress in 1986, attempted to extend the legal restrictions on wiretapping of phone conversations to electronic communications, providing both criminal and civil penalties for illegal interception and disclosure of protected communications (Perritt, 1996). However, the legislative history of the ECPA suggests that Congress did not intend to restrict employers from reading the e-mail of employees (Cole, 1997; Cozzetto & Pedeliski, 1997).

First of all, the ECPA does not prohibit the monitoring of communications in situations where one of the parties consents (Cole, 1997; White, 1997). Thus, if an employer has a policy stating that e-mail will be monitored, which employees are aware of and at least implicitly agreed to, that alone would appear to provide legal justification for reading of employee e-mail. However, there are employment situations where the implied consent might be questioned. For example, in a case involving telephone monitoring (*Watkins v. L. M. Berry & Co.*, 1983), the court ruled that the employee had not given consent for the monitoring of personal calls; the employer could intercept the calls to determine if they were personal or business calls, but once it was determined that they were personal, monitoring was no longer permitted (Cole; Cozzetto & Pedeliski, 1997; White). Still, the absence of clear consent on the part of the employee does not guarantee e-mail privacy, as other provisions of the ECPA may provide employers with broad rights in this area.

The ECPA includes a system provider exemption, which essentially states that the provider of the communications system being used by employees in the ordinary course of business (which almost always will be the employer in the case of e-mail systems) has a right to intercept messages, at least if there is a legitimate business purpose for doing so (Cole, 1997; White, 1997). Moreover, the ECPA also distinguishes between intercepting a communication and accessing a stored communication, providing more latitude for the latter action. This is an important distinction for e-mail, which almost always would be accessed by the employer from stored files rather than during transmission. Title II of the ECPA created the Stored Communications Act (SCA), which states that restrictions on the reading of stored files do not apply to the provider of the electronic communications service (Bramsco, 1997; Cole, 1997; Perritt, 1996). Thus, any employer that owns the computer network on which the employee e-mail messages reside could cite both the general employer-owned-system exemption and the stored-communications provision. The combination of these would appear to confer a very broad right to read stored e-mail, at least as far as federal law is concerned.

Interpretation of the stored-communication provision is evolving, however. For example, in *Fraser v. Nationwide Mutual Insurance Co.* (2001), a federal district court ruled that the ECPA protected stored e-mail messages that had not been read by the recipient just as strongly as e-mail messages in transmission; the lesser level of protection for stored e-mail messages applied only to those that had already been read by the recipient (Starr & Lippner, 2001). The distinction between read and unread stored e-mail messages is one that had not been made in earlier cases. Of course, even if the SCA is interpreted as not applying to some or even all stored e-mail messages, other aspects of the legislation, such as the system-provider provision, may provide the employer with protection for the monitoring of employee e-mail. Furthermore, it should be noted that the application of the ECPA and the SCA to workplace e-mail is complicated by the fact that this legislation was written in 1986, when e-mail was in its infancy, as well as by the dense and confusing aspects of parts of the legislation. Future court cases may well alter current interpretations of this law.

While many observers have noted the limitations of the ECPA in this area, there has been little subsequent action at the federal level to provide privacy protection for employee e-mail. One bill, entitled the Privacy for Consumers and Workers Act (PCWA), was introduced into Congress in the early 1990s (Schnaitman, 1999; Wilborn, 1998). The PCWA attempted to protect employees from secret electronic monitoring in the workplace; it would have allowed employers to monitor employee e-mail, but only after informing employees of the scope and form of the monitoring and the intended use of the information collected in the course of monitoring (Schnaitman). The legislation failed to pass either house of Congress, however. Another attempt, the Notice of Electronic Monitoring Act (NEMA), was introduced into Congress in 2000. The proposed legislation would have required employers to provide their employees with prior notice of intended monitoring of e-mail, including the means and frequency of monitoring, the type of information to be obtained, and the intended use of the obtained information (Watson, 2001). The NEMA was opposed by employer groups and, like the PCWA, languished in Congress.

While the employer may be able to read the employee's e-mail, the ECPA clearly prohibits outside third parties from doing so. The same rule generally applies to state wiretap statutes. A private third party who surreptitiously accessed an e-mail conversation between two individuals would be guilty of criminal conduct, just as if that person tapped the phone conversation between two individuals. However, many experts feel that even in these cases the ECPA provides little real protection against such actions. Computer hackers are capable of breaking into the computer network of an organization and accessing stored files, including stored e-mail, perhaps far more so than people suspect (Behar, 1997). Many individuals undoubtedly feel that such a threat is extremely remote, in large part because they have nothing in their stored e-mail files that would be of interest to an outside individual. But as one expert puts it, a hacker "has to learn how to hack, and will start off by breaking into a computer system that is relatively easy to break into" (Garcia, 1996). Organizations often feel that e-mail messages do not contain the kind of sensitive information that requires elaborate security measures, so e-mail files may be left more vulnerable to external hackers than other computer files. Hackers may often operate without great fear, either because they believe there is little chance that they will be caught, or because they do not believe that they will be severely penalized if apprehended.

The National Labor Relations Act (NLRA) may also provide some protections for employee use of e-mail, at least in situations related to protected activity, although the law in this area is not settled at this time. There have been findings that employee e-mails are protected concerted activity under the NLRA. For example, in *Timekeeping Systems, Inc.* (1997), the National Labor Relations Board (NLRB) found in favor of an employee who was fired for misconduct after he sent an e-mail critical of the employer's vacation policy to other employees. In another case, the NLRB found that an employer's prohibition on use of its e-mail system to distribute union-related material (while allowing its use for a wide variety of other reasons) violated the NLRA (*E.I. DuPont de Nemours & Company and Chemical Workers Assn.*, 1993). There have been decisions that an employer's business-use-only policy with regard to e-mail was acceptable, as well as decisions that a business-use-only e-mail policy constitutes improper interference with employees' rights to discuss unionization (see, e.g., *Guard Publishing Co.*, 2002, where the employer's business-use-only policy for e-mail was acceptable, and *Prudential Insurance Co. of America*, 2002, where the employer's business-use-only policy for e-mail was improper interference with employees' right to organize). The NLRB's Office of the General Counsel Division of Advice has stated that a ban on all nonbusiness use of e-mail on workplace computers is unlawful (*Pratt & Whitney*, 1998). It appears likely that there will continue to be some doubt as to whether employers may completely restrict e-mail use to business purposes only without running afoul of the NLRA, and employers should take this into consideration when drafting an e-mail policy. The NLRA applies only to private-sector employees, although some states have statutes providing the right to unionize to public-sector employees.

State Protections

State and local public employees also can claim privacy rights under applicable state constitutional provisions or state statutes, which also might apply to private-sector employees. Ten states have privacy clauses in their constitutions, and most states have statutes that protect privacy in personal communications (Barsook & Roemer, 1998; Fitzpatrick, 1997; Prysby & Prysby, 1999, 2000; White, 1997). For example, the Florida constitution states that every person "has the right to be let alone and free from governmental intrusion into the person's private life except as otherwise provided herein" (Rodriquez, 1998, p. 1446). Provisions such as this may provide public employees with some protection against employee monitoring of their e-mail. In *Hill v. NCAA* (1994), the California Supreme Court stated that in interpreting the California constitution, a balancing test should be applied in which the privacy interest of the individual should be weighed against governmental interests and public concerns, which is similar to the Fourth Amendment reasonableness test discussed above.

In recent years, some states have added legislation to deal specifically with electronic communications and/or unauthorized access to computer systems (for a state-by-state summary of such legislation, see Perritt, 1996; Rodriquez, 1998). Several states have legislation prohibiting unauthorized access to computer systems or data, which presumably would cover e-mail (for example, see Cal. Penal Code § 502; Conn. Gen. Stat. Ann. § 53a-251; Iowa Code Ann. § 716A.2; S.C. Code Ann. § 16-16-20). However, these statutes are not specifically written for the protection of employees, and it is unclear if they could even be enforced against an employer. The statutes generally refer to the owner of the system as the enforcer of the law; because the employer is almost certain to be the system owner, employees probably would not be able to use these statutes.

Several states have enacted specific employer-employee monitoring statutes. Under Nebraska law, an employer, on the business premises, may intercept, disclose, or use electronic monitoring in the normal course of employment. The employer may not randomly monitor employees unless notice of the policy is given to the employees (Neb. Rev. Stat. § 86-702). Similarly, Colorado employers who maintain an e-mail system are required to adopt a policy on the monitoring of e-mail, including the circumstances under which monitoring may take place (Dawes & Dallas, 1997). State law in Connecticut and Delaware also requires public and private employers to give prior written notice to employees, informing them of the type of monitoring of their e-mail that may take place. The Connecticut and Delaware laws both provide for penalties for violations. Under the Delaware law, a civil penalty of $100 per violation may be imposed; in Connecticut, for the first violation, a civil penalty of $500 may be imposed, with penalties up to $1,000 for a second offense and $3,000 for third and subsequent offenses (Conn. Gen. Stat. Ann § 31-48d; Del. Code Ann. tit. 19, § 705).

Adding statutory protections for employee e-mail privacy has been controversial in many states. For example, the California legislature passed a bill in 1999 that would have prohibited employers from inspecting or reviewing any personal electronic mail without the consent of the employee, but Governor Gray Davis vetoed the bill, stating that it would be an undue regulatory burden on businesses (Watson, 2001). Davis argued that there was likely to be undesirable controversy over whether proper notification was provided to the employee and that employees should understand that computers provided by the employer are for business purposes and may be monitored by the employer, particularly because employers have a legitimate need to monitor e-mail to prevent unauthorized, inappropriate, or even illegal behavior. In 2004, the California legislature passed a bill that would have required clear and conspicuous notice before an employer could monitor the e-mail of an employee, but Governor Arnold Schwarzeneggar vetoed the bill.

In addition to or in lieu of statutory requirements, many states have formulated policies that apply to public employees in the state. For example, the Arkansas state policy recommends that state agencies inform employees that users should not consider e-mail to be private or secure, as state agencies retain the right to monitor e-mail communications, with or without notice, although monitoring will not occur on a regular basis (Arkansas Office of Information Technology, 2006). The California state policy in 2002 included similar language and also included a consent form to be signed by all employees, acknowledging that all e-mail may be monitored, with or without notice (California Department of Information Technology, 2002).

The Virginia state policy applicable to all state agency employees clearly states that no employee should have an expectation of privacy in "any message, file, image or data created, sent, retrieved or received by use of the Commonwealth's equipment and/or access." The policy also specifies that the state has the right to monitor any computer system, including all e-mail sent or received by state employees, and that the monitoring may take place at any time and with or without notice (Virginia Department of Human Resource Management, 2001). The Idaho governor issued an executive order in 2001 that allowed state employees to use e-mail at work for occasional personal use, but with the stipulation that all e-mail messages may be subject to monitoring (Kempthorne, 2001).

Tort Law Protections

Employees, public or private, may be able to assert a violation of privacy under common law if their e-mail is monitored by their employer. Tort law varies from state to state, but in most states, it is now recognized that violation-of-privacy torts do apply to employment situations (Fitzpatrick, 1997). Invasion of privacy is a tort with four possible causes of action, two of which might apply to e-mail. The first would be an intrusion by a method objectionable to a reasonable person into affairs that an individual has a right to keep private; the second would be a public disclosure of private matters, where the disclosure is highly offensive and not of public concern (Perritt, 1996; Watson, 2001).

The reasonable expectation of privacy is an essential element in common-law tort claims, just as it is in cases asserting constitutional (federal or state) protections. An employee cannot claim that his or her privacy has been invaded if there was no reasonable expectation of privacy in the employment situation. With regard to e-mail at work, privacy expectations have been very narrowly defined by the courts. First of all, if the employer has a written statement that e-mail messages may be monitored, which many employers do have, the employee will find it very difficult to claim an expectation of privacy. Even in the absence of a written statement, it may still be difficult for an employee to assert a reasonable expectation. In fact, in one case involving a private company, *Smyth v. Pillsbury Co.* (1996), the court ruled against an employee who sued his employer for wrongful discharge following his firing, which was based on information gathered by the employer through reading his e-mail. The court decided that the employee had no reasonable expectation of privacy despite the fact that the employer had assured employees that their e-mail messages were considered confidential items (White, 1997).

Several important cases have reaffirmed the employer's right to inspect employee e-mail. In a well-known case, *Shoars v. Epson* (1992), an employee claimed that her employer wrongfully terminated her after she complained that her supervisor was reading employee e-mail (Adelman & Kennedy, 1995). Her attorney also filed a companion class-action invasion-of-privacy lawsuit, *Flanigan v. Epson* (1992). The California Superior Court rejected the invasion-of-privacy claim, stating that the California statute protecting privacy of communications did not extend to e-mail at work (Adelman & Kennedy). In *Bourke v. Nissan Motor Co.* (1991), two employees sued their employer because they were terminated after their supervisor read their e-mail messages and discovered that they had been making fun of him, but the court rejected their argument that the company had violated their privacy (Cozzetto & Pedeliski, 1997). In *Bohach v. City of Reno* (1996), the court rejected any claim of a reasonable expectation of privacy on the part of police personnel using the departmental computer system (Bramsco, 1997).

However, at least one court has recognized a cause of action for invasion of privacy based on the employer's reading of employee e-mail (*Restuccia v. Burk Technologies, Inc.*, 1996). The *Restuccia* case involved a supervisor who read subordinates' e-mail and discovered that they had a variety of nicknames for him and that they were aware of his extramarital affair. The employees were terminated, and they then filed a lawsuit claiming, among other things, invasion of privacy and wrongful discharge in violation of the public policy against invasion of privacy. The court found that there were issues of fact as to whether the employees had a reasonable expectation of privacy in their e-mail messages and whether the reading of the messages constituted an unreasonable, substantial, or serious interference with the employees' privacy. The ruling was based on a Massachusetts statute that guarantees a right to privacy and contains language analogous to that found in a common-law cause of action. To date, other courts have not followed the *Restuccia* decision.

From the limited number of cases that have been decided, it appears that employees may have fewer privacy rights for e-mail than for other forms of communications. This difference may be explained in part by the fact that the employee is using the employer's computer system, and therefore the employer may be regarded as having a right to monitor what is transmitted on the system. The argument is that if the employer owns the computer, then the employer should be able to read what is stored on the computer. In most cases, the employer can do so very easily and surreptitiously, making the action perhaps seem less intrusive than opening personal mail, for example. Moreover, since the e-mail messages generally are stored in a common location, such as on a network server, the employee cannot regard them in the same light as personal items stored in a desk drawer in his or her office. In addition, actual knowledge that the system is not private may be inferred. For example, the *Bourke v. Nissan Motor Co.* (1991) court cited the fact that the plaintiffs knew that their e-mail was occasionally read by individuals other than the sender and recipient as evidence that the plaintiffs had no reasonable expectation of privacy.

It also may be that the relative newness of e-mail means that social conventions have not firmly developed in this area. While most people would regard listening in on another person's phone conversation as very impolite, perhaps the same social stigma is not attached to the reading of someone else's e-mail (Doss & Loui, 1995). Societal perceptions are important because for recovery under most privacy common law, an intrusion, even if regarded as outside usual social norms, must usually be shown to be "highly offensive" to a reasonable person.

Concluding Points on Employee Privacy

While recent court cases involving the monitoring of workplace e-mail have almost always upheld the right of the employer to engage in such monitoring, it would be a mistake to conclude that there are no restrictions on the extent and nature of this monitoring. The cases that have been decided have usually involved monitoring that was in some way job related. The employer was capable of making an argument that the monitoring was necessary for business purposes. Furthermore, it should be noted that in many of these cases, the employee had engaged in some sort of inappropriate or illegal behavior and thus would not be seen in a sympathetic light by most people. It is not certain that the courts would rule that an employer had the right to routinely read the personal e-mail messages of an individual when there was no reason to believe that the contents of the messages were in any way job related (Cozzetto & Pedeliski, 1997).

Another unsettled area involves the use of Web-based e-mail systems, such as Hotmail, by employees while at work. In this situation, the employee would use the employer's computer to access the e-mail via the Internet, but the messages would be not be stored on the employer's server, as they would be if the employee was using the employer's e-mail system, and this would undercut some of the employer's legal justifications for monitoring that have been discussed above. In one case, *Fischer v. Mt. Olive Lutheran Church* (2002), the employer read an employee's Hotmail messages (including ones that had not yet been read by the employee) without the employee's knowledge, which probably would violate the ECPA, although factual disputes in the case prevented the court from making this determination (Chapman, 2003). In this case, the employer accessed the employee's Hotmail account by correctly guessing the individual's password, something that probably would be difficult to do in many cases, so this case might have limited applicability to most workplace situations. However, employers might be able to monitor employee use of Web-based e-mail systems through various monitoring software ("spyware") or potentially by accessing the hard drive of the employee's computer. The legal status of such actions seems unclear at this time, but if the employer has notified employees that Internet use may be monitored, it seems unlikely that an employee could claim a reasonable expectation of privacy.

Although employers have broad power to read the e-mail of their employees, there may be political restrictions on such behavior when it comes to the public sector. What might be tolerated if carried out in a private company could be seen as objectionable if it occurred in a public organization. There may be more concern on the part of the public, the media, or elected officials regarding the monitoring of e-mail in public agencies. Even though the courts might rule otherwise, there may be widespread perceptions that the Fourth Amendment prohibits e-mail spyware of public employees, except in carefully defined situations, and these beliefs may impose practical limits on such monitoring.

Public universities in particular tend to stress the privacy of individual e-mail (Doss & Loui, 1995). For example, the University of North Carolina at Chapel Hill's Policy on the Privacy of Electronic Information specifically states that the university respects the privacy of users and does not monitor e-mail routinely (University of North Carolina at Chapel Hill, 2002). The policy does state that an employee's use of e-mail should not burden the system or interfere with the employee's job. The policy provides that e-mail will be read only for listed purposes, which include: (a) troubleshooting of hardware or software problems, (b) investigation of system misuse, (c) retrieving university-related information, (d) investigating reports of violations of law or university policy, (e) investigating reports of employee misconduct, (f) complying with legal requests for information, and (g) retrieving information in emergency services. For reading e-mail in any of the above situations other than the first two, the system administrator will need approval from the Provost and the General Counsel (University of North Carolina at Chapel Hill, 2002).

Other universities have similar policies. For example, the University of Arizona's Electronic Mail Policy contains similar restrictions on employees' use of e-mail for personal reasons and outlines situations in which an employee's e-mail may be monitored. These situations include: (a) when required by law or policy, (b) when there is a reasonable suspicion that violations of law or university policy have occurred, and (c) when required for time-dependent critical operational needs. The university will normally try to inform e-mail users about the monitoring, except when notification might be detrimental to an investigation (University of Arizona, 1998).

It should, of course, be understood that nothing prevents the recipient of an e-mail communication from divulging its contents. An individual has no reasonable expectation of privacy regarding information communicated to another person, except in some narrowly defined situations of privileged communications, such as between an individual and his or her attorney (Garcia, 1996). The recipient of the message is free to tell anyone, including law-enforcement officials, about its contents (Sundstrom, 1998). In the case of e-mail, this could include forwarding the message to others, a common practice. Moreover, if the recipient of an e-mail message chooses to divulge its contents to third parties, those parties are equally free to disseminate the message to others. Even if the third parties received the message from the original recipient with the understanding that they would keep it confidential, they are not bound by statutory or tort law to maintain secrecy. Again, with the power to easily forward e-mail to many individuals, the divulgence of messages is a very real possibility.

Employer Obligations to Monitor

Under some circumstances, the employer may have not only the right but also the responsibility to read or monitor e-mail communications. When one employee is suspected or accused of harassment by sending offensive e-mail messages to another employee, the employer's obligation to ensure that employees do not have to work in a hostile environment may require the monitoring of the employee's e-mail. If, for example, an employee is known to have sent harassing e-mail to a coworker previously, some monitoring might be considered necessary to shield the employer from further liability. Also, employers may be concerned about potential illegal behavior on the part of some employees, and monitoring e-mail messages may be necessary to investigate this possibility. Since employers can in some circumstances be held liable for the illegal behavior of their employees, the employer has a responsibility to monitor possible illegal actions.

Sexual Harassment and Other Civil Rights Claims

Under federal and state civil rights laws, employers may be liable for harassment occurring in the workplace unless they exercise reasonable care to prevent and correct promptly any harassing behavior. Sexual harassment in the workplace is of two types: (1) quid pro quo behavior, in cases in which a supervisor pressures an employee for sexual favors, using either the promise of rewards or the threat of punishment, and/or (2) a hostile workplace environment, in which unwelcome sexual conduct, including communications, creates an abusive climate for an employee. E-mail might be a component of either type of harassment, but it can be particularly significant in creating a hostile workplace. Employees in office settings could be subjected to offensive e-mail messages, which might be directed either at a single employee or distributed to most or all employees in the office. Furthermore, a hostile workplace environment does not have to be based on sex or gender. The Civil Right Act also prohibits harassment based on other protected classes, such as race, ethnicity, religion, and perhaps even disability and age (Towns and Johnson, 2003).

E-mail has been used to support claims of discrimination in several reported cases to date. In one case (*Peterson v. Minneapolis Community Development Agency*, 1994), a court overturned summary judgment for the employer by the lower court and cited the e-mail from a supervisor to a subordinate, pressing for a relationship, as one of the facts in support of the decision (Towns & Girard, 1998). In a case involving posting to an electronic bulletin board, the New Jersey Supreme Court found that postings to the board could create a hostile work environment and that if the employer had notice of the harassing postings, it had a duty to take action to stop the harassment (*Blakely v. Continental Airlines, Inc.*, 2000). In *Strauss v. Microsoft Corp.* (1995), the plaintiff alleged that she had not been promoted because she was female. Microsoft tried to exclude evidence of e-mail messages, sent by her supervisor, that contained comments of a sexual nature. The court found that while the e-mail messages did not prove sexual discrimination, they were relevant evidence for the jury to consider (Towns & Johnson, 2003).

It is unclear at this point exactly what e-mail will need to contain to support a claim of a hostile environment. Several courts have found that a few offensive e-mail messages do not create a hostile environment (*Curtis v. DiMaio*, 1999; *Harley v. McCoach*, 1996; *Owens v. Morgan-Stanley Co.*, 1997). However, Michael and Lidman (2000) report an increasing tendency for courts to hold employers responsible for offensive e-mail at work, so the legal standards in this area may be evolving. Significantly, some courts have been willing to allow the introduction of e-mail evidence to demonstrate general behavior at an office. Even if the e-mail messages in question were not targeted at a particular employee, they still might be considered as evidence of a hostile workplace. Thus, distribution of sexist or racist jokes on the workplace e-mail system might qualify as creating a hostile climate, as mentioned above (see, e.g., *Autoliv ASP, Inc. v. Department of Workforce Services*, 2001; *Greenslade v. Chicago Sun-Times, Inc.*, 1997; *Yamaguchi v. U.S. Dept. of the Air Force*, 1997).

An employer may escape liability if an appropriate response is taken following a complaint of harassment. In a case decided under Texas law, an employer received a complaint that racially harassing e-mail was being sent over the company's e-mail system. The employer reprimanded the employees responsible for the e-mails and had two company-wide meetings to discuss appropriate use of company e-mail. The court found that the employer's prompt action relieved it of liability (Towns & Girard, 1998, citing *Daniels v. Worldcom Corp.*, 1998). Perhaps in an attempt to shield themselves from legal action, some employers are following an even stricter and more punitive policy with respect to senders of inappropriate e-mail by firing those who are found to have violated company policy. In December 1999, the *New York Times* fired more than 20 staff employees for the sending of e-mail that contained offensive material (Kurtz, 1999). In addition, a St. Louis brokerage firm, Edward Jones & Company, fired 19 employees over e-mail containing off-color jokes and pornography (Seglin, 1999). In both of these cases, the offensive e-mail was discovered following a complaint to the employer.

Some employers have adopted a more proactive policy for preventing offensive e-mail in the workplace by employing software for the routine monitoring of e-mail for offensive content. For example, PornSweeper analyzes e-mail attachments for pornographic images and blocks transmission of inappropriate messages (Swanson, 2001). Other software screens e-mail messages for obscene, racist, or otherwise offensive language (Adams, Scheuing, & Feeley, 2000). Although it is unlikely that an employer would be held liable for failing to monitor employee e-mail when there was no reason to suspect inappropriate behavior, some employers apparently have decided to take no chances in this matter. However, while these employers may feel that routine monitoring of e-mail is necessary to shield them from potential civil rights lawsuits, such monitoring might make them more vulnerable to an invasion-of-privacy claim (Ciapciak & Matuszak, 1998). So far, blanket monitoring of e-mail for offensive content appears to occur primarily in the private sector, perhaps because public organizations are more politically sensitive to charges of invasion of privacy, as discussed earlier.

While the obligation to prevent a hostile workplace environment applies to both public- and private-sector employees, governmental organizations may feel more vulnerable to problems in this area. In a case known to one of the authors, the head of a local government department intended to send an e-mail message containing a joke about the Catholic Church to a few friends while at work but accidentally sent it to everyone in the department. When he realized very shortly what he had done, he issued an apology. Since this was an isolated incident, and since no employee in the department had charged the department head with harassment, the matter most likely would have ended if this had been a private organization.

However, one of the e-mail recipients forwarded the message to the local media, which then publicized the incident. The considerable media attention resulted in the department head feeling that he had to resign. Had this incident occurred in a private organization, the local media undoubtedly would not have regarded it as newsworthy.

Misuse of Government Property

In addition to potential liability for sexual harassment, public employers have another reason to monitor e-mail usage of their employees: the misuse of resources. Because the computers and software used by state employees are publicly owned, excessive use of e-mail for personal use could be considered a misuse of state property. Moreover, excessive personal e-mail activity probably means that the employee is not devoting sufficient attention to his or her job, which would be grounds for dismissal (McEvoy, 2002). While these considerations apply to both the private and the public sector, there may be greater concern with public employees not spending sufficient time on the job as this would be regarded as wasting taxpayers' money. Some state employee e-mail use policies address this problem. For example, the Virginia policy allows for limited personal use when the use does not interfere with productivity or adversely affect the efficient operation of the computer system, among other reasons (Virginia Department of Human Resource Management, 2001). The Texas state policy provides the following:

Personal e-mail should not impede the conduct of state business ... Incidental amounts of employee time—time periods comparable to reasonable coffee breaks during the day—should be used to attend to personal matters via e-mail or other telecommunications, similar to personal telephone calls ... Personal e-mail should not cause the state to incur a direct cost in addition to the general overhead of e-mail. Consequently, employees, upon receiving personal e-mail, should read it and delete it. (Texas Department of Information Resources, 2006)

The North Carolina Administrative Office of the Courts policy has similar provisions, stating that e-mail may be used for personal reasons where the direct measurable cost to the public is none or is negligible and there is no negative impact on employee performance of public duties (North Carolina Court System, 2006).

A recent New York case highlights a more restrictive policy for public employee use of e-mail for non-work-related reasons. In that case, *Benson v. Cuevas* (2002), the state education department had an e-mail policy that prohibited the use of the department's e-mail system for unofficial purposes, except for an unwritten agreement between the department and the union that allowed limited use by union officers to schedule meetings. The employee, Darcy, used it for other union business, including mass distribution of meetings summaries and a newsletter. The department sent out an e-mail reminding all employees that e-mail was not to be used for union business and also had a meeting with Darcy in which he was informed of the department's position. When Darcy continued to use the department e-mail system to e-mail union members, the department disconnected his access to the system. Darcy then filed suit, claiming that the department had interfered with participation in union activities in violation of state law. While an Administrative Law Judge found that the department had violated the law, the Public Employee Relations Board (PERB) reversed the decision and the PERB's findings were upheld on appeal to the New York appellate court. The court stated that the department had an e-mail policy, Darcy was in violation of the policy, and therefore the termination of e-mail access did not violate state law.

Perhaps the most significant misuse of resources in a public organization or agency occurs when an employee uses resources to conduct private, for-profit, business activity. Federal, state, and local laws and policies routinely prohibit employees from using public resources for private gain. While most private-sector employers would not want their employees using company resources for external business activities, such inappropriate use of company resources would not raise any legal questions. In a public organization, however, the matter is different. Public disclosure of governmental employees using their workplace for private business activities could well result in politically damaging charges of misuse of taxpayer funds. Thus, governmental agencies may feel more of a responsibility to guard against such misuse, including making sure that the organizational e-mail system is not used to conduct personal, for-profit business, even if the cost of using the system and the time involved in these business activities are minimal. Similar concerns apply to partisan political activities.

It is possible, though, for a policy to be too restrictive. The Connecticut state policy states that state systems must be used solely to conduct state business and that any and all personal use is unacceptable, including creating or forwarding any non-work-related e-mail as well as checking personal Web-based e-mail via the Internet (Connecticut Department of Information Technology, 2006). This policy appears draconian, and it is likely that most state employees who have their own office computer violate the policy. Indeed, it is doubtful that the state attempts to vigorously enforce this policy.

Public Access

Public organizations must contend with one important problem involving e-mail that their private-sector counterparts do not have to face. The public has some right to see the e-mail messages in a public agency. At the federal level, the Freedom of Information Act (FOIA), first passed in 1966 and subsequently amended in 1974 and 1986, outlines the public's right to access written materials and documents (Garson, 1995; Hammett, 2000). In 1996, Congress passed the Electronic Freedom of Information Act (EFOIA), which more specifically addresses electronic documents (Lewis, 2000). E-mail communications qualify as documents covered by the FOIA and the EFOIA. At the state level, public access depends on the specifics of state legislation. Many states have freedom-of-information acts that provide for even more public access to e-mail than does the federal FOIA.

The Federal Freedom of Information Act

The FOIA stipulates that federal agencies must make public records available when requested. Government reports, notices, bulletins, newsletters, publications, and policy statements are considered public documents. Exemptions are made for information defined as private or confidential, which would include certain aspects of personnel records, proprietary data contained in government contracts or proposals, certain criminal records, and so on. Most of the exemptions are the usual and familiar ones. Also included in the exemptions are internal communications prior to a policy decision (Garson, 1995). Thus, while a policy is being discussed within an agency, the employees are free to send messages to each other, either written or via e-mail, expressing thoughts about various aspects or implications of some possible policy, and those messages are not considered part of the public record. The exemption is designed to protect only those materials bearing on the formulation or exercise of policy-oriented judgment; peripheral documents are not protected (*Ethyl Corporation v. United States E.P.A.*, 1994). To meet the exemption, the agency would need to show that the document in question is both predecisional and deliberative (*City of Virginia Beach v. United States Dept. of Commerce*, 1993).

Depending on how strictly this exemption is interpreted, it potentially could exclude a considerable amount of e-mail communication. However, in one case, *Armstrong v. Executive Office of the President* (1993), the court ruled that the National Archive could not refuse to provide a computer tape of certain e-mail created during the Reagan administration as these e-mail messages could qualify as public records (Hunter, 1995). Furthermore, the court ruled that the e-mail messages had to be retained in electronic form; printing and saving the hard copies was not equivalent to saving the electronic versions. The court explained that the printed version might not contain all of the information found in an electronic file, such as the headers stating who sent the message, the recipients, and any attachments.

The FOIA provides that individuals requesting information must be specific about the content of the information being requested. The request must "reasonably describe" the records sought (5 U.S.C. § 552(a)(3)(A)). Little guidance is provided as to any more exacting requirements. However, case law indicates that although the size of the request is not the test for specificity, blanket requests for all of a certain type of record are generally considered not to be sufficiently descriptive and that broad requests that would require unreasonably burdensome searches need not be fulfilled (*Sears vs. Gottschalk*, 1973/1974). The agency may charge for providing such information, although if the information is provided to news media, only the actual duplication costs may be charged. Other organizations or individuals can be charged for the costs involved in searching for and reviewing the information that has been requested.

The 1996 EFOIA was supposed to require federal government agencies to make records more readily available in electronic format. However, it appears that this legislation had little effect. For one thing, agencies were not required to search for and report electronic records when doing so would not be technically feasible or when doing so would interrupt agency information systems, loopholes which appear to have significantly limited the impact of the EFOIA (Garson, 2006; Lewis, 2000). Additionally, enforcement of the provisions of this act has been limited. Garson reports that 6 years after the Department of Justice required all federal agencies to include an EFOIA link on their Web sites, many agencies had failed to do so. The fact that agencies were not provided with additional funding to implement the EFOIA probably played a large role in the lack of impact of this legislation (Lewis, 2000). It also is the case that following the September 11, 2001, terrorist attacks, the Bush administration followed a policy of limiting public access to government information to the greatest degree possible, thus making the FOIA and the EFOIA weaker than they otherwise might have been (Garson, 2006).

State Freedom-of-Information Laws

A review of state legislation in 2000 found that 14 states specifically defined public records to include electronically stored information, and another 24 states had statutes that did not specifically identify electronically stored information as public records but had legislation that was broadly worded so that such information would likely be covered (Bush & Chamberlin, 2000). However, few states specifically mention e-mail in their statutes. Four states (Arizona, Colorado, Florida, and Maryland) explicitly include e-mail in the definition of public records, and one state (Mississippi) specifically excludes e-mail (Bush & Chamberlin). Thus, in many cases, decisions on whether e-mail records have to be disclosed must be based on more general statutory language.

State laws sometimes provide for more public access than does the federal FOIA. For example, the North Carolina Public Records Act defines public records far more broadly. Included in the definition of public records are internal memos and messages, electronic or otherwise, that are part of the consideration of policies or actions, not just the statements of the policies and actions (L. Capone, personal communication, July 8, 2002). A considerable amount of e-mail might fall into this category. Colorado's open public records law contains a similar provision, including electronic mail used for public business as correspondence subject to the law. The Colorado law also states that e-mail is included under the law whether transmitted locally or globally and whether read, printed, or stored (Colo. Rev. Stat. Ann. §§ 24-6-402.2, 24-72-202). Therefore, as in North Carolina, it would appear that e-mail in Colorado could be broadly accessed. On the other hand, Michigan law under certain conditions, discussed below, exempts communications that are of an advisory nature and are preliminary to a final policy decision (Hunter, 1995). California law specifically provides that "writings" subject to the state inspection of public records act include any means of recording information, including material on a disk, but that law also contains an exemption for preliminary drafts and notes of a public agency (Cal. Gov't. Code §§ 6252, 6254).

The question of whether e-mail messages that are preliminary discussions of possible policies or actions qualify as public records seems especially important. E-mail frequently substitutes for the type of conversation that individuals often have in person or by telephone. Various ideas or thoughts may be put forth just to stimulate thought or explore various possibilities. In some cases, the messages may be little more than an employee thinking out loud. There is a valid interest in encouraging open and honest discussions of various decision options, and e-mail often is a convenient way of doing so. If government employees fear that all their e-mail messages dealing with some topic may be publicly disclosed, open and honest discussions may be discouraged, at least through e-mail. In fact, Michigan law recognizes this point, as advisory or preliminary discussions are exempt from public disclosure if the public agency shows that the public interest in encouraging frank communications outweighs the public interest in disclosure (Hunter, 1995). Connecticut law contains a similar exception (Conn. Gen. Stat. Ann. § 1-210). Other states, however, may not have such exemptions. The Virginia Freedom of Information Act does not contain an exclusion for such preliminary types of documents, and the Colorado law contains an exemption only for the work product of an elected official (Va. Code Ann. §§ 2.1-3700 *et seq.*; Colo. Rev. Stat. Ann. § 24-72-202).

State law routinely excludes a variety of government records from public disclosure. Commonly excluded information includes individual tax records, certain law-enforcement and criminal justice records, personnel information of a confidential nature, and communications made within the scope of the attorney-client relationship. Regarding e-mail, it is worth noting that in most cases, a government agency could not refuse to publicly disclose e-mail communications on the grounds that personal and confidential information is commingled with the requested information. The agency would have to separate the confidential information from the public information. Moreover, if the agency has a policy that the e-mail system is to be used for business purposes, an employee may not have a valid objection to the release of his or her e-mail messages if they fall within the scope of a public record, even if personal information is contained in some of the communications. From this perspective, a public employee may have less of a privacy right regarding his or her e-mail than a private-sector employee.

In most cases, a purely personal e-mail communication sent or received by a public employee using the workplace e-mail system would not qualify as a public record and thus would not be covered by freedom-of-information statutes. This point is illustrated by a recent Florida case, which is particularly significant because Florida law provides for very broad access to public records. In the case, a local newspaper sought copies of all e-mail communications between two city employees. The city allowed the employees to divide their e-mail messages into personal and public communications, and only the latter were released to the newspaper, which then sued for all of the records. A Florida court ruled that a newspaper was not entitled to all of the e-mail communications between two city employees, even though they had used the city computer system to send and receive the messages, because state law only provided for disclosure of messages that involved official government business (*Times Publishing Co. v. City of Clearwater*, 2002).

In a similar case under the Ohio public records statute, a court found that internal e-mails by county employees communicating racial epithets were not public records because the e-mails were not documenting the activities of the county (*State ex rel. Wilson-Simmons v. Lake County Sheriff's Department*, 1998). In *Tiberino v. Spokane County* (2000), a public employee who was dismissed for excessive personal e-mail use sued for reinstatement. While the Washington Court of Appeals upheld the employee's discharge, it ruled that the public did not have the right to see the content of the personal messages (McEvoy, 2002). The public did have the right to know that a government employee was spending excessive time on personal business, but the public did not have a right to know the details of that personal business.

As is the case with federal law, state laws generally restrict the charges that can be collected from those who request public records. Most states specify that only the actual cost of providing the information can be imposed on those requesting the information, although what that means can vary from state to state (Bush & Chamberlin, 2000). Under North Carolina law, the government agency cannot charge more than the actual cost of providing the information, regardless of whether or not the request is made by a news organization, a provision that is more generous than the federal FOIA (N.C. Gen. Stat. § 132-6.2). This cost cannot include the time involved in complying since that is considered part of the job definition of state employees (L. Capone, personal communication, July 8, 2002). The agency also must supply the information in the medium chosen by the requestor if the agency is reasonably capable of providing the records in that format (N.C. Gen. Stat. § 132-6.2). If an individual requested e-mail messages in an electronic format, then presumably the only cost that could be charged would be for the disks given to the individual. Other states, however, are able to impose higher fees, including charges for labor and for the amortization of computer facilities (Bush & Chamberlin, 2000).

If state law regards public records as the property of the people, it logically follows that state and local government agencies have certain obligations to retain these records, including electronic ones. This does not mean that every e-mail message must be retained. For example, a North Carolina document that established guidelines for state employees distinguished between transitory messages and those of lasting value (North Carolina Office of Archives and History, 2002). A message setting up or confirming a meeting time is clearly transitory and can be deleted after the meeting has been held. A message of lasting value must be retained for a longer period of time, either in electronic or print form. In either case, the files can be reviewed periodically, and out-of-date information discarded.

By contrast, Michigan law is far less specific regarding e-mail, and it appears that very little of the e-mail communications in a government agency would have to be retained (Hunter, 1995). It should be noted that the Michigan freedom-of-information act does not impose an obligation to keep records, only an obligation to make public records that have been retained; the management and budget act specifies record-keeping requirements. Thus, if e-mail communications are retained, they must be provided on request unless they fit into one of the specifically exempted categories; but if the communications are not required to be retained, they can be destroyed and therefore are unavailable for public access (Hunter, 1995).

Toward a Model \ Policy

We conclude this discussion with an examination of possible components of an e-mail policy for an organization. All authorities in this field recommend that organizations, pubic or private, develop clear policies regarding e-mail and communicate those policies to their employees. Differences of opinion exist on what features an e-mail policy should contain, however, and we attempt to outline some of the thinking in this area.

Barsook and Roemer (1998), writing specifically for governmental organizations, recommend that employers tell their employees that the workplace e-mail system is for business use only, that e-mail may be monitored for any reason, and that employees should not expect any privacy regarding their e-mail. Similar recommendations are made by others for both public and private employers (Adams et al., 2000; Fitzpatrick, 1997; White, 1997). Such recommendations frequently appear in state employment law newsletters (for example, see Coie, 1999; Walker, 1999). Some attorneys even recommend that the policy appear on the computer monitor screen every time the user logs on and that the user be required to acknowledge consent in order to proceed. Others suggest that employees be required to sign a statement acknowledging the employer's e-mail policy. As earlier discussions make clear, such a policy provides the employer with almost complete protection in the event employee e-mail is monitored. Of course, many of those who recommend a policy of this nature would not recommend that employee e-mail be regularly monitored; they simply are recommending a course of action that is likely to minimize potential legal problems for an employer.

Another opinion is that employees should be given some privacy rights regarding their e-mail (Doss & Loui, 1995; Etzioni, 1997). First of all, limiting the workplace e-mail system to business use only is a substantial restriction. Many employees find it extremely convenient to use their e-mail for personal use, just as they use their workplace telephone for personal calls. While a stronger argument can be made for permitting personal telephone calls at work—urgent or even emergency calls might be necessary, for example—prohibiting personal e-mail would be a significant inconvenience that would raise several questions. What happens if an employee receives an external e-mail message of a personal nature? What if a communication between two colleagues contains both business and personal items, just as the two can easily be contained in one phone call? Should an employee be prohibited from using the workplace e-mail system for personal messages during a lunch hour or after work? Rather than completely prohibiting personal e-mail, public employers might consider appropriate guidelines for allowable personal messages. For example, personal e-mail might be permitted if it (a) is not excessive in amount and/or does not overburden the e-mail system, (b) is not used for private business (i.e., profit-making) purposes, and (c) is not in violation of agency policies or federal or state laws.

It also may be desirable to provide employees with some privacy rights regarding their e-mail, especially if personal communications are allowed. Rather than stating that all e-mail is subject to monitoring, possible conditions for monitoring could be spelled out. Monitoring naturally would be conducted when there was some legitimate business reason for doing so, such as a situation in which there was reason to believe that the employee was violating organizational policies. In such cases, however, monitoring of e-mail should be limited in duration and scope (Fitzpatrick, 1997). Moreover, the results of the monitoring should be kept confidential. Thus, employees should expect that their e-mail communications would not be randomly monitored without cause, but that they could be monitored under appropriate situations.

There are possible benefits from a policy that provides employees with some privacy expectations, as opposed to a policy that provides maximum legal protection for the employer. First of all, employee morale is likely to be higher. Employees do not like being distrusted. They are likely to resent monitoring of their e-mail, even if perfectly legal, just as they would object to their employer routinely searching their office. Recognizing and respecting these privacy desires undoubtedly contributes to a healthier atmosphere in the organization (Doss & Loui, 1995). Second, employees may feel freer to use e-mail for honest and open discussions when they have some privacy expectation. Encouraging such interchange is desirable, especially for public organizations. A free flow of ideas should contribute to better decisions and policies. Thus, establishing an e-mail policy that provides employees with some privacy rights can be defended on pragmatic as well as ethical grounds, and these considerations might outweigh the desire to have maximum legal protection. Some employers agree with these arguments and have established policies that permit reasonable personal use of e-mail and provide for employee privacy rights (Sipor, Ward, & Rainone, 1999).

There is almost universal agreement that an e-mail policy should contain a prohibition of obscene, hostile, threatening, or harassing communications (Cozzetto & Pedeliski, 1997; Fitzpatrick, 1997). An employee should not have any expectation of privacy for communications that contain offensive language. As we discussed earlier, the employer has certain responsibilities to maintain a workplace that is free of harassing or hostile behavior, which mandates that certain language be restricted. Suspicion of violating this policy naturally would be a valid reason for monitoring an employee's e-mail. As discussed previously, the employer's ability to monitor e-mail for prohibited content has been enhanced by the development of software designed just for such a purpose. Several programs will scan all e-mail messages and identify those that appear to contain a significant number of questionable items (Etzioni, 1997). These messages can then be scrutinized by supervisors to determine if they truly are inappropriate communications. However, such blanket monitoring of e-mail may not be well received by employees, so there is a real question as to whether this is a desirable practice.

Another important part of the e-mail policy for a public organization is a clear statement of what types of e-mail messages qualify as public records and what requirements exist to retain qualifying e-mail communications. As we have noted above, there are two competing public interests here: the desire to encourage open discussion of alternatives and the desire to inform the public. Of course, government organizations must operate within established legal parameters, which may not allow much flexibility on the part of the organization. To the extent that flexibility in interpreting and implementing the laws exists, the agency or organization has to determine the proper balance between keeping e-mail private, especially communications that are advisory or preliminary to an action or decision, and providing the public with the information that legitimately should be available for scrutiny. However this is decided, an important part of the policy is to communicate to employees what the legal requirements are and what constitutes acceptable implementation of the legislation. What employees need are clear guidelines that allow them to make appropriate decisions regarding the retention of their e-mail. Given the lack of long-standing conventions in this area, there is considerable potential for confusion over how to treat e-mail communications.

To conclude, there are important legal questions surrounding e-mail in the public workplace. These questions involve issues that are not unique to e-mail, such as questions of employee privacy rights, employer obligations to provide a nondiscriminatory workplace, and public access to information. It is useful for managers to understand current legal thinking and to consider the issues involved in order to develop an appropriate and useful e-mail policy.

References

Adams, H., Sheuing, S. M., & Feeley, S. C. (2000). E-mail monitoring in the workplace: The good, the bad, and the ugly. *Defense Counsel Journal, 67*, 32-47.

Adelman, E., & Kennedy, C. (1995). *The right to privacy.* New York: Knopf.

Arkansas Office of Information Technology. (2006). *Sample Internet use policy.* Retrieved July 20, 2006, from http://www.oit.state.ar.us/AgPlan/Policies/InternetExample.doc

Barsook, B., & Roemer, T. (1998). Workplace e-mail raises privacy issues. *American City and County, 113*, 10.

Behar, R. (1997). Who's reading your e-mail? *Fortune, 3*, 56-61.

Bramsco, J. W. (1997). *Employee privacy: Avoiding liability in the electronic age* (Litigation and administrative practice course handbook series, No. H-562). New York: Practicing Law Institute.

Bush, M., & Chamberlin, B. F. (2000). Access to electronic records in the states: How many are computer friendly? In C. N. Davis & S. L. Splichal (Eds.), *Access denied: Freedom of information in the information age.* Ames, IA: Iowa State University Press.

California Department of Information Technology. (2002). *Statewide Internet usage policy.* Retrieved June 20, 2002, from http://www.doit.ca.gov/SIMM/policy/usagepolicy.asp

Chapman, K. W. (2003). I spy something read! Employer monitoring of personal employee Webmail accounts. *North Carolina Journal of Law and Technology, 5*, 121-154.

Ciapciak, J. J., & Matuszak, L. (1998). Employer rights in monitoring employee e-mail. *For the Defense, 40*, 17-20.

Coie, P. (1999). Does your employee handbook have an e-mail policy? *Washington Employment Law Letter, 6*.

Cole, W. S. (1997). *E-mail: Public records and privacy issues.* Paper presented at the Annual Meeting of the National Association of College and University Attorneys.

Connecticut Department of Information Technology. (2006). *Acceptable use of state systems policy.* Retrieved July 24, 2006, from http://www.ct.gov/doit/cwp/view.asp?a=1245&Q=314686

Cozzetto, D. A., & Pedeliski, T. B. (1997). Privacy and the workplace: Technology and public employment. *Public Personnel Management, 26*, 515-527.

Dawes, S. J., & Dallas, S. E. (1997). Privacy issues in the workplace for public employees: Parts I and II. *Colorado Lawyer, 26*, 61-85.

Dixon, R. (1997). Windows nine-to-five: Smyth v. Pillsbury and the scope of an employee's right to privacy in employer communications. *Virginia Journal of Law and Technology, 2*, 4-27.

Doss, E., & Loui, M. C. (1995). Ethics and the privacy of electronic mail. *The Information Society, 11*, 223-235.

Etzioni, A. (1997, November 23). Some privacy, please, for e-mail. *New York Times*, p. C12.

Fitzpatrick, R. B. (1997). Technology advances in the information age: Effects on workplace privacy issues. In *Current developments in employment law* (ALI-ABA course of study). Washington, DC: Fitzpatrick and Associates.

Garcia, E. C. (1996). *E-mail and privacy rights.* Retrieved April 9, 1998, from http://wings.buffalo.edu/ academic/... /Complaw/CompLawPapers/garcia.html

Garson, G. D. (1995). *Computer technology and social issues.* Harrisburg, PA: Idea Group Publishing.

Garson, G. D. (2006). *Public information technology and e-governance.* Sudbury, MA: Jones and Bartlett.

Greengard, S. (1996). Privacy: Entitlement or illusion? *Personnel Journal, 75*, 74-88.

Hammitt, H. A. (2000). The legislative foundation of information access policy. In G. D. Garson (Ed.), *Handbook of public information systems.* New York: Marcel Dekker.

Hunter, D. F. (1995). Electronic mail and Michigan's public disclosure laws: The argument for public access to governmental electronic mail. *University of Michigan Journal of Law Reform, 29*, 977-1013.

Jenero, K. A., & Mapes-Riordan, L. D. (1992). Electronic monitoring of employees and the elusive "right to privacy." *Employee Relations Law Journal, 18*, 71-102.

Kempthorne, D. (2001). *Executive Order No. 2001-12: Establishing statewide policies on computer, the Internet, and electronic mail usage by state employees.* Retrieved June 20, 2002, from http://www2.state.id.us/gov/mediacenter/execorders/eo01/eo_2001_12.htm

Kurtz, H. (1999, December 1). Not fit to print—or transmit: *New York Times* fires 20 workers for sending offensive e-mail. *The Washington Post*, p. C3.

Lewis, J. R. T. (2000). Electronic access to public records. In G. D. Garson (Ed.), *Handbook of public information systems.* New York: Marcel Dekker.

McEvoy, S. A. (2002). E-mail and Internet monitoring and the workplace: Do employees have a right to privacy? *Communications and the Law, 24*, 69-83.

Michael, A. R., & Lidman, S. M. (2000). Monitoring of employees still growing. *National Law Journal, 23*, B9.

North Carolina Court System. (2006). *AOC policy and guidelines on Internet use.* Retrieved July 20, 2006, from http://www.nccourts.org/Employees/HRPolicies/InternetPolicy.asp

North Carolina Office of Archives and History. (2002). *Electronic mail as a public record in North Carolina.* Retrieved June 21, 2002, from http://www.as.der.state.nc.us/sections/ archives/rec/email.htm

Perritt, H. J. (1996). *Law and the information superhighway.* New York: John Wiley.

Prysby, C., & Prysby, N. (1999). Legal aspects of electronic mail in public organizations. In G. D. Garson (Ed.), *Information technology and computer applications in public administration.* Harrisburg, PA: Idea Group Publishers.

Prysby, C., & Prysby, N. (2000). Electronic mail, employee privacy, and the workplace. In L. Janczewski (Ed.), *Internet and intranet security management: Risks and solutions.* Harrisburg, PA: Idea Group Publishers.

Rodriquez, A. I. (1998). All bark, no byte: Employee e-mail privacy rights in the private sector workplace. *Emory Law Journal, 47*, 1439-1462.

Schnaitman, P. (1999). Building a community through workplace e-mail: The new privacy frontier. *Michigan Telecommunications and Technology Law Review, 5*, 177.

Seglin, J. L. (1999, July 18). You've got mail, you're being watched. *The New York Times*, p. C4.

Sipor, J. C., Ward, B. T., & Rainone, S. M. (1999). A strategy for ethical management of e-mail privacy. *Information Systems Security, 8*, 64-73.

Starr, M., & Lippner, J. (2001). Monitoring employee e-mail. *National Law Journal, 23*, B8.

Sundstrom, S. A. (1998). You've got mail! (and the government knows it): Applying the Fourth Amendment to workplace e-mail monitoring. *New York University Law Review, 73*, 2064.

Swanson, S. (2001, August 20). Beware: Employee monitoring is on the rise. *Information Week*, p. 57.

Texas Department of Information Resources. (2006). *Acceptable use of the Internet, e-mail/messages and peer-to-peer* (Document 06-120, draft). Retrieved July 17, 2006, from http://www.dir.state.tx.us/pubs/pfr/06-120/appendixB.htm#contentBody

Towns, D. M., & Girard, J. (1998). Superhighway or superheadache? E-mail and the Internet in the workplace. *Employee Relations Law Journal, 24*, 5-29.

Towns, D. M., & Johnson, M. S. (2003). Sexual harassment in the 21st century: E-harassment in the workplace. *Employee Relations Law Journal, 29*, 7-24.

University of Arizona. (1998). *Electronic mail policy.* Retrieved July 21, 2006, from http://web.arizona.edu/ ~records/efinal.htm

University of North Carolina at Chapel Hill. (2002). *Office of the Vice Chancellor and University Counsel: Policy on the privacy of electronic information.* Retrieved July 21, 2006, from http://www.unc.edu/campus/policies/elec_info.htm

Virginia Department of Human Resource Management. (2001). *Use of Internet and electronic communications systems.* Retrieved July 24, 2006, from http://www.dhrm.state.va.us/hrpolicy/policy/pol1_75.pdf

Walker, A. N. (1999). Monitoring of electronic communications. *Connecticut Employment Law Letter, 7*.

Watson, N. (2001). The private workplace and the proposed "Notice of Electronic Monitoring Act": Is "notice" enough? *Federal Communications Law Journal, 54*, 79-104.

White, J. J. (1997). E-mail@work.com: Employer monitoring of employee e-mail. *Alabama Law Review, 48*, 1079-1104.

Wilborn, S. E. (1998). Revisiting the public/private distinction: Employee monitoring in the workplace. *Georgia Law Review, 32*, 825-858.

Court Cases and NLRB Cases

Armstrong v. Executive Office of the President, 877 F. Supp. 690 (D.D.C. 1993).

Autoliv ASP, Inc. v. Department of Workforce Services, 29 P.3d 7 (Utah App. 2001).

Benson v. Cuevas, 2002 N.Y. Slip Op. 03207, 2002 WL 722877 (N.Y.A.D. 3 Dept. April 25, 2002).

Blakely v. Continental Airlines, 751 A.2d 538 (N.J. 2000).

Bohach v. City of Reno, 932 F. Supp. 1232 (D.Nev. 1996).

Bourke v. Nissan Motor Co., No. YC 00379 (Cal. Sup. Ct., Los Angeles 1991).

City of Virginia Beach v. United States Dept. of Commerce, 995 F.2d 1247 (4th Cir. 1993).

Curtis v. DiMaio, 46 F. Supp. 2d 206 (E.D.N.Y. 1999).

Daniels v. Worldcom Corp., 1998 WL 91261 (N.D.Tex. February 23, 1998).

E.I. DuPont de Nemours & Company and Chemical Workers Assn., 311 N.L.R.B. 893 (1993).

Ethyl Corporation v. United States E.P.A., 25 F.3d 1241 (4th Cir. 1994).

Fischer v. Mt. Olive Lutheran Church, 207 F.Supp.2d 914 (W.D.Wis. 2002).

Flanigan v. Epson, No. BC 007036 (Cal. Sup. Ct., Los Angeles 1992).

Fraser v. Nationwide Mutual Insurance Co., 135 F. Supp. 2d 623 (E.D. Pa. 2001).

Greenslade v. Chicago Sun-Times, Inc., 112 F.3d 853 (7th Cir. 1997).

Guard Publishing Co., 2002 N.L.R.B. LEXIS 70 (2002).

Harley v. McCoach, 928 F. Supp. 533 (E.D. Pa. 1996).

Hill v. NCAA, 865 P.2nd. 633 (Cal. 1994).

Leventhal v. Knapek, 266 F.3d 64 (2nd Cir. 2001).

Muick v. Glenayre Electronics, 280 F.3d 741 (7th Cir. 2002).

O'Conner v. Ortega, 480 U.S. 709 (1987).

Owens v. Morgan Stanley & Co., 1997 WL 403454 (S.D.N.Y. July 17, 1997).

Peterson v. Minneapolis Community Development Agency, 1994 WL 455699 (Minn. App. August 23, 1994).

Pratt & Whitney, 1998 N.L.R.B. GCM 40 (1998).

Prudential Insurance Co. of America, 2002 N.L.R.B. LEXIS 551 (2002).

Restuccia v. Burk Technologies, Inc., No. 95-2125 (Mass. App. Ct. August 12, 1996).

Sears v. Gottschalk, 357 F. Supp. 1327 (E.D.Va. 1973), *aff'd* 502 F.2d 122 (4th Cir. 1974).

Schowengerdt v. United States, 944 F.2d 483 (9th Cir. 1991); *cert denied* 503 U.S. 951 (1992).

Shoars v. Epson, No. SWC 112749 (Cal. Sup. Ct., Los Angeles 1992).

Smyth v. Pillsbury Co., 914 F. Supp. 97 (E.D.Pa. 1996).

State ex rel. Wilson-Simmons v. Lake County Sheriff's Dept., 693 N.E.2d 789 (Ohio 1998)

Strauss v. Microsoft Corp., No. 91 civ. 5928 (SWK), 1995 WL 324692 (S.D.N.Y. June 1, 1995).

Tiberino v. Spokane County, 13 P.3d 1104 (Wash. Ct. App. 2000).

Timekeeping Systems, Inc., 323 N.L.R.B. 244 (1997).

Times Publishing Co. v. City of Clearwater, No. 2D01-3055 (Fla. Dist. Ct. App. 05/10/2002).

United States v. Slanina, 283 F.3d 670 (5th Cir 2002).

Watkins v. L.M. Berry & Co., 704 F.2d 577 (11th Cir. 1983).

Yamaguchi v. U.S. Dept. of the Air Force, 109 F.3d 1475 (9th Cir. 1997).

Chapter XVII

World Wide Web Site Design and Use in U.S. Local Government Public Management

Carmine Scavo, East Carolina University, USA

Jody Baumgartner, East Carolina University, USA

Abstract

The World Wide Web has been widely adopted by local governments as a way to interact with local residents. The promise and reality of Web applications are explored in this chapter. Four types of Web utilizations are analyzed: bulletin board applications, promotion applications, service delivery applications, and citizen input applications. A survey of 145 municipal and county government Web sites originally conducted in 1998 was replicated in 2002, and then again in 2006. These data are used to examine how local governments are actually using the Web and to examine the evolution of Web usage over the 8-year span between the first and third survey. The chapter concludes that local governments have made progress in incorporating many of the features of the Web but that they have a long way to go in realizing its full promise.

Introduction

Not long ago, a group of "netizens"—permanent residents of cyberspace—posed a question for themselves, Technology, Yea or Nay? In a self-congratulating tone, one of them responded, "Technology is wonderfully liberating. I don't need my stockbroker or travel agent anymore. I may choose to use them for a variety of reasons, but I don't NEED them anymore. Multiply that by millions of people and you have an entire industry that could be irrelevant in the Information Age. ... Take this even further—maybe technology at some point will make government irrelevant ..." ("Technology, Yea or Nay," 1998)

Late last fall, Detective Chris Hsiung of the Mountain View, Calif., police department began investigating a suspicious pattern of surveillance against Silicon Valley computers. From the Middle East and South Asia, unknown browsers were exploring the digital systems used to manage Bay Area utilities and government offices. ... Working with experts at the Lawrence Livermore National Laboratory, the FBI traced trails of a broader reconnaissance. A forensic summary of the investigation...said the bureau found "multiple casings of sites" nationwide. Routed through telecommunications switches in Saudi Arabia, Indonesia, and Pakistan, the visitors studied emergency telephone systems, electrical generation and transmission, water storage and distribution, nuclear power plants and gas facilities. Some of the probes suggested planning for a conventional attack ... But others homed in on a class of digital devices that allowed remote control of services such as fire dispatch and of equipment such as pipelines. More information about those devices—and how to program them—turned up on al Qaeda computers seized this year ... (Gellman, 2002, p. A01)

The newly improved Durham County [NC] Web site launched this week offers so much useful information that a county commissioner thinks there ought to be a charge for using it. In addition to offering taxpayers the chance to pay their bills online, the site now allows Internet users to browse the recent sale prices and some characteristics of all homes in a given neighborhood. ... The data is useful for real estate agents and appraisers performing market analyses to estimate where to set the price of a home going on the market. County Commissioner Becky Heron said that the site will do a lot of those professionals' work for them and that there should be a small charge. "It has cost the taxpayers a few bucks," Heron said. "People who are using this to make a profit, yes, they should pay." (Phillips, 2006, p. C3)

We used the first anecdote to open the original version of this chapter. It illustrated the vision that some had about what the World Wide Web could do and perhaps what they hoped that it would do: reduce the costs of doing business in such spheres as travel, stock brokerage, or government. Coupled with this comes the capacity for making institutions more transparent and accessible—posting governmental information on Web sites so that citizens can use that information to make more informed policy choices, to interact more closely with policy makers, and to become more empowered.

The second anecdote illustrates some of the concerns associated with the post-September 11 use of the Web by government. Posting more information on the Web and using the Web to conduct internal and external business increases the possibility that government services can be disrupted through denial of service, Trojan horses, viruses, or worm-type attacks (Lo, 2000, 2002). Before 2001, such disruptions were often considered almost innocent intrusions by hackers exercising their skills with nefarious but not malicious intentions. Attacks were often viewed as temporary disruptions to the normal conduct of Web site business—annoying but not deadly. After 2001, however, the situation may be much more serious: Terrorists may be attempting to infiltrate government Web sites to bridge the gap between cyberattacks and physical attacks, to use the very features of government Web sites that Web designers publicize—ease of use, widespread availability, and so forth—to stop the delivery of emergency services in time of physical terrorist attacks and to increase the number of casualties caused by the physical attack itself.

The third anecdote brings us back to more traditional concerns of government. Is the taxpayer getting what he or she paid for? Who pays for public services? Are services provided to the public also being used by private enterprise to increase its profits? The three anecdotes show the constant state of flux the Web represents: from early, almost naive enthusiasm about the potential use of the Web, to security concerns, to issues of who pays and who benefits when local government decides to invest in e-government.

By 2000, the assumptions made by many authors about the future of government use of the Web involved monotonically upward curves, offering more sophisticated services, hosting more users, moving into more efficient and effective modes of delivering services to citizens, and so forth (Layne & Lee, 2001; Musso, Weare, & Hale, 2000; Waisanen, 2002). However, the wholesale move to e-government is not without some kinks in these upward sloping curves. Increased security concerns after the terrorist attacks of 2001 coupled with the economic slowdown that began in 2000 brought reduced governmental revenues, increased attention to costs, downsizing of personnel, and other belt-tightening measures, some of which have had negative impacts on the adoption and use of high technology by local governments.

It is therefore uncertain that progress in the use of Web technology by local government can be assumed. The diffusion of such technology may depend both on economic and security concerns. Rather than a monotonic upward sweeping curve, we may see a hill-and-valley curve with significant advances followed by plateaus and even periods of retrenchment.

This chapter explores the potential of the Web as a method of information and service delivery for local governments. The chapter updates 1998 and 2002 surveys of local-government Web sites with data from a third survey conducted in early 2006. The three waves of survey data allow us to examine the assumption of progress that is often made by those who research such matters. Based on the results of three waves of this survey, we can discuss some major issues involved in Web application: how to establish and maintain a service delivery system on the Web, and what impacts such a system would have on the personnel and structure of a government organization and its users.

What is the Web?

The Web was originally introduced in 1991 by Tim Berners-Lee, a computer engineer associated with CERN, the European Particle Physics Laboratory in Geneva, Switzerland. The Web was designed to be a place where people all over the world could come together to work by combining their knowledge in a web of documents formatted in a nonsequential manner known as hypertext. Berners-Lee originally envisioned the Web to be much more of a collaborative environment than it became. In fact, his original term for a Web browser was a Web *client*, meaning that the individual could write as well as read in cyberspace and information would not only flow from the server to the user but also from the user to the server. As Berners-Lee said in an interview with David Plotnikoff of the *San Jose Mercury*,

That collaborative environment didn't take off ... But I'm not settling for it ... When we have a collaborative medium, where people are building things together on the Web, people won't call it a Web revolution, but that might lead to a change in how we do politics, a change in how we govern, how we run organizations. That would trigger a social revolution. ("Meet the Man," 1999, p. D3)

Berners-Lee's vision of the Web as a collaborative medium and his belief in a Web-led social revolution is shared at least in theory by a large number of those who write on cyberspace. Much has been made of the ways in which IT and the Web are transforming institutions (Fountain, 1999; Reschenthaler & Thompson, 1996; Shi & Scavo, 2000; for views that differ with this conclusion, see Kim & Lee, 2006; Scavo & Shi, 2000). Certainly, the potential of the Web to do so is evident. The Web has only had a short history, but since it was made available to the Internet community, it has proven to be the most powerful and exciting innovation in information technology. The Web offers a promising future for local governments to deliver services more quickly and conveniently. First, as a menu-driven Internet tool, the Web is easy to use, even for a person with little or no computer background. By clicking a mouse on certain elements in a Web page, a viewer can jump from one document to another without knowing where the next document is located and in an order that makes most sense for him or her. Second, the information distributed on the Web can be accessed at anytime and from anywhere. Third, with the assistance of plug-ins or application helpers, information transmitted via the Web can be displayed in the form of text, graphics, audio, and video. Fourth, Web browsers (Microsoft Internet Explorer, Netscape Communicator, Mozilla, etc.) have bundled all other Internet tools, including electronic mail, remote operation, file transfers (FTP, file transfer protocol), and the Adobe reader (for PDF [portable document format] documents) into one package. Such incorporation has facilitated simple straightforward access to all online information, not merely Web documents. With the common gateway interface (CGI) and Java technologies, the Web was equipped with interactive capacities, allowing a person browsing a Web page to interact with the person who authored the Web page. With the addition of features such as Web logs (blogs), podcasting, really simple syndication (RSS), XML (extensible markup language), and so forth, the Web is approaching the collaborative environment envisioned by Berners-Lee.

The Web and Public Management

The dominant philosophy in public management for many years has centered on the notion of efficiency. Public managers have been disposed to raise the productivity within the organization and reduce the costs of operation. This has compelled public managers to seek new technologies that can help them increase managerial control and productivity of individual employees. Studies show that as the computer has penetrated into every area of government operation, ranging from accounting to personnel, it has been utilized to beef up office automation and data processing. This includes word processing, spreadsheet, record keeping and retrieval, graphics, scheduling, financial calculation, statistics, engineering design, expert systems, and geographic information systems (GISs; Hadden, 1986; Hurley & Wallace, 1986; Kraemer & Northrop, 1989; Lan & Cayer, 1994; McGowan & Lombardo, 1986). The computer has become indispensable to any public agency seeking a high level of efficiency.

The advent of the Web, however, has added a completely new dimension to computer applications in the public sector. In our view, this technology's most important potential is that it gives public managers an opportunity to practice a management philosophy that emphasizes more the quality of service than efficiency. Such a philosophy, introduced under names like total quality management (TQM), entrepreneurial government, and customer-driven government, considers customer satisfaction rather than productivity as the top priority of good management and encourages public managers and employees to find the ways to achieve this goal (Osborne & Gaebler, 1992). Computer-based communication such as the Web holds great promise for this new preaching for at least two reasons. The first is that it permits a public agency to deliver its services more quickly and conveniently. The second is that it gives customers the opportunity to evaluate the services and enter their opinions back into some monitoring mechanism. As many scholars in public administration have long recognized, by improving government communication with the public, computer networks will help government become more open and responsive (Gorr, 1986; Kraemer & King, 1986; Vasu, 1988; Weitzman, Silver, & Brazill, 2006). As far as the Web is concerned, there are at least four specific areas of application:

1. **Bulletin board:** This type of utilization refers to the case where a government agency maintains a Web site to publish online the information regarding the services it provides; the programs it runs; the policies or regulations it enforces; the names, telephone numbers, and e-mail addresses of its key personnel; job opportunities; the important events associated with the organization; and so forth. The information posted on the bulletin board can be used by both the agency's employees and citizens.
2. **Promotion:** This type of utilization refers to the case where a government agency maintains a Web site to offer a wide range of information about the city, county, or region in which it is located. This includes population, economy, education, recreation, history, tourist attractions, business opportunities, and cultural events in the community. The purpose of such application is to raise the profile of the area, increase the spirit of community, and attract outsiders. Regional economic development is often the most important aspect of this use of the Web. Another important part of promotional Web site is a linkage to many public and private organizations—libraries, schools,

hospitals, and so forth—located in the area. Still another feature is the application of GIS software that allows data to be overlaid on maps of the community.

3. **Service delivery:** While both bulletin board and promotion are information dissemination, service delivery represents a government agency maintaining a Web site to provide services to citizens online. For example, specific information on local taxes and payment for local taxes, applications for city permits and licenses, and so on can be programmed into the Web site, allowing the local government to deliver services virtually. This utilization requires that the Web site be interactive: The viewer must be able both to view information on the computer screen and to input information to the Web site.
4. **Citizen input:** This type of utilization refers to the case where a government Web site allows citizens to communicate with local government officials and personnel directly. This often reflects the degree of willingness on the part of government personnel to tailor the services to the needs of citizens. The highest form of such solicitation will be some kind of public forum in which citizens can express their opinions on various policy issues. Today, local governments have to decide on a wide range of controversial issues concerning community development, environment protection, local taxes, cable television rates, traffic management, public health, zoning regulations, and so forth. Like many private companies soliciting consumers' preferences via the Web, government agencies can also use the Web to invite citizens to contribute their opinions on policy issues facing the community. However, unlike private companies (whose primary goal is to increase sales no matter to whom they are made), government needs to be concerned about the representativeness of those who utilize a Web page. We will take this point up later in this chapter.

One can see internal and external dimensions in these four utilizations. The latter two utilizations, service delivery and citizen input, are largely directed at residents within the city. Citizens would most likely make use of the citizen input utilization. While there is no reason that a local government would (or could) limit a discussion group only to residents or would not allow nonresidents to comment on city services, Web sites, and so forth, there is also no overwhelming reason why nonresidents might be interested in becoming involved in a discussion of a local issue unless that local issue had national implications. Residents, business operators, and property owners are those who would be most interested in city services, and so they would most likely make use of the service delivery utilization. The items posted on a city's bulletin board would also be of most interest to city residents, but public hearings concerning zoning changes, for example, might be of interest to business operators and property owners outside the city. Other announcements, such as street closings, schedules for public buildings, and so forth, would also have their own external constituencies. The promotion utilization would most likely be of the greatest interest to external constituencies. External constituencies would be interested in entertainment events, tourist attractions, business opportunities, history, recreational opportunities, and other information a city's promotional utilization would maintain. External researchers might also find the links to other institutions and the inclusion of GIS software useful for their own purposes.

Survey Results

To investigate the four areas of Web application described above, we conducted surveys of local government Web sites listed on the Public Technology, Inc. Web page (http://www.pti.nw.dc.us/)[1] in summer 1998, spring 2002, and then again in spring 2006. A one-fifth stratified sample of local governments in each state was identified in 1998 and the Web sites of those governments were coded to ascertain the extent that each Web site used components of the four areas of Web application identified above. The resulting sample of 145 local governments (see Table 1) contained 114 (79%) municipalities and 31 (21%) counties. There were no statistically significant differences between cities and counties in the overall Web site score or the scores on any of the four application scales, so we combined the scores of the two forms of local government in the discussion below. In order to maintain strict comparability across time, the same sample identified in 1998 was used in both the 2002 and 2006 Web site surveys.

There were 30 separate indicators in the four areas of application that we sought to code in the first two waves of the surveys of Web sites. While no local governments received perfect scores of 30 in 1998, two (Palmdale, CA, and Roanoke, VA) received perfect scores in 2002. In 1998, the highest score of 23 was obtained by Roanoke, while scores of 22 were obtained by Virginia Beach, VA; Annapolis, MD; Torrance, CA; Cape Girardeau, MO; Des Moines, WA; and San Bernardino County, CA. Other high scores (29) in 2002 were given to Volusia County, FL; Richardson, TX; and Virginia Beach. Low-scoring Web sites in 1998 were Riverton, WY (0); Stillwater, MN (1); Blackfoot, ID (3); and Wakefield, MA (3). In 2002, low-scoring Web sites included Embarrass, MN (4); and San Pedro, CA (7). In 2005, we added additional indicators to the survey, increasing the potential maximum score to 37. One local government—Virginia Beach—received the maximum score while two others—Volusia Count, FL, and Honolulu—received scores of 36. The mean, maximum, and minimum scores for the overall Web site index and the four utilization scores for the 1998, 2002, and 2006 surveys are included in Table 2.

In 1998, there were 10 indicators of the bulletin board type of utilization; the highest score that any local government attained was 10. Highly ranked governments on the bulletin board utilization were Tumwater, WA (10); St. Charles County, MO (9); and seven municipalities tied at 8. In 2002, we added an 11th indicator to the bulletin board utilization—whether the Web site listed the e-mail addresses of various government officials. Five Web sites received perfect scores of 11: Virginia Beach; Reno; Clearwater, FL; Volusia County, FL; and Tompkins County, NY. In 2006, there were also five Web sites with perfect scores of 11 on the bulletin board utilization: Virginia Beach; Roanoke; Winston-Salem; Pima County, AZ; and Dane County, WI.

In 1998, there were 13 indicators of the promotion-type application; the highest score that any local government attained was 13. This was attained by five local government Web sites: Roanoke; Volusia County, FL; Simi Valley, CA; Anderson, IN; and Aurora, NE. These five were followed by 14 Web sites tied with scores of 12. By 2002, there were 19 Web sites tied with perfect scores of 13 (Roanoke; Virginia Beach; Annapolis; Tuscaloosa; College Station, TX; Lufkin, TX; Catawba County, NC; Renton, WA; Winston-Salem, NC; Tompkins County, NY; Rocky Mount, NC; Oak Ridge, TN; Kitsap County, WA; Texas City, TX;

Table 1. Listing of municipal and county Web sites in sample

Municipalities				Counties
Birmingham AL	Byron GA	Morganton NC	Dumfries VA	Fairbanks AK
Tuscaloosa AL	Thomasville GA	Rocky Mount NC	Poquoson VA	Maricopa AZ
Hot Springs AR	Storm Lake IA	Winston-Salem NC	Roanoke VA	Pima AZ
Chandler AZ	Blackfoot ID	Aurora NE	Virginia Beach VA	Orange CA
Alameda CA	Barrington IL	Claremont NH	Burlington VT	San Bernardino CA
Benicia CA	LaSalle IL	Keene NH	Des Moines WA	Arapahoe CO
Calimesa CA	Mundelein IL	Monroe NJ	Renton WA	Sarasota FL
Colton CA	Pekin IL	Princeton NJ	Tumwater WA	Volusia FL
El Cerrito CA	Skokie IL	Willingboro NJ	Vancouver WA	Honolulu HI
Fremont CA	Westmont IL	Roswell NM	Muskego WI	Cook IL
Hayward CA	Anderson IN	Reno NV	Clarksburg WV	Jefferson KY
La Mirada CA	Seymour IN	Clifton Park NY	Riverton WY	Grand Traverse MI
Livermore CA	Hesston KS	Rochester NY		St. Charles MO
Manhattan Beach CA	Lindsborg KS	Dublin OH		Cascade MT
Monterey CA	Belmont MA	Maumee OH		Catawba NC
Murrieta CA	Marblehead MA	Sidney OH		Forsyth NC
Palmdale CA	Wakefield MA	Tulsa OK		Renville ND
Poway CA	Annapolis MD	Eugene OR		Bergen NJ
Ridgecrest CA	Greenbelt MD	Medford OR		Queens NY
San Jacinto CA	Westminster MD	Edinboro PA		Tompkins NY
San Pedro CA	Veazie ME	Norristown PA		Auglaize OH
Santa Barbara CA	Brownstown MI	Woonsocket RI		Carter OK
Simi Valley CA	Mt. Pleasant MI	Myrtle Beach SC		Benton OR
Torrance CA	Owosso MI	Chattanooga TN		Washington OR
Victorville CA	Barnesville MN	Oak Ridge TN		Beaver PA
Aurora CO	Chanhassen MN	Arlington TX		Harris TX
Littleton CO	Embarrass MN	Brownwood TX		Utah UT
Superior CO	Maple Grove MN	College Station TX		Accomack VA
Meriden CT	Owatonna MN	Denton TX		Floyd VA
Clearwater FL	Stillwater MN	Lufkin TX		Kitsap WA
Crystal River FL	Winona MN	Nassau Bay TX		Dane WI
Margate FL	Cape Girardeau MO	Richardson TX		
Naples FL	Independence MO	Texas City TX		
Quincy FL	Vicksburg MS	Cedar City UT		

Table 2. Comparison of Web site features on overall scale and four subscales (1998, 2002, 2006)

	OVERALL			BULLETIN BOARD			PROMOTION			SERVICE DELIVERY			SOLICIT FEEDBACK		
	1998	**2002**	**2006**	**1998**	**2002**	**2006**	**1998**	**2002**	**2006**	**1998**	**2002**	**2006**	**1998**	**2002**	**2006**
Mean	13.9	21.4	25.3	4.8	7.4	7.6	7.5	10.6	13.0	0.5	2.2	3.1	1.1	1.2	1.6
Median	14	22	26	5	8	8	8	11	14	0	2	3	1	1	2
Maximum	23	32	37	10	10	11	13	13	18	3	6	6	3	3	3
Minimum	0	4	8	0	0	1	0	2	4	0	0	0	0	0	0
N	145	140	139	145	140	139	145	140	139	145	140	139	145	140	139

Richardson, TX; Palmdale, CA; Maricopa County, AZ; Floyd County, VA; and Muskego, WI. These were followed by some 48 Web sites with scores of 12. In 2006, we added five new indicators to the promotion utilization. Some 80 Web sites received scores of 13 or above on this utilization, but only three, Roanoke, Volusia County, FL, and Honolulu, HI, received perfect scores of 18 on the new index.

In 1998, there were six indicators of the service delivery type of application; no government Web site attained a score higher than 3 on this scale. Roanoke; College Station; Renton, WA; and Norristown, PA, all attained that score. By our 2002 survey, there were eight Web sites with perfect scores of 6: Roanoke; Palmdale, CA; Volusia County, FL; Cape Girardeau, MO; Thomasville, GA; San Jacinto, CA; Murietta, CA; and Honolulu County, HI. These were followed by seven local-government Web sites with a score of 5. In 2006, there were 15 Web sites with perfect scores of 6 on this utilization with another 18 showing scores of 5.

Last, there were three indicators for the feedback solicitation application. In 1998, San Bernardino County; Manhattan Beach, CA; and Roswell, NM, all scored at this highest level of feedback solicitation. By 2002, none of these three were still at this high level; two other Web sites, those of Birmingham, AL, and Hot Springs, AR, had replaced them. The Web sites of these two local governments were followed by some 30 others tied at a score of 2. In 2006 there were 13 Web sites with scores of 3, and none of these appeared on either the 1998 or 2002 list.

It is useful to compare the results of these three waves of surveys with results from other Web evaluation efforts. In particular, the Center for Digital Government (http://www.centerdigitalgov.com) has given "Best of the Web" awards to government Web sites since 2000. Table 3 presents a summary of the five highest scoring municipal and county Web sites for the time period of 2003 through 2005, and more generalized measures for the 2000 through 2002 time period. While our sample of Web sites cannot truly be compared with the Center for Digital Government's universe of local-government Web sites, it is reassuring to see how some of our highest scoring sites show up as winners in the "Best of the Web" awards competition. In particular, Honolulu and Virginia Beach—high scorers on our four utilizations—are mentioned often in the Center for Digital Government rankings.

More interesting than the absolute scores of the local governments on these four scales are the interrelationships of the scales themselves (see Table 4). Are local governments specializing in one single type of utilization or do local governments that do well on one utilization also do well on the other three? In 1998, the scales were interrelated with low to moderate correlations. The bulletin board utilization was significantly correlated with all three other scales (Pearson's $r = 0.31$ with promotion; 0.36 with service delivery; and 0.25 with feedback solicitation), supporting the argument made above that bulletin board applications are the baseline of Web utilizations. Promotion was moderately correlated with service delivery (0.25) and with feedback solicitation (0.20). Feedback solicitation was insignificantly correlated with service delivery (0.07) and so was only significantly correlated with the bulletin board application. The hypothesized structure in the four areas of Web page design is therefore somewhat supported by empirical observation: Bulletin board uses appear to be the most common and are the base of any hierarchy of uses that Web pages show. Above bulletin board uses would be promotion, then service delivery; feedback solicitation would be at the top of the hierarchy, being the least common and most specialized of any of the four uses.

Table 3. Center for Digital Government "Best of the Web" local government Web site winners (2000-2005)

	2005	2004	2003	2002	2001	2000
	CITIES			OVERALL		
1st	Washington, DC New Orleans, LA	Fort Collins, CO	Washington, DC	Tampa, FL	New York	Seattle OR
2nd	Virginia Beach, VA	Phoenix, AZ	Virginia Beach, VA	Miami-Dade, FL	Montgomery, MD	
3rd	Seattle, WA	New York	Boston	Marion/Indianapolis, IN	Conyers, GA	
4th	Tampa, FL	Chicago New Orleans, LA	Phoenix AZ	Honolulu, HI	Miami-Dade, FL	
5th	New York	Dallas, TX	Dallas, TX	Dallas, TX	Chicago	
	COUNTIES					
1st	Hennepin, MN	Miami-Dade, FL	Montgomery, MD			
2nd	Fairfax, VA	Santa Clara, FL	Santa Clara, CA			
3rd	Orange, FL	Fulton, GA	Fairfax, VA			
4th	King, WA	King, WA	Marion/Indianapolis, IN			
5th	Fulton, GA	San Diego, CA	King, WA			

Note: The Center for Digital Government introduced separate awards for cities and counties in 2003. In 2000, only one general local-government Web site award was given. Other awards were given to subcategories of local-government Web sites.

Table 4. Pearson's R correlations of four subscales across three survey dates

Note: * $p \leq .05$

	Bulletin Board 1998	Promotion 1998	Service Delivery 1998	Citizen Input 1998	Bulletin Board 2002	Promotion 2002	Service Delivery 2002	Citizen Input 2002	Bulletin Board 2006	Promotion 2006	Service Delivery 2006	Citizen Input 2006
1998 Measures												
Bulletin Board		.31*		.25*								
Promotion			.25*									
Service Delivery		.25*										
Citizen Input		.20*	.07									
2002 Measures												
Bulletin Board	.25*	.19*	.18*	-.04		.53*		.24*				
Promotion	.29*	.24*	.12	-.02			.50*					
Service Delivery	.28*	.19*	.23*	.10		.48*						
Citizen Input	-.05	-.07	.09	.00		.24*	.26*					
2006 Measures												
Bulletin Board	.42*	.14	.20*	-.03	.44*	.28*	.27*	.04		.55*		.46*
Promotion	.21*	.20*	.15	-.03	.34*	.32*	.31*	.06			.42*	
Service Delivery	.32*	.13	.25*	.25*	.30*	.28*	.27*	.00		.51*		
Citizen Input	.12	.10	-.06	-.06	.23*	.10	.15	.04		.46*	.36*	

The structure of the four utilizations in 2002 is somewhat different: All four utilizations are significantly related to each other. The bulletin board utilization was significantly correlated to all three other scales, as above (Pearson's r=0.53 with promotion; 0.48 with service delivery; and 0.26 with feedback solicitation). Promotion is significantly related to service delivery (0.50) and to feedback solicitation (0.26). Although all of the correlations are statistically significant, they decline as one moves from the bulletin board utilization to feedback solicitation, thus displaying some evidence supporting the hypothesized hierarchical structure described previously.

By 2006, all six of the correlation coefficients were statistically significant and none was below 0.35. This higher level of intercorrelation suggests that local governments are learning how to design Web sites; the sophisticated sites are sophisticated across the board. In particular, these higher intercorrelations suggest that there are fewer specialists: local-government Web sites that do a great job on one utilization but do not on the other three.

It should be noted, however, that there is little across-time consistency in the individual scores for the four utilizations. So, the two bulletin board scores (1998 and 2002) are correlated only at the 0.25 level; the two promotion scores are correlated at the 0.24 level; the two service delivery scores at the 0.23 level (all statistically significant), while the two feedback solicitation scores are not related to each other at all (0.00). The 2002 scores are only minimally more highly correlated, with the 2006 scores—bulletin board at 0.44, promotion at 0.32, service delivery at 0.27, and feedback at 0.04.

The pattern that emerges from all these data is one with a great deal of individual variation from 1998 to 2002 to 2006 (low over time correlations) with general improvement across all four scales (increase in mean scores) and somewhat more consistency across the four utilizations (higher intercorrelations); those local governments having Web sites that have a great deal of bulletin board features also have Web sites with promotion features, service delivery features, and (to a lesser extent) feedback solicitation features.

The administration of local-government Web sites is the other issue that can be addressed through our survey data. In 1998, some 25% (37) of the 145 Web sites were administered by local government on its own. Some 32% (47) sites were administered by organizations outside of local government acting on their own. Most of these were local nonprofit organizations or businesses, but one city's Web site was maintained by a local university and another by the local Chamber of Commerce. Another 9% (13) Web sites were administered by such organizations acting in conjunction with local government. By 2002, this pattern had changed completely: Some 53% (78) of the Web sites were administered by local government, a finding that is in tune with the 2001 ICMA (International City/County Managers Association) survey of e-government (Norris, Fletcher, & Holden, 2001). The remaining 47% of Web sites were administered by either a local business or nonprofit organization (35%, *n*=52) or by a national or regional Web hosting company (8%, *n*=11). Apparently, some of these outsourcing experiences were not very successful since 4 years later some 94% (*n*=131) of the Web sites were coded as being administered by local government itself.

Designing and Managing a Web Site

In recent years, a number of guides for local-government managers who want to establish a Web presence have been published (Center for Technology in Government, 2001; Stark, 2001). In addition, a great deal of technical assistance is available on the Web itself for local governments who want to improve their Web sites. Since these guides and Web sites do an excellent job of addressing Web design issues, and since some 83% of local governments in the recent ICMA survey reported already having a working Web site (Norris et al., 2001), our discussion here will be devoted to Web management issues.

Planning

Past research on computer applications in public and private organizations has demonstrated that simply focusing on the technical aspect of a new technology will not achieve the results as expected (Cats-Baril & Thompson, 1995; Davenport, 1993; Hammer & Champy, 1993; Overman & Loraine, 1994). Any gains from investment in computer technology depend on (a) management decisions that shape the path of technology diffusion, (b) resources that are available, (c) government employees who use computers, (d) the population that is served by the organization, and (e) past computer use in the organization (Brudney & Selden, 1995; Caudle, 1990; King & Kraemer, 1988; Martin & Overman, 1988; Northrop, Kraemer, Dunkle, & King, 1990).

In deciding what information to place on the Web, we have delineated two common rules. First, all the information that can be made public and of importance to the public can be put online. This includes information on the agency's services; programs; policies or regulations; the names, telephone numbers, and e-mail addresses of its key personnel; job opportunities; important events associated with the organization; and so forth. A Web site can also link to other Web sites that provide information about the local community such as schools, hospitals, news media, recreation, business, demographics, environment, and so forth. Second, all the services that only involve paper work can be delivered through the Web. An example will be an application for a home business license. To accomplish this in person, the process will start with a visit by an applicant to city hall to obtain all the application forms and instructions from city clerks or the department granting the license. However, such a trip can be saved if the forms and instructions are put on a Web server for anyone to download.

The choice of exactly what information to place online is, of course, a much more complicated matter than stated here. Only a large group of decision makers can develop a comprehensive list of such information; leaving the decision to one person (or a small group of people) risks omitting key pieces of information in which the public may be interested. Most importantly, however, it is not necessary to decide once and for all what information to put online. Web sites can be designed to solicit feedback for users as to what information they would like to see on the Web site that is currently not provided. As our three surveys showed, however, few local-government Web sites solicit such feedback.

A second issue concerning information placed on a Web site has to do with security. In the past, the rule was (as noted above) the more the better. However, in the more recent security-conscious environment, governmental agencies are rethinking what information should be placed online or made freely available to the public. Gellman (2002) reports that information gleaned from Web sites in Silicon Valley was found on al Qaeda computers in Afghanistan after U.S. troops seized those computers. National Public Radio (NPR) reporter Allison Aubrey reported on *All Things Considered* that the U.S. Environmental Protection Agency removed information about the dangers of accidents at chemical plants around the United States because such information might supply terrorists with knowledge about where potentially dangerous chemicals are stored and how secure storage of the chemicals actually is ("Government Off the Web," 2001). Also, NPR's Adam Hochberg has reported that states have been reconsidering portions of their public records laws that mandate public access to certain government documents (and also determine what information can be placed on Web sites) in light of terrorist incidents and the possibility that terrorists are getting potentially damaging information on such subjects as water supplies, nuclear power plants, and so forth ("Terrorists and Libraries," 2002) from state and local Web sites (Kriz, 2001; Toner, 2001).

Much more than simple posting of information can be done with the Web if a government agency has the necessary technical expertise and its Web server is equipped with interactive capability. Again, take the application for a home business license, for example. Typically, after getting the application forms and instructions, an applicant must fill in the forms and mail them to the licensing agency. But in many of our sample cities and counties, not only is there no need to visit city hall or the county courthouse for the forms or instructions, an applicant need not send the forms back via the mail at all. He or she can supply all the information on the government's Web server and send it to the relevant office by simply clicking a submit button. Residents that have this type of interactive Web site might also be able to request other services as well via the Web; they can complain about shady businesses, airport noise, traffic conditions, discrimination, or the state of the city in general. They can request graffiti removal, street maintenance, bus itineraries, or library cards, and they can also report petty thefts, sign up for recreation classes, register for volunteer activities, and pay some taxes or fees.

A common error in many Web applications is the overemphasis on formats. The Web is attractive to a large part as a result of its multimedia interface, which has offered a variety of means for data presentation and service delivery. Creative use of multimedia enhances the functionality of Web pages and reduces boredom. It brings attention to important information conveyed via the Web. However, overuse of multimedia can be counterproductive. A Web page garnished with elegant graphics or animated pictures, for instance, often takes a long time to be loaded on a user's computer (unless the user happens to have a high-speed Internet connection), forcing the user to stare at his or her computer monitor waiting for something to happen. In the case of audio and video, not only does transferring files take a long time, but additional software is often required for the program to display its full effect. Designers must be aware that long waiting time and frequent technical troubles can easily turn away the browsing public. The city of Greenville, NC, Web site in the mid-1990s included audio greetings from the mayor. The greeting message included a short description of the city, its geographical location in North Carolina, the fact that it was the home of East Carolina University, and so forth. While this greeting was an attractive feature of the Web site, it was discontinued (and the Web site simplified) after users complained about the amount of time the message took to download. A recent study in the United Kingdom has shown that the public sector is learning this point faster than is the private sector. Muncaster (2006) reports that the Web performance specialist Site Confidence (http://www.siteconfidence.co.uk) reported that public sector Web sites downloaded in an average of 16.1 seconds, between 7 and 34 seconds faster than commercial-sector Web sites. According to Site Confidence's Bill Kirkwood, the public sector has "got the message that getting to the information quickly and easily is the most important thing. There's a strong message that if you have uncluttered pages without loads of Flash they will download easily" (Muncaster, 2006, p. 16). In an era of relatively slow-speed dial-up connections, users might attribute a slow downloading Web site to their own slow connection. However, as broadband connections have become more common, users are getting more accustomed to virtually instantaneous downloads; a site that downloads slowly may thus be less tolerated in this quicker moving virtual world than it would have been previously.

There is no universal rule as to what and how much multimedia should be used in a particular case. The rule of thumb is that if a format does not serve any functional purpose, then it should not be used. A government Web site is not for entertainment purposes. Another point to bear in mind is that a functional Web site should have a clean and short front page that provides a general overview of the information available on the Web site. The requirement of being short also applies to other components of a Web site. A Web page that forces a viewer to scroll down and up or left and right many times will turn away many people. Extensive text also requires users to read in a way that they are not accustomed to doing on a computer. Research has shown that users read Web sites similar to how they read newspapers, scanning headlines and perhaps reading a few of the beginning sentences in an article before moving to other headlines and articles (Morkes & Nielsen, 1997). Thus, providing long-text articles on a Web site makes it unlikely that the user will read the entire article. A CGI or Java program that takes a long time to execute can also be frustrating. The solution is to break a long document or program into small and sequentially related segments and to use hyperlinks to allow the user to jump to other parts of the Web site (or other Web sites) when he or she wishes.

Hyperlinks embedded in Web site text can take the user to a variety of related sites. Often, Web designers will highlight uncommon or jargon words in a document and link these to definitions so that the naive or inexperienced user can jump to the definition, read how the Web site is using the word, and then jump back to the document to continue browsing. The hyperlink need not be textual; Web designers can use PowerPoint slides to illustrate points they are attempting to make, or insert audio explanations or short streaming video to address issues that are better explained visually rather than textually. However, while all of this is technically possible, the Web designer needs to remember that each time a high-bandwidth application is included in a Web site, it will slow down the access of users who have relatively slow connections. While many users today are linked to the Internet via fast cable modems or DSL (digital subscriber line) lines, many still rely on slower dial-up services with 56K modems. Attempting to access streaming video, for example, from a 56K modem dial-up service is virtually an exercise in frustration, one that Web designers (most of whom abandoned their dial-up services years ago) might not readily appreciate. As Stark (2001) puts it,

Portions of the local government site geared to residential users should definitely be keyed to minimal technical capacity. It may be wise to also offer a text-only version for users with slow connections or individuals who prefer textual presentations, including many senior citizens. Sections intended primarily for business users, however, could be developed with an assumption of higher technical capability.

The point is that the local-government Web designer needs to know the potential market for each of the features incorporated in the Web site, and the only way for the designer to know this is to consult broadly with those both inside and outside of local government (see the discussion on personnel).

Usability

To ascertain how user-friendly Web sites were, we focused on the front (or home) page of the site. Of the sites we surveyed, 79 (54.5%) had contact information (address and telephone number) for the government offices displayed on the front page. This means that almost half did not. This was surprising. Displaying this information seems to be a low-cost and fundamental element for any government Web site. For example, many people may find it necessary or desirable to phone their government directly rather than spend time searching for the information on the Web site. Seventy-three sites had a search box on their front page as well. While not necessary, this feature certainly adds a certain convenience to many Web site visits. On the other side of the coin, 26 sites (17.9%) had no search capability anywhere on their Web sites.

A number of form- or design-related aspects of Web development are directly related to how user friendly a given site is. Several design elements seem decidedly user unfriendly, and only a very few of the sites we surveyed included these elements in their site. For example, only six (4.1%) had a design that employs frames. Frames are a little-used HTML (hypertext markup language) design feature that divides a page into two or more sections, allowing each section of the page to display independently of the other. It is quite tempting to use frames in many cases, since frames allow, for example, for a menu to be statically displayed in the left-most column (frame) while the user scrolls through right-column text. This provides for an easy navigation solution. Similarly, a header (e.g., a logo) could be displayed in a top frame while the user scrolls through text in a bottom frame. As tempting as it might be to employ frames, their use is generally not a good idea since many older browsers have trouble displaying them. Some sites that use frames will explicitly tell users that they do so, and that if the page does not display correctly, the user should upgrade their browser. However, this seems to stray a great deal from an ideal of being user friendly. Web sites should be written for the user: Web developers should adapt to them.

So, for example, government Web pages should be written to display in 775 pixels wide or less. This is a fairly well-accepted Web development standard. A Web page written to approximately 775 pixels wide will display in 800×600 screen resolution without a horizontal scroll bar appearing on the bottom of the browser window. Alerting users that the "site is best viewed at 1024×768 screen resolution" is not a good alternative to a 775-width standard. While bigger and better quality monitors have made screen resolutions of 1024×768 far more common, many users continue to set their monitor resolution to 800×600. A lower resolution makes fonts display larger, and this is important to many elderly or occasional computer users. Moreover, many (perhaps most) users do not know how to change their screen resolutions. Only 15.2% (22) of the sites we surveyed were written to display wider than 800 pixels.

Relatedly, good Web developers never assume that a user will stay at their site longer than a few seconds and will design front pages accordingly. When users are forced to scroll down several pages in order to find what they are looking for, they often get frustrated and leave. A good rule of thumb is to try and have the page height written to display in 600 pixels or less. We used a more generous standard in our evaluation. Of the sites we surveyed, only 46.2% (67) displayed in less than two pages. In all, 41.4% (60) of the sites displayed properly in 800×600 mode, while 10.3% (15) failed both of these design display tests.

Another design aspect of Web sites to be avoided is to present users with an introductory front page that forces them to click again to actually get to the site. This assumes again that users will actually have the interest or patience to do so, which is not a safe assumption. Even if we could assume that users would continue past this introductory page, we are left with the question of why such a design is necessary in a government Web site, whose function it is to provide information and services, not to entertain. Fortunately, only five of the sites we surveyed (3.4%) had these nonfunctional introductory front pages.

A full 38.6% (56) of the sites we examined employ cascading or pop-out menus somewhere on their front page. Cascading or pop-out menus are written with Javascript or other non-HTML scripting languages. When the user hovers their mouse over a menu item (either text or image), a submenu appears presenting more choices. Like frames, this design strategy provides a convenient solution for navigation problems, especially when there are many pages or sections in a Web site. However, unless done properly, the appearance of the submenus can be distracting. Moreover, they often appear and cover other important information on the Web site. This would not be problematic, except that users often do not know that the page has these cascading or pop-out menus, meaning that the user can cause the submenus to appear inadvertently by unwittingly hovering their mouse over a menu item. In other words, beyond being distracting, they can be annoying to some people. One solution to this is to use expanding rather than cascading or pop-out menus. With these, when users click an expanding menu choice, the submenu appears below the menu choice. This has the advantage of avoiding the chance appearance of submenus.

A final design aspect deals with layout. Virtually all Web sites employ a header of some sort, consisting of a logo, place name, and so forth—what Web designers commonly call branding (the logo carries over to Web pages past the front page). This header, incidentally, should rarely be greater than 100 to 125 pixels high. Below this header, pages are typically divided into columns. Usually this translates into a narrower left-hand column that is reserved for navigation (menus) and one or more columns to the right that have information. Of our sample, only 17 sites (11.7%) were not divided into columns. Forty-one sites (28.3%) employed a two-column layout, while 72 (49.7%) used a three-column layout. The three-column layout has a right-hand column that is roughly the same size as the left (leaving the middle column as the largest) and is used for announcements, special links, and other information—features (e.g., weather) or links the host wants to prominently display. The two- or three-column designs are probably the best choices. Having more than three columns becomes rather confusing to the user. Only 13 (9%) sites that we examined used a four-column layout, while 2 used five columns.

Beyond design elements, we also looked at how well the site was linked to other sites. We used a free software package called Web Link Validator (Version 4.5) to test this. The sites we surveyed had a wide variation in the number of external links contained on their front pages. Because of this, the mean number of external links is not a good measure of central tendency (69.6, with a standard deviation of 41). The median number of external links on the front pages of our sample was 61.

We also examined how many sites were linked to each of our sample sites. To do this, we used the popular free software package Link Popularity Check (Version 3.0). This software polls the major search engines (AllTheWeb, AltaVista, Google/Tiscali, MSN Search, and Yahoo) and counts the number of Web sites that are linked to the sample site. The variation in this respect was even greater than in the case of external links on the front page. The number of sites linked to our sample sites ranged from a minimum of 18 (Vicksburg, MS) to 444,088 (Honolulu County, HI). The median number of sites linked to our sample sites was 3,543.

The linked popularity of a site is greatly enhanced by the use of META tags. These are lines of HTML code that are hidden from the user by the browser, functioning as key words for search engines' databases. Without META tags, search engines are forced to use the content of a page itself in order to categorize the site, a practice that is often not very precise. There are two types of META tags: key words or description. Key words are simply words or phrases, while the description is a short sentence. An example of both, taken from the town of Superior, CO, is given as follows:

```
<META NAME="description" CONTENT= "Welcome to the Town of Superior, Colorado">
<META NAME= "keywords" CONTENT= "superior, superior colorado, superior co">
```

We tested for both types of META tags and found that only 67 (46.2% of 145) sites in our sample used either or both. This means that 78 (53.8%) had no META tags whatsoever. Perhaps most interesting is the fact that of these sites, 15 were sites that were managed by outside firms, who should understand the value of using this simple coding device.

A final aspect of the Web sites in our sample that we examined was how fast the site loaded in the browser. Download speed is contingent on a number of factors. One of these factors is how large the page is. There are at least three elements that make up how large a given Web page is: the amount of actual text on the page, the number of images on the page, and the amount of code that is written to display the page. To measure these elements, we used a well-known Web site application named Dr. Watson (Version 5.1, found at http://watson.addy.com/). The application, provided by Addy & Associates, has analyzed over 2 million Web pages since the mid-1990s for almost as many visitors. The top half of Table 5 presents the median, range, mean, and standard deviation of each of the above elements, as well as the total size of the front page, measured in bytes.

Table 5. Download speeds in seconds by connection speed

	Median	Range	Mean	St. D.
Number of words (text)	351	8 - 4425	517	606.7
Number of images	16	0 - 88	19	14.6
Size of HTML code in bytes	23,201	1,483 - 224,475	28,056	25,403.4
Total size of front page in bytes	113,207	1,664 - 628,849	135,152	97,148.6
Download speed, 28.8k (seconds)	45	0.7 - 252	54	38.9
Download speed, 56k (seconds)	32	0.5 - 130	38	25.6
Download speed, T1 (seconds)	5.7	0.1 - 31	7	4.9

Notes: Download speeds correlate at 0.945 or greater, $p \leq 01$.

Easily, the most obvious conclusion that can be drawn from the data in Table 5 is that these sites varied widely with respect to all of these elements. This said, we should probably note some fairly obvious points about size and Web site design. First, while minimal file sizes are no longer necessary with faster technologies, it is probably not wise to make files too large. Even with faster connection speeds, it remains true that the larger the file is, the longer it will take to download. So, for example, it is not necessary to have a great deal of text on a front page, and the more text there is, the longer the file will take to download. Relatedly, a site with more images will probably load slower than one with fewer images. Finally, advancements in HTML coding (Javascripts, cascading style sheets, and more) have made the size of the HTML code itself larger. A number of the front pages we surveyed were larger than one quarter of a megabyte in size. In short, overkill is clearly possible.

Finally, we measured actual download speeds of the front pages of the sites in our sample. We relied again on Dr. Watson to compute these speeds. The bottom half of Table 5 presents various measures of the speed of our sample, according to 28.8k and 56k modems, as well as a T1 line.

File size relates, of course, directly to the issue of download speed. For example, in our sample, "download speed, T1" and "total number of bytes of front page" correlate at 0.995, significant at the 0.01 level (two-tailed).

Funding

While the initial investment for a Web site is manageable for most government agencies, the cost of maintenance could be a serious problem. Running a Web site requires a level of technical expertise often beyond what most local governments are accustomed to and their employees are ready for. The cost of hiring computer specialists can be prohibitive for some local governments even in a slack job market. According to the research on technological adoption in the public sector, government units of small size or located in the less developed areas tend to lag behind large governments in keeping up with technological innovation because they are either constrained by their limited resources or by the small returns from investment in technology (Agnew, Brown, & Herr, 1978; Brudney & Selden, 1995; Smith & Taebel, 1985; Walker, 1969).

From our three surveys, we have found several possible solutions to this problem. The first is to solicit donations from the private sector in exchange for advertisement space on an agency's Web pages. The city government of Greenville, North Carolina, exemplifies this approach as its Web site is financially sponsored in part by Pepsi Cola, Inc. The second solution is a kind of cost-sharing arrangement among several government units. In the San Francisco Bay area of California, local governments share the cost of one Web site where each contributor has access to its share of space. Milward and Snyder (1996) have made a convincing case for cost sharing from a service perspective. They argue that a cooperative arrangement will help integrate clients with the services they need. A homeless mentally ill person, for example, will need food and shelter from the social-service department and a medical evaluation and care from the mental-health unit. While Milward and Snyder's interest is mainly in computer networking, the argument they make offers a useful insight into the funding problem of Web applications.

A third possible solution is to attempt to reduce costs through outsourcing Web development and maintenance. However, there is little empirical evidence that contracting for Web-related services actually reduces costs in comparison to own-source provision of such services. Outsourced provision may make Web services possible for local governments that do not have such capacity to develop such services on their own (and may do so in a cost-effective manner), but there is little if any research that directly compares outsourced services with own-source services.

Even this simple own-source (internal) vs. contracted (external) dichotomy is difficult to ascertain in the survey data. In attempting to examine this, we coded the Web site as internally managed if we found any of the following conditions: (a) The domain name was included in the e-mail address to contact the Webmaster, (b) the lack of professionalism made it obvious that an outside firm did not design and/or administer the site, or (c) the site lacked any obvious signs that it was outsourced (e.g., credit given to the outside firm). We were able to determine with a fair amount of certainty whether 118 sites in our sample were managed internally or externally. Of these, 90 (76.3%) were administered internally. A caveat, however, is in order. Of these internally managed sites, many were either initially designed by an outside firm, used a content-management software package, or both. Content-management systems allow employees to enter updates and information into a database; information in the database is then served to the Web server. This allows for a Web site to be more frequently and easily updated since it is not necessary to train employees to write HTML code in order for them to update information about their departments. So, while initial development may have been outsourced, management, maintenance, and updating the site are internally accomplished.

Personnel

A successful Web presence must involve employees in every functional department of a government from the very beginning, especially those who have direct contacts with the public. This is because these employees know more about what information and services ought to be posted online than do computer technicians. Another benefit of increasing employee participation is to boost the level of responsiveness to citizens requesting information and services. There is evidence that in public organizations, employees in functional departments are more interested in new technologies than computer specialists and top managers (Bugler & Bretschneider, 1993; Caudle, 1990). According to Kraemer and Dedrick (1997), this is because program and middle-level managers are more likely to be engaged with people outside the organizations and understand the problems and challenges facing the organizations. Users also need to be consulted about a Web site's performance, and this consultation needs to be conducted on an ongoing basis. Given this, it is reassuring that a percentage of Web sites in our sample that have provisions for user feedback incorporated into the site increased from 79% in 1998 to 95% in 2002 to 97% in 2006.

In any case of a Web presence, a centralized coordination must be maintained as to who has access to a Web site (access here means that a person can alter the contents of a Web page, and enter and delete Web pages) and how the Web pages are connected. While much of the early centralization-decentralization debate in computer technology centered on questions of economies of scale (Danziger et al., 1993; Reschenthaler & Thompson, 1996), we find such questions to be somewhat old fashioned. The technology of computer design has changed enough that a decentralized computing environment can now show the same kinds of economies that a centralized, mainframe system showed in the past. Using content-management systems, local governments can provide both the benefits of a decentralized system and the controls of a decentralized system. Thus, we support a balanced control model of computer management. This is based on the recognition that a government agency must present a unified image to the browsing public (branding) and that the data on a Web server need to be updated, integrated, and safeguarded. What is more, managing a Web site involves some difficult technical problems that require a certain degree of technical expertise. In general, designing and writing a simple Web page is a relatively easy task. Anyone who can do word processing on a computer can master the vocabulary and grammar of HTML with some basic training. When the application moves toward a higher level, involving database searches, interactive citizen input, and service delivery, Web pages must be integrated with some database, CGI programs, and Java applets. Such integration, along with security and other networking needs, would require a certain degree of technical expertise in database management, programming, and network engineering, even though Web design software programs such as Dreamweaver, ColdFusion, or FrontPage have made this much simpler than it was in the past.

From the management point of view, a balanced model is more difficult to manage than a purely decentralized or purely centralized model. A major problem is how to cultivate a cooperative relationship between network specialists and other employees when a Web site is run in house. A number of surveys have found that computer technicians in the public sector tend to attribute problems in computer application to employees who they believe lack a good understanding of the role of information technology (Caudle & Gorr, 1991; Swain, White, & Hubbert, 1995). Conversely (but not surprisingly), nontechnical employees tend to put the blame squarely on computer technicians (Danziger et al., 1993). Successful communication and the solution of problems thus require strong leadership from individuals who can move back and forth from the information technology to the nontechnical environments.

Outreach

Finally, we want to address the access problem. Current estimates of the number of users on the Web range upward of 200 million in the United States alone and some 1 billion worldwide, up from only 3 million in 1994. While the rate of expansion of Web users simply must slow down (as we approach virtually 100% saturation), the number would seem to certainly keep growing in the years to come. This suggests that online government services would have great potential in meeting public demand. But how can a government agency inform those customers who are equipped with Web access of the services provided online? There are two approaches to this. One approach, targeted mainly at local residents, is to utilize all the local media, such as newspaper, radio, and television, to bring about some publicity for a government Web site. Internet service providers in the local community can also provide information about local Web access. Government Web site addresses can be posted in telephone books, on mass transportation vehicles and facilities, on utility bills, and so forth. The other approach, targeted at the general online public, is to have a government Web site catalogued by such online search engines as Yahoo, Excite, or the meta-search engines such as Google.

A more important question for a public agency is how to reach those customers who have little online access. The data on Internet use show that Web access has not been randomly distributed historically, but rather correlated with socioeconomic and demographic differences, something that has been termed the digital divide. Those who have lagged behind in catching the new technology tend to be minorities and have a low level of socioeconomic status (Novak & Hoffman, 1998; Scavo, 2005). The implication of this is both ethical and technical. On the ethical side, with the different levels of the Web, online government services would benefit the affluent more than the poor, thus reinforcing the socioeconomic divide in the information era between the haves and the have-nots. On the technical side, the cost-benefit ratio in setting up and maintaining a government Web site may vary from one community to another. If the residents of a community have little Web access, the benefits may not justify the cost, unless local government is willing (as many have been) to devote resources to increasing Web access by lower income and minority populations. For example, programs to make free computer access available in public libraries have become quite popular. Several foundations, including the Bill and Melinda Gates Foundation, have been providing funding for such projects. Most recently, Microsoft itself has introduced prepaid computers using FlexGo technology that allow the user to pay a small percentage (typically 50%) of the price of a computer up front and then pay the remainder of the fee through the purchase of prepaid cards. Microsoft based this technology on the prepaid mobile-phone model, citing the popularity of prepaid mobile phones in developing countries such as India, Brazil, China, Vietnam, and Russia (*A Computer You Can Afford*, 2006).

A second aspect of the digital divide is not so easily addressed. Relatively excluded populations of Web users include large percentages of the handicapped, those of lower intellectual or reading abilities, and (in the United States) non-English speakers. Local government Web sites tend to fare worse than either federal or state government Web sites on such indicators as the Bobby test (http://www.watchfire.com), which tests for compliance with a variety of legal requirements of Section 508 of the U.S. Rehabilitation Act of 1973 and guidelines of the World Wide Web Consortium (W3C), and the Flesch-Kincaid test used by the U.S.

military as a test of readability. For example, some 90% of city Web sites in the United States were shown to have a 12th-grade reading level while the median reading level of the American public is at the 8th grade (West, 2003).

Conclusion

The Web represents a widely used technology for local government. The potential for the Web to reduce the costs of service delivery, promote civic pride, increase awareness of the types of programs local governments offer, and so forth cannot be doubted by even the most casual observers of Internet operations. However, this potential is not being completely realized by current local government use of the Web. The surveys summarized in this chapter show that while there have been significant increases in the sophistication of local-government Web sties, for some local governments, the Web is no more than an elaborated telephone directory, allowing those who access the Web page to see an online listing of city officials, along with titles, addresses (perhaps e-mail addresses), and telephone numbers. Fewer local governments have taken the additional steps to take fuller advantage of Web technology and move into e-government, by making their Web pages interactive, for example.

Our observations on the findings from our three surveys and our review of the literature on both Web adoption and technology diffusion are that this situation will continue as long as those in charge of Web technology at the local-government level persist in thinking of the Web in the same terms as they think of more traditional avenues of service delivery, citizen input, and local promotion. When the Web was first introduced, several observers noted that it had the potential to remake virtually every type of human interaction. Web technology has been cited as a major component of the possible reforming of local government in several recent studies (Musso et al., 2000). Several years of experience with the Web, coupled with the recent downturn in the economy and the terrorist attacks of 2001, have caused some rethinking of this position. Many users of the Web who may refuse to be seduced by the promises of entirely new modes of interpersonal and intergovernmental exchanges may instead not recognize that the Web can fundamentally change some of the ways people interact with each other and with government. While the exaggerated claims of developers and pundits that the introduction of virtually any new technology should be treated with caution, one should also not underestimate the change that new technology can cause in an existing system. Local governments have been tempted to fit Web technology into their preexisting notion of what the relationship between itself and the population is, and thus may be squandering the more innovative or progressive aspects of the Web. When local governments act in this way, they demonstrate the underlying business assumptions that much of the e-government movement is based on (Abramson & Means, 2001). All of this can result in underuse of the more sophisticated capabilities of the Web. As more and more people become accustomed to going online to conduct such things as their personal correspondence, banking, travel plans, and so forth, local government cannot afford to be left behind. The process is not a simple or quick one.

However, progress is being made. Our survey results show progress on virtually every indicator of local-government Web activity. Our conclusion is thus more positive than in

2002: Local-government Web sites are much more sophisticated than they were when we began this project in 1998, and the progress shown between 1998 and 2002 has accelerated in the 2002-2006 time period. Nagging questions of cost, equity, and access remain, but these are the same questions that local governments strive to answer in the delivery of more traditional services such as sanitation, public safety, and recreation.

References

Abramson, M., & Means, G. (2001). *E-government 2001.* Blueridge Summit, PA: Rowman & Littlefield Publishers.

A computer you can afford. (2006). Retrieved June 10, 2006, from http://www.microsoft.com/presspass/features/2006/may06/05-21EmergingMarket.mspx

Agnew, J. A., Brown, L. A., & Herr, P. (1978). The community innovation process: A conceptualization and empirical analysis. *Urban Affairs Quarterly, 14*, 3-30.

Brudney, J., & Selden, S. (1995). The adoption of innovation by smaller local governments: The case of computer technology. *American Review of Public Administration, 25*, 71-86.

Bugler, D., & Bretschneider, S. (1993). Technology push or program pull: Interest in new information technologies within public organizations. In B. Bozeman (Ed.), *Public management: The state of the art* (pp. 275-293). San Francisco: Jossey-Bass.

Canales, M. (2001, July 16). Service to citizens is focus of e-government. *Federal Times*, p. 16.

Cats-Baril, W., & Thompson, R. (1995). Managing information technology projects in the public sector. *Public Administration Review, 55*, 559-566.

Caudle, S. (1990). Managing information in state government. *Public Administration Review, 50*, 515-524.

Caudle, S., & Gorr, W. (1991). Key information systems issues for the public sector. *MIS Quarterly, 15*, 171-188.

Center for Technology in Government. (2001). *A World Wide Web starter kit.* Retrieved July 25, 2006, from http://www.ctg.albany.edu/publications/guides/www_starter_kit

Chen, Y., & Gant, J. (2001). Transforming local e-government services: The use of application service providers. *Government Information Quarterly, 18*, 343-355.

Danziger, J., Kraemer, K., Dunkle, D., & King, J. (1993). Enhancing the quality of computing service: Technology, structure, and people. *Public Administration Review, 53*, 161-169.

Davenport, T. (1993). *Process innovation: Reengineering work through information technology.* Boston: Harvard Business School Press.

Fountain, J. (1999). The virtual state: Toward a theory of bureaucracy for the twenty-first century. In E. Kamarck & J. Nye (Eds.), *Democracy.com: Governance in a networked world* (pp. 133-156). Hollis, NH: Hollis.

Gellman, B. (2002, June 27). Cyber-attacks by Al Qaeda feared: Terrorists at the threshold of using Internet as tool of bloodshed, experts say. *Washington Post*, p. A01.

Gorr, W. (1986). Use of special event data in government information system. *Public Administration Review, 46*, 532-539.

Government off the Web. (2001, October 19). *Morning Edition.* Retrieved May 20, 2002, from http://search.npr.org/cf/ cmn/segment_display.cfm?segID=131723

Hadden, S. (1986). Intelligent advisory systems for managing and disseminating information. *Public Administration Review, 46*, 572-578.

Hammer, M., & Champy, J. (1993). *Reengineering the corporation: A manifesto for business revolution.* New York: Harper Business.

Ho, A. (2002). Reinventing local government and the e-government initiative. *Public Administration Review, 62*(4), 434-441.

Holden, S., & Ha, L. (2002). Do the facts match the hype? Public demand for, and government attitudes about, e-government. *PA Times*, 3.

Hurley, M., & Wallace, W. (1986). Expert systems as decision aids for public managers: An assessment of the technology and prototyping as a design strategy. *Public Administration Review, 46*, 563-571.

Kaylor, C., Deshazo, R., & Van Eck, D. (2001). Gauging e-government: A report on implementing services in American cities. *Government Information Quarterly, 18*, 293-307.

Kim, S., & Lee, H. (2006). The impact of organizational context and information technology on employee knowledge-sharing capabilities. *Public Administration Review, 66*(3), 370-385.

King, J. L., & Kraemer, K. (1988). Information resource management: Is it sensible and can it work? *Information & Management: The International Journal of Information Systems Management, 15*, 7-14.

Kraemer, K., & Dedrick, J. (1997). Computing and public organizations. *Journal of Public Administration Research and Theory, 7*, 89-112.

Kraemer, K., & King, J. L. (1986). Computing and public organizations. *Public Administration Review*, 488-496.

Kraemer, K., & Northrop, A. (1989). Curriculum recommendations for public management education in computing: An update. *Public Administration Review, 49*, 447-453.

Kriz, M. (2001, October 20). Vanishing Web pages. *New York Times*, pp. 33, 42.

Lan, Z., & Cayer, J. (1994). The challenges of teaching information technology use and management in a time of information revolution. *American Review of Public Administration, 24*, 207-222.

Layne, K., & Lee, J. (2001). Developing fully functional e-government: A four stage model. *Government Information Quarterly, 18*, 122-136.

Lo, J. (2000). *Trojan horse or virus?* Retrieved May 20, 2002, from http://www.irchelp.org/irchelp/security/trojanterms.html

Lo, J. (2002). *Denial of service or "nuke" attacks.* Retrieved May 20, 2002, from http://www.irchelp.org/irchelp/nuke/

Martin, J., & Overman, S. (1988). Management and cognitive hierarchies: What is the role of management information systems? *Public Productivity Review, 11*, 69-84.

McGowan, R., & Lombardo, G. (1986). Decision support systems in state government: Promises and pitfalls. *Public Administration Review, 46*, 579-583.

Meet the man who wove the Web: A chat with Tim Berners-Lee. (1999, November 29). *The News and Observer*, p. D3.

Milward, B., & Snyder, L. (1996). Electronic government: Linking citizens to public organizations through technology. *Journal of Public Administration Research and Theory, 6*, 261-275.

Moon, M. (2002). The evolution of e-government among municipalities: Rhetoric or reality. *Public Administration Review, 62*(4), 424-440.

Morkes, J., & Nielsen, J. (1997). *Concise, scannable, and objective: How to write for the Web.* Retrieved June 7, 2005, from http://www.useit.com/papers/webwriting/writing.html

Muncaster, P. (2006, July 3). Government Web sites beat the private sector. *IT Week*, p. 16.

Musso, J., Weare, C., & Hale, M. (2000). Designing Web technologies for local governance reform: Good management or good democracy? *Political Communication, 17*(1), 1-19.

Norris, D., Fletcher, P., & Holden, S. (2001). *Is your local government plugged in? Highlights of the 2000 electronic government survey.* Washington, DC: International City Management Association.

Norris, D., & Moon, M. (2005). Advancing e-government at the grassroots: Tortoise or hare? *Public Administration Review, 65*(1), 64-75.

Northrop, A., Kraemer, K., Dunkle, D., & King, J. L. (1990). Payoffs from computerization: Lessons over time. *Public Administration Review, 50*, 505-514.

Novak, T., & Hoffman, D. (1998, April 17). Bridging the digital divide: The impact of race on computer access and Internet use. *Science.*

Osborne, D., & Gaebler, T. (1992). *Reinventing government.* New York: Addison-Wesley.

Overman, S., & Loraine, D. (1994). Information for control: Another management proverb? *Public Administration Review, 54*, 193-196.

Phillips, G. (2006, July 14). Official wants to charge for using county Web site. *Durham Herald-Sun*, p. C3.

Reschenthaler, G., & Thompson, F. (1996). The information revolution and the new public management. *Journal of Public Administration Research and Theory, 6*(1), 125-143.

Safai-Amini, M. (2000). Information technologies: Challenges and opportunities for local governments. *Journal of Government Information, 27*(4), 471-479.

Saia, R. (2002). E-government has potential to deliver on Web promises. *Computerworld, 34*(24), 30-31.

Scavo, C. (2005). Citizen participation and direct democracy through computer networking: Possibilities and experience. In D. Garson (Ed.), *Handbook of public information systems* (2nd ed., pp. 255-280). Boca Raton, FL: Taylor and Francis.

Scavo, C., & Shi, Y. (2000). The role of information technology in the reinventing government paradigm: Normative predicates and practical challenges. *Social Science Computer Review, 18*(2), 166-178.

Shi, Y., & Scavo, C. (2000). Citizen participation and direct democracy through computer networking. In D. Garson (Ed.), *Handbook of public information systems* (pp. 247-264). New York: Marcel-Dekker.

Smith, A. C., & Taebel, D. A. (1985). Administrative innovation in municipal government. *International Journal of Public Administration, 7*, 149-177.

Stark, N. (2001). *Getting online: A guide to the Internet for small town leaders.* Retrieved May 20, 2002, from http://www.natat.org/ncsc/Pubs/Getting%20Online/getting_online.htm

Swain, J., White, J., & Hubbert, E. (1995). Issues in public management information system. *American Review of Public Administration, 25*, 279-296.

Technology, yea or nay? (1998, May 13). *Web culture.* Retrieved May 14, 1998, from http://form.netscape.com/ directory/community/html/pc_unreg_mian.htm

Terrorists and libraries. (2002, April 11). *All things considered.* Retrieved May 20, 2002, from http://search.npr.org/cf/cmn/ segment_display.cfm?segID=141525

Toner, R. (2001, October 27). Reconsidering security, US clamps down on agency Web sites. *New York Times*, p. B4.

Vasu, M. (1988). Information utilities and telecommunications networks for public administration. *Public Productivity Review, 11*, 219-227.

Waisanen, B. (2002). The future of e-government: Technology-fueled management tools. *Public Management*, 6-9.

Walker, J. (1969). The diffusion of innovations among the American states. *American Political Science Review,* 63, 880-899.

Weitzman, B., Silver, D., & Brazill, C. (2006). Efforts to improve public policy and programs through data experience: Experiences in 15 distressed American cities. *Public Administration Review, 66*(3), 386-399.

West, D. (2003). *Achieving e-government for all: Highlights from a national survey.* Retrieved July 25, 2006, from http://www.benton.org/publibrary/egov/access2003.html#disability

Endnote

[1] PTI, Inc. is "the non-profit technology research, development and commercialization organization for all cities and counties in the United States. The National League of Cities (NLC), the National Association of Counties (NACo), and the International City/County Management Association (ICMA) provide PTI with its policy direction, while a select group of city and county members conduct applied R&D and technology transfer functions" (http://www.pti.nw.dc.us).

Chapter XVIII

An Information Technology Research Agenda for Public Administration

G. David Garson, North Carolina State University, USA

Abstract

Research questions are outlined, forming the dimensions of a research agenda for the study of information technology in public administration. The dimensions selected as being the most theoretically important include the issues of the impact of information technology on governmental accountability, the impact of information technology on the distribution of power, the global governance of information technology, the issue of information resource equity and the "digital divide," the implications of privatization as an IT business model, the issue of the impact of IT on organizational culture, the issue of the impact of IT on discretion, the issue of centralization and decentralization, the issue of restructuring the role of remote work, the issue of implementation success factors, the issue of the regulation of social vices mediated by IT and other regulatory issues.

Introduction

If anything has mushroomed faster in the past quarter century than information technology, it is perhaps literature about information technology. One large chunk of this literature is theoretically descriptive of various IT projects, another large segment is on the order of how-to manuals, and yet more centers on policy guidelines about computer security, privacy, access, and other management concerns. When one seeks theoretical social-science literature on information technology, one is apt to find empirically disconnected speculation infused with utopian optimism or dystopian cynicism. There remains, however, a growing body of work which is empirically grounded yet raises issues of broad theoretical importance to the public administration community. In this chapter, an attempt has been made to identify some of the primary dimensions of this body of work and to outline the research agenda which it poses.

Political Issues

The Issue of How Information Technology Affects Governmental Accountability

The increasing importance of accountability in public administration was signaled by the change in the name of the Government Accounting Office to the Government Accountability Office (GAO) in 2004. This change is related in part to information technology. A decade earlier Helen Nissenbaum (1994) had argued that "accountability is systematically undermined in our computerized society...it is the inevitable consequence of several factors working in unison—an overly narrow conceptual understanding of accountability, a set of assumptions about the capabilities and shortcomings of computer systems, and a willingness to accept that the producers of computer systems are not, in general, fully answerable for the impacts of their products. If not addressed, this erosion of accountability will mean that computers are 'out of control' in an important and disturbing way." Given the centrality of accountability to democracy, surprisingly little research has been undertaking in the years since Nissenbaum's warning.

For their part, federal managers are aware of the importance of accountability if only in terms of avoiding IT project failure. For example, in early 2006, a GAO report criticized the Department of Defense's Global Information Grid (GIG) project, a massive attempt to weld a variety of programs and systems, including communications satellites and next-generation radio, into an Internet-like but secure worldwide network. The GAO found GIG was "being managed in a stovepiped and bungled manner," with no one ultimately accountable in spite of its estimated cost of $34 billion over 5 years (Onley, 2006). A clear line of accountability is a classic Weberian tenet of ideal-typical bureaucracy, but a variety of forces, including outsourcing and decentralization, operate to undermine it. Public administrationists may seek a resolution by advocating a return to centralization or by seeking some more creative response to the problem of accountability in a decentralized world. Unfortunately, many simply advocate reform in other dimensions (e.g., privatization) while sweeping the accountability issue under the rug.

In principle, information technology brings greater transparency to government operations, and through transparency brings greater accountability, which is a cornerstone of democracy. Greater transparency comes about because of more access to more information by more people, and greater accountability comes because access promotes citizen participation, and participation promotes demands for responsiveness and accountability. The logic is impeccable at an abstract level, but is it empirically supported by real-world data?

If one defines government accountability in terms of Web site openness, then there is evidence that accountability has increased over time, not only in the United States, but also in other countries (Welch & Wong, 2001). The research community has well documented the linkage between demands for accountability on the one hand, and agencies' greater use of information technology, including the Internet, on the other. With respect to academia itself, Wells, Silk, and Torres (1999) for instance have shown how legislative demands for accountability have been the driving force for expansion of university institutional research offices and their associated information databases.

Although there is agreement that governments all over the world are developing more extensive and "open" Web sites, this does not meet the ordinary criteria by which accountability is defined. The dictionary definition of accountability centers on willingness of government to report on actions taken so that the process of being called to account can take place. Accountability requires that government respects the rights of citizens to know, informs citizens of actions, provides avenues for participation in change processes, and responds to demands arising from participation.

The access-participation linkage: Social capital theory suggests that increased networking will lead to increased participation, and a number of scholars have tied the rise of the Internet to a rise in social capital and hence to accountability. However, the evidence in support of this access-participation linkage is mixed at best. In the United States, the higher rates of voter turnout in the 2000 and 2004 presidential elections have been partly attributed to the advent of Internet campaigning, but others argue that this increase is merely the normal effect of close contests when the electorate is evenly split, as it has recently been in the United States. Wilhelm's (1999) study of online Usenet groups found little evidence of the "deliberative discussion" required for participatory democracy. Abroad, Villanueva (2003) has noted the "high expectations" that Internet-supported information access would lead to greater participation in seven Latin American countries studied, but found that in practice a variety of barriers disrupt the access-participation linkage. On a grander scale, Frost (2006) examined the hopes that networked relationships might fulfill the postnationalist hope that new Internet-based social and political bonds can ground transnational projects such as the European Union. Rather, Frost found that the Internet fails to support the commitment and cohesiveness needed to reorient traditional commitments to identification with an emerging transnational organization such as the EU.

- **The access-accountability linkage:** While increased demands for accountability lead to greater IT-based access, the reverse need not be true. Barata and Cain (2001), for instance, have noted that in Sub-Saharan Africa, financial functions were automated early on but hoped-for accountability results did not materialize, at least as measured in terms of levels of corruption and theft of state assets. One reason for failure is simply poor record-keeping infrastructure: greater access to bad information gives the appearance of greater accountability but not the reality. Another reason is that people engaging in impropriety do not post their activities to the Internet.

There is little evidence that greater transparency has brought a reduction in corruption in the United States either. In the most recent presidential administration, there have been no fewer than 34 major ethics scandals (for a listing, see Dizikes, 2005). In 2005, seven members of Congress were indicted or investigated for improper conduct, including conspiracy, securities fraud, and improper campaign donations. A host of lawmakers were reported to have been involved with shady connections to Jack Abramoff, a lobbyist who billed Indian tribes for $82 million in fees that may then have been used for improper political purposes. Among those forces to step down was House Speaker Tom DeLay, long a leading spokesperson for the Bush policy agenda. Though not corruption, other scandals of 2005 related to accountability were false claims about alleged weapons of mass destruction in Iraq, torture in Abu Ghraib, allegations of abuses in secret contract prisons abroad, and the illegal outing of a CIA (Central Intelligence Agency) operative married to a diplomat who had criticized the administration. Although all of these set off waves of Internet discussion, none were exposed due to greater transparency brought by the Internet.

- **The access-responsiveness linkage:** Do administrators actually respond to increased access brought about by information and communications technology? One of the few scholars to examine this had been Darryl West, who has conducted an e-mail responsiveness test since 2000. This test takes a minimalist definition of responsiveness, requiring only that the agency answer a simple e-mail request asking for their hours of operation. Whereas in 2000, when the launch of e-government Web sites stirred interest, some 91% of 286 agencies surveyed were responsive by this minimal criterion, by 2003 the rate had dropped to under 70% bothering to respond. When the request was for simple but specific information, the response rate was lower (West, 2005). One may speculate that responsiveness defined in terms of policy influence would be far lower yet.

  Moreover, there is the matter of the quality of responsiveness under ICT. When input is digital, whether in citizen e-mail or online survey responses, responsiveness can be digital also. That is, rather than some step toward electronic deliberative democracy, computerization can actually be closer to automated marketing, with citizens receiving form responses tailored to what will politically advantage the agency or official.

  Even seemingly major advances in responsiveness, such as the Bush administration's e-regulation initiative, may lessen rather than augment substantive responsiveness. The e-regulation initiative allows citizens to read proposed regulations online, then provide e-mail comments as policy input, seemingly an excellent example supportive of the access-responsiveness linkage. But is there actually more responsiveness to this policy input? Empirical studies have yet to be done, but there is reason to raise the issue. Compare e-regulation with the traditional format for handling the same regulations: holding live hearings around the country. The hearings method gives a forum for activist groups and attracts media coverage, whereas e-regulation individualizes the process and isolates it from media attention, making it plausible to think that ICT may actually diminish the substance of the access-responsiveness linkage.

The Research Need

There is a need to resurrect accountability to the forefront of the public administration research. In terms of information technology, serious empirical questions arise regarding the validity of each of three alleged linkages: that electronic access leads to political participation, that Internet-based increases in transparency lead to greater accountability as measured by reduced wrong-doing, and that increases in access and transparency lead to public decision makers actually being more responsive as measured by changing policy decisions in the direction of the preponderance of digital input.

The Issue of How Information Technology Impacts Power Distribution

Enthusiasts for the information age have seen technology as an irresistible force for the democratization of institutions and societies. It is an open empirical research question, however, whether the rise of public information technology and e-government does, in fact, increase participation levels or even affect the distribution of power. A number of public administrationists, such as the authors associated with Heeks (1999), argue that properly implemented, information technology will "reinvent government" by decentralizing bureaucracy and empowering communities and citizens. This is in line with a number of authors who have argued that the Internet empowers individuals and communities by providing a new, democratic forum for both civic debate and civic action, thereby increasing social capital, as is, for example, the hope of the "smart communities" movement in Canada (Coe, Paquet, & Roy, 1999) and the "digital places" movement in the United States (Horan, 2000). Many authors over a long period of time have articulated this optimistic view of information technology and democratization (e.g., Alexander & Grubbs, 1998; Becker & Scarce, 1984; Blanchard & Horan, 2000; Cohill & Cavanaugh, 1997; Hiltz & Turoff, 1978; Jones, 1995; Loper, 2001; Mantovani, 1996; Negroponte, 1995; Rheingold, 1994; Sackman & Nie, 1970; Schiller, 1996; Schneider, 1996; Warren & Wechsler, 1999).

Protestors from Chinese dissidents at Tiananmen Square to terrorists of Ben Laden's Al Qaeda network have used the Internet extensively, as have a host of environmental, welfare, and assorted other activist groups. Advocates for women have routinely extolled the Internet for its capacity to empower women (Harcourt, 1999). The labor movement has made extensive use of the Internet (Lee, 1997). The Internet was critical to organizing recent World Trade Organization demonstrations (Smythe & Smith, 2006). The Internet makes possible a wide range of activist tactics, from mass e-mail letter-writing campaigns to cyberterrorist damage to the computer networks of enemies. Numerous studies emphasize the capacity of the Internet in particular to promote political organizing and mobilization, even when organizers have few resources by traditional standards (Bimber, 1998; Bonchek,

1995; Grossman, 1995; Gurak, 1997; Tsagarousianiu, Tambini, and Bryan, 1998; Wittig & Schmitz, 1996). Postmes and Brunsting (in press) attribute the global rise in activism in part to resources provided by the Internet. Some argue that citizens are taking a more active role and demanding more information, knowing that IT makes open government more possible (King, Feltry, & O'Neil, 1998).

Nonetheless, thus far the bulk of scholarly opinion is on the side of viewing IT primarily as a tool like any other tool, conferring power on the wielder of the tool. As the wielder with the greatest resources to exploit computers and the Internet tends to be a member of the powers that be, IT functions primarily as a tool to consolidate the status quo, of little overall political consequence and certainly not a harbinger of revolution. Gattiker (2001) notes, for example, that the Internet does offer additional avenues for political participation, but mostly functions simply as one more communications medium, in addition to print, radio, and television, equally prone to manipulation by lobbyists and well-funded interest with the resources and motives to invest in the medium. Vadén and Suoranta (2004) argue that the digital era has brought media concentration and monopoly inimical to digital democracy. Other researchers who have found that IT implementation does not disrupt the persistence of existing political structures within organizations include Danziger, Dutten, Kling, and Kraemer (1982), Fuchs (2004), Kolleck (1993), Kraemer (1991), Pinsonneault and Kraemer (1993), Robertson & Seneviratne (1995), and Riedel, Wagoner, Dresel, Sullivan, & Borgida (2000). Moreover, it should be noted that studies of other technologies, such as the telephone, have likewise found a lack of substantial impact of these other technologies on the distribution of power (Fischer, 1992).

Control systems can be programmed to reflect the political priorities of those in power. As systems design is not normally covered in the press, this comes to light only occasionally, but the practice is routine. A recent example was the attempt of the Bush administration to embed antiunion features in both the proposed merit-based personnel system of the defense department's National Security Personnel System (NSPS) and in a similar human-resources system of the Department of Homeland Security. Antilabor aspects of both were ruled illegal in federal court decisions in 2006 and 2005, respectively. The NSPS software system was found, for instance, to fail to ensure that employees could bargain collectively, did not provide for congressionally required third-party review of labor relations decisions, and did not provide a fair appeals process for employees. In commenting on the case, Colleen M. Kelley, president of the National Treasury Employees Union (NTEU), said it vindicated "NTEU's consistent argument that the White House has clearly overstepped its authority in attempting to take away longstanding federal employee rights" (Walsh, 2006). Normally, however, embedding political controls in ostensibly neutral software goes unchallenged and unnoticed by the public at large. Even in the NSPS case, the Department of Defense did not accept the court ruling, but continued development of the system as it was contested in the courts.

Within organizations, there is a propensity of top officials to use information systems to enhance their own control. For instance, when the U.S. military implemented MCS2 (electronically networked maneuver-control systems) in the Gulf war, an expected result was the centralization of information in the hands of top-ranking officers. What was less expected was an increase in micromanagement, with many commanders unwilling to use new information systems to unleash the abilities of lower level staff to make decisions (Turner, 2003). The desire to not relinquish decision-making power made available by centralization arising from information-systems implementation reinforced and enhanced existing bureaucratic structures in spite of the vision of some that these systems would allow flexible, informed, decentralized decision making in the field.

That IT either reinforces the status quo and increases political inequities or at least does not challenge the status quo is the theme of the dozen authors contributing to *Access Denied* (Lax, 2001). Likewise, critical social theorists such as Robins and Webster (1999) argue that the "information revolution" is not revolutionary at all, but is simply an extension of capitalism, governed by the same profit-based rules. This point is reinforced by empirical studies (cf. Riedel et al., 2000) that find that market forces left to their own tend toward information inequality and undermine attempts to use the Internet for purposes of democratization.

In an empirical study of some 40 community-based information networks, Tonn, Zambrano, and Moore (2001, p. 201) concluded bluntly, "It does not appear that, either individually or in combination, the websites are working to strengthen the social capital of the communities they serve." In fact, information technology may be working in the opposite direction in American democracy. Novotny's (2000) study of presidential election campaigns and IT noted, for instance, that technology was furthering political disengagement of the citizen: "Where candidates once coveted relationships with voters in their districts, they now purchase lists of these same voters on CD-ROM and data files on the World Wide Web as a part of the new campaign technologies" (p. 65).

The Research Need

The research need in relation to information technology and power distribution is to understand better the extent to which the Internet and information technology represent an additional tool of manipulation from the top, or as other authors (e.g., Pruijt, in press; World Bank, 1999) argue, an empowering vehicle for the increase of social capital at the bottom. Although the preponderance of research to date on information technology and democratization tends to cast strong doubt on the enthusiasm of cyberdemocracy advocates, it is nonetheless true that some efforts at using IT to extend democratic values are more successful than others and this variance merits empirical explanation. Research is needed on the prerequisites and success factors associated with greater success (Rocheleau, 1999) whether or not the overall picture is that IT reinforces the status quo. More broadly, research is needed to uncover how the Internet and related information technologies affect how issues are framed in public discourse, and whether there is an effect on national agenda setting different from mass media in general.

The Global Governance of Information Technology

Since 1998, Internet domain names have been under the authority of the Internet Corporation for Assigned Names and Numbers (ICANN). ICANN, in turn, operates under the authority of the U.S. Department of Commerce and has some representation from foreign governments as well. Referring to Microsoft, Verisign, and other multinational firms, one commentator on the domain-name issue recently wrote,

The Internet has become over the past quarter century so darn expensive that the huge tech multinationals that control the internet have become the world's new governments in fact. With war chests of billions of dollars in reserve and little debt on their balance sheets, certain ones are wealthier than the countries in which they are domiciled ... It is governed by a few major multinationals which report to no one but their shareholders. (T1, 2004)

A host of issues ranging from trademark law to enforcing security features of the Internet—even the weak and largely privatized regulatory structure of ICANN has brought it into sharp conflict with multinationals like Verisign, which prefer unregulated markets.

Because of the U.S. origins and composition of ICANN, there has been mounting international pressure to place coordination of the domain-name system under some multilateral body such as the International Telecommunications Union (ITU), which is a UN body headquartered in Geneva that serves to coordinate global telecom networks and services. In 2003 this issue arose at the UN World Summit on the Information Society, where it became clear that many nations, including China and several Middle Eastern and African nations, want more of a say in governance of the Internet. At this meeting, the United States was successful in arguing that multilateralization would imperil technological innovation and free speech on the Internet, a none-too-veiled allusion to the poor record of multilateralization proponents on human rights and transparency in government.

A second World Summit on the Information Society was help in Tunis in 2005, where the governance of the Internet again became a contentious issue. The 2005 Tunis Summit reached a compromise, in which a new governance structure was created: the Internet Governance Forum (IGF). The IGF structure embraced governments, nongovernmental organizations, and the business community (multinationals pressed for a direct role and supported the U.S. position). As such, it represented a partial accession to the demands of many countries, led by Brazil, Russia, China, and Iran, to loosen U.S. control of ICANN, but left some key aspects of ICANN decision making remaining with the U.S. Department of Commerce. Critics saw the IGF as a token structure and the Tunis Summit outcome as a victory for U.S. control, while others defended the IGF structure as a victory for civil society over pure governmental dominance of the Internet. Two key facts emerged from the Tunis process: (a) at least until 2010, U.S. dominance of Internet governance will continue based on existing institutional arrangements, with the IGF having no powers of control, and (b) increasing numbers of nations object to U.S. dominance.

The Research Need

Research on the governance of the Internet calls for political science and international relations studies related to interest-group theory, principal-agent theory, and critical social theory. While there has been considerable coverage of the technical issues associated with governance of the Internet, studies of its politics have largely been left to journalistic accounts (for an exception, see Koppell, 2005). Instead, the research agenda should return to the Lasswellian roots of political science as the study of who gets what, when, and how. A central contribution of social science to the study of information technology has been the demonstration that sociopolitical factors eclipse technological factors in the explanation of

IT implementation. The research need is to examine whether (and if so, how) this is so in the case of Internet governance as well.

The Issue of Information Resource Equity and the Digital Divide

One of the most discussed topics in information-technology literature has to do with assessing the import of the digital divide, the gap separating largely high-socioeconomic-status (SES) technology haves from largely low-SES technology have-nots (Schiller, 1996; Stevens, 2006). The digital divide separates the well-off from the poor, whites from blacks, technology corridors from old-economy regions, and rich nations from developing nations. To the extent that the digital divide exists, it is becoming critical for public administration because the rise of e-government means information-technology inequality is directly related to unequal access to governmental services, not just unequal access to recreational cybermedia.

In its 1999 report, *Falling Through the Net: Defining the Digital Divide* (National Telecommunications and Information Administration [NTIA], 1999), the U.S. Commerce Department studied the extent of the digital divide for the United States for the 1988 to 1996 period. This study document found significantly lower computer ownership and Internet access among Hispanics, blacks, rural individuals, central city residents, and those with low income. As one might expect, income was strongly related to computer ownership. A subsequent commerce department report (NTIA, 2000a) found minorities and rural residents were closing the digital divide. A third report (NTIA, 2000b) found that the digital divide had disappeared with respect to location (discounting lack of high-speed access in rural areas) and with respect to gender. Further optimism was found in a UCLA (University of California, Los Angeles) Center for Communication Policy study that found Internet access was well on the way to becoming universal: 72.3% of Americans went online in 2001, up from 66.9% in 2000 (Lyman, 2001). Other studies concurred that in the United States, the digital divide had sharply diminished or disappeared by the turn of the century (Murdock, 2000; Thierer, 2000).

The digital divide, by the most recent data, showed areas of improvement but also of persistence. By summer 2005, men were only slightly more likely to use the Internet than women (69% to 67%). Racial divisions in Internet use by race persist. Data from summer 2005, show that while 70% of non-Hispanic whites and English-speaking Hispanics used the Internet, only 57% of blacks did so. Household income is also a persistent differentiator of Internet use. Pew survey data showed that in summer 2005, some 93% of those with incomes over $75,000 a year used the Internet, compared to only 49% of those with incomes of $30,000 or less. Some 73% of lower and middle-income bracket ($30,000-$50,000) households and 87% of upper middle bracket ($50,000-$75,000) households used the Internet. This paralleled trends by education, which is highly correlated with income: 29% of those with less than high school education used the Internet, 61% of those with high school, 79% of those with some college, and 89% of those with a college degree used the Internet (Pew Internet and American Life Project, 2005).

Within the United States, there are several developments mitigating the digital divide. By the mid-1990s, inner city and rural public libraries and schools routinely had computers and Internet connections for public access (Henderson & King, 1995). There has been a large number of small-scale experiments, such as the community computing movement, which have sought with varying degrees of success to bring technological knowledge and even equipment into the hands of inner city residents, minorities, and the technologically dispossessed (for examples, see O'Dubhchair, Scott, & Johnson, 2000; Schön, Sanyal, & Mitchell, 1999; Schuler, 1996; Shi & Scavo, 2000; Tonn et al., 2001). And even broadband connectivity has become more affordable and accessible (NTIA, 2004).

When one considers the digital divide between developed nations and poorer nations, the picture is less benign. A 1999 United Nations report concluded, for instance, that "new information and communications technologies are driving globalization but polarizing the world into the connected and the isolated" (United Nations Development Program [UNDP], 1999, p. 5). Hedley (2000, p. 278) found the information revolution was limited primarily to the developed nations of the North. Ferguson (2000) has argued that absent of an effective international communication policy (a less than likely prospect), the world may well divide into the information rich and the information poor. Prospects for Africa, in particular, seem pessimistic (Pfister, 2000). Numerous other authors have asserted that the international digital divide is likely to grow, not recede, in the future (Brody, 2005; Gasco, 2005; Salman, 2004). While developing countries often have some minimal Web presence, with exceptions, frequently the sites are lacking in regular content updates and are poorly designed as well as lacking in network security or any form of certification authority for transactions (Bhatnagar, 2004).

The Research Need

The issue of information technology's impact on resource equity raises two sets of research questions: (a) continued tracking, for the United States and internationally, of the actual extent of the digital divide and whether trends are toward convergence or divergence, and (b) prescriptively, strategies that would be required to successfully diminish the digital divide. Are existing strategies of the UN and other world bodies likely to reduce the international digital divide, significantly? Which ameliorative strategies are more successful than others, and why? Digital divide research will be more meaningful when defined in systemic terms involving all those human as well as physical prerequisites to allowing an individual to be an effective user of information technology, not just yes/no access to hardware.

Organizational Behavior Issues

The Impact of IT on Organizational Culture

Information technology, intranets, and the Internet have the capacity to reinforce or to undermine organizational culture (Dudley, 1994; Hakken, 1993; Orr, 1997; Schoenberger, 1997; Shields, 1997; Wilkinson, 1983). On the side of reinforcement theory is research such as that by Sproull and Kiesler (1991, p. 90), in which they found, "Respondents seem to believe that sharing information enhances the overall electronic community and leads to a richer information environment. The result is a kind of electronic altruism quite different from the fears that networks would weaken the social fabric of organizations." In fact, technology may reinforce existing organizational culture more than is desirable, as the Department of Defense's MCS2 case cited earlier demonstrated, where new information systems conferred on top officials broad powers to micromanage to the detriment of overall effectiveness (Turner, 2003).

Erosion of organizational culture may be a particular problem in public organizations, where fiscal austerity often means information systems are introduced in a more top-down manner with inadequate time for employee participation, testing, and buy-in (Caudron, 1997). In general, research on IT success has pointed to the paramount importance of organization development, leadership, training, and other human factors, yet at least in the public sector, these soft aspects of information systems are precisely the ones most likely to be neglected (Lynn, 2000). When they are neglected, the result may be "loss of control, social isolation and employee disappointment" (Brod, 1994, p. 39). Information technology requires greater collaboration and partnering skills in public management compared to the private sector (Kernaghan & Gunraj, 2004). Management of information technology, especially given the increased importance of public-private partnerships, requires a team-based collaborative organizational culture to motivate and retain knowledge workers. Classical approaches to technology management do not take politics and organizational culture into account adequately (Haynes, 2005).

At the same time, those who are in control of technology implementation may represent a technocratic class within organizations, seeking to displace organizational culture with their own technocratic subculture. Vallas (2000) found that new technologies provide new opportunities for professionally educated experts to enjoy new opportunities for the expansion of their power and autonomy in a form of technocratic undermining of organizational culture. Vallas is in a long tradition of scholarship that argues that new technology reinforces elite culture within organizations, frequently at the expense of the less educated (Aronowitz & DiFazio, 1995; Burris, 1993; Curtis, 1988; Derber & Schwartz, 1991; Michaelson, 1996; Reich, 1993; Rifkin, 1995). Some authors have even gone so far as to see IT as a catalyst for a new form of class warfare within organizations (Perelman, 1998). In addition to the problem of the possible subversion of organizational culture to a technocratic subculture, there is also the problem of cultural corruption through cyber-slacking and Internet addiction (Lavoie & Pychyl, 2001; Marron, 2000).

The Research Need

The traditional Marxian analysis of technocrats cast them as hired guns of the capitalist class. Is there really a technocratic subculture in large organizations, and do technocrats seek to make this subculture the general organizational culture? Or in the end, are technocrats simply still hired tools of more powerful organizational interests and not an organizational force in their own right? Do intranets, systems approaches to work process, and other aspects of IT better socialize workers into acceptance of existing organizational goals and norms, or does this mode of management encourage disengagement from cultural commitment to the organization and instead promote either (a) an ethos of organizationally alienated pragmatic self-interest or (b) the kind of reinvention of organizations along more effective lines, as envisioned by systems reformers? An example of the kind of research needed is a study of three cases of IT implementation using a systems analysis paradigm (SPRINT, the Salford Process Reengineering Involving New Technology methodology). Kawalek and Wastall (2005) found that in every case, organizations in the end opted for outcomes that were the less radical of a set of alternative proposals. In spite of a proclaimed ethos of participatory planning, the systems process in fact was systematically biased against original visions of radical organizational change.

The Impact of IT on Employee Discretion

Discretion is the employee-level view of the issue of power distribution. Will information technology mean that employees will have increased discretion, resting on better information shared more freely through networks? Zuboff (1988) represents the optimistic side of empirical research on this question. Her study of drug company employees' use of a computer conference system concluded that "knowledge displayed itself as a collective resource; non-hierarchical bonds were strengthened; individuals were augmented by their participation in group life; work and play, productivity and learning, seemed ever more inseparable" (p. 386). Other authors in this optimistic tradition include Piore and Sabel (1984) and Heckscher (1994).

An important opposing point of view appeared in the work of Bovens and Zourides (2002). These articles contended that information and communication technology was rapidly restructuring large public agencies from ones in which street-level bureaucrats exercised wide discretion, into a new form in which street-level bureaucrats have diminished discretion or even have vanished altogether. The force for this has been the embedding of rules and even of political dictates into software systems used to track and control the flow of work processes. In the terms of Bovens and Zourides, IT moves bureaucracy from being street level to being system level. The theme of how IT embeds rules and reduces employee discretion was also a major aspect to Jane Fountain's book, *Governing the Virtual State* (2001).

Cynthia West's (2001) work is indicative of findings of researchers who have taken a similar view. Studying the "digerati," information-technology workers, West found a general diminution of discretion, freedom, and privacy on the job as employers discovered and implemented new forms of electronic surveillance and accountability and, more important, as the digerati themselves embraced IT's immersive culture, willingly sacrificing even social and family freedoms to organizationally defined needs. Other researchers have likewise found that while information technology may add conceptually complex content to jobs, this does not mean that the discretion of employees over this content is increased (Burris,1998).

The Research Need

Information technology has the capacity to increase the access of public employees to various forms of expertise, enabling them to make better decisions across a wider array of topics, warranting increasing their level of discretion and freedom. Information technology also has the capacity to increase the intensity and closeness of supervision in ways hierarchically oriented "Theory X" managers only dreamed of in the last century, severely eroding professional discretion and freedom. Again, which pole in this duality will prevail in any given setting, or when considered globally, is an empirical research question. Research is needed to understand the dynamics of forces that under some contingencies tend toward increased discretion, and under other contingencies toward less.

Organizational Design Issues

The Issue of Centralization vs. Decentralization

That the information revolution of the computer era has reduced information costs drastically is the starting point of an argument made by Reschenthaler and Thompson (1996) about the impact of information technology specifically on public administration. They theorized that the effects of reduced information costs are the following:

1. Increased efficacy of market mechanisms vis-a-vis governmental services or regulation
2. Increased efficacy of decentralized allocation of resources and after-the-fact control vis-a-vis centralized and before-the-fact controls (e.g., pushing decision making as low in the organization as possible)
3. Increased efficacy of job-oriented process structures vis-a-vis broad functional structures (e.g., multidisciplinary teams working holistically rather than functional departments working sequentially)

These authors argued that information explosion in the financial sector lessens the need for governmental financial regulations to protect investors, that advances in information processing allow such things as pollution rights or airport landing rights to be securitized and bought and sold on the market, and that technology-based improvements in costing and pricing transportation and telecommunications services justify the deregulation of these industries. More generally, Reschenthaler and Thompson argued that information technology eroded economies of scale in administration, production, and marketing, giving even smaller organizations first-class management tools in each of these functional domains.

Authors such as Reschenthaler and Thompson (1996) saw the technological tide pushing administrators, public ones included, toward increased span of control, reduced layers of management, and delegation of more power to computer-equipped frontline employees, and toward decentralization if not actual contracting out of governmental services. There is an important empirical research question in determining if this dynamic does, in fact, exist. While the advent of desktop computing undeniably has brought striking decentralization in information systems, other researchers have seen a more complex picture characterized by a mix of centralization and decentralization, with the center maintaining critical controls over the decentralized periphery (Heeks, 2000).

Yet other researchers see technology as a primarily centralizing force, reducing autonomy (Burris, 1998; Harris, 2000). Most researchers who have looked at the governmental trends that Reschenthaler and Thompson (1996) cite have attributed them to a combination of the rise of conservative political forces in combination with an era of fiscal austerity economically, largely unrelated to trends in information technology (Liner, 1989). The world of practice provides more telling evidence. Perhaps the foremost thrust of the managerial ideology of chief information officers (CIO) in the last decade has been to overcome department-based stovepipe applications in favor of centralizing IT authority under the CIO, who then promoted enterprise-wide software systems, consolidating previously decentralized ones. This has been the official doctrine of the U.S. Office of Management and Budget, at least since the Clinger-Cohen Act of 1996, and it finds its counterparts at the state level as well (Mosquera, 2005). One observer noted recently, "At the state level, consolidation seems to be a religion of sorts" (Hanson, 2006, p. 27).

The Research Need

Research is needed not just to better describe how information technology increases the power of both the center and the periphery, but also to understand where decentralization does occur, what the several forces behind it are, and whether information technology is a determining factor or merely one of many supporting factors. If the latter, then change in the primary determinants (e.g., prevailing political ideology, state of the economy) may find information technology is not an independent dynamic in its own right but rather supports whatever direction, centralization or decentralization, which these primary determinants dictate. This in turn should help researchers understand whether recent marked trends toward re-centralization are part of a pendular cycle or are a predominantly unidirectional tendency innate to systems logic.

The Implications of Privatization as an IT Business Model

For public administrationists, privatization may be the extreme extension of decentralization. As in other areas of governance, the issue of possible privatization of public information systems is a controversial one deserving more research. Outsourcing is now common and has many longstanding examples, including the noteworthy rise of EDI and other firms handling health care transactions for both the private and public sector. Since then, outsourcing has become a routine reform favored by state legislatures. The state of Connecticut even made an attempt to privatize all governmental IT functions (Daniels, 1997).

Inadequate contract management in relation to privatization due to stripping agencies of internal IT staff needed for policy formation and contract management came back to haunt the British government during the Y2K crisis of the late 1990s. This crisis had to do with legacy software encoding years in only two digits and thus not being able to distinguish 2000 from 1900, for instance. UK administrative reforms in the 1980s and 1990s had hollowed out the civil service in favor of relying on a much larger network of providers. This in turn made the government vulnerable to IT opportunists who exploited media alarmism over Y2K, reaping lucrative contracts at the expense of the public interest (Quigley, 2004).

Republicization may occur when privatization fails. A notable example is the U.S. Postal Service's Strategic Transformation Plan 2006-2010, approved in September 2005, setting forth a plan to reduce costs and streamline operations by transitioning its IT infrastructure from "high-cost IT contractors" to in-house staff (Thormeyer, 2005a). At the state level, in 2003 the Alaska State Legislature authorized a pilot program for outsourcing the state procurement function. The pilot, running from 2004 to 2006 for the Alaska Department of Transportation, suspended the state's existing procurement code, including its lowest-cost requirements (Frisch, 2005). The National Institute of Governmental Purchasing (NIGP) and the National Association of State Procurement Officials (NASPO) lobbied against the plan, calling for reform of state code to facilitate state e-procurement as an alternative to privatizing the procurement function. NASPO argued that privatization would undermine public faith that the procurement process was free of fraud, waste, and favoritism, and that private procurement tended to minimize the number of vendors with whom the state had a supplier relationship, depriving the state of some of the benefits of competition. Toward the end of the pilot, an audit firm was retained to compare procurement records in the year prior to privatization with the first year of privatized procurement. The audit reviewed tens of thousands of procurements and concluded that per-unit costs increased 9% under privatization. Where units and quantities were exactly the same, costs increased by 16% (Elton, 2006).

The Research Need

While privatization and outsourcing are proposed as efficiency measures "to put government on a business-like basis," in truth there are only a few empirical research studies (such as Cohen, 2000) about when and under what circumstances this organizational design strategy actually improves effectiveness and when it actually works in the opposite direction. Moreover, as Rocheleau (1999) noted, in the public sector strategies of privatization and outsourcing are subject to the additional threats of political corruption and favoritism, not to mention the more generalized problems of goal displacement and lack of accountability in the contracting out process. Goal displacement and even outright corruption in privatization and outsourcing are themselves important issues that also are presently underresearched.

The Issue of the Restructuring Effect of Remote Work

Information technology holds the potential to have profound effects on the way organizations, including government agencies, are designed. One of the most commonly mentioned possibilities is telework and telecommuting. Remote work and telecommuting promise efficiencies in terms of nonmonetary job incentives for employees and in some cases major savings in the form of ability to forego capital construction as increased office space is no longer needed (Fulk & deSanctis, 1995). The reality of these promised developments remains largely unresearched.

Telework is persistently if slowly transforming the public workplace. In its 2005 annual survey of federal telework in 82 agencies, the Office of Personnel Management (OPM) found that teleworking by federal employees increased 37% in 2004, meaning that over 140,000 employees participated in telecommuting programs. This is almost double the 73,000 when the survey started in 2001. The OPM attributed the increase to efforts to promote telework, such as the portal Telework.gov (Thormeyer, 2005a). A 2006 report from CDW Government Inc., based on a survey of 500 federal workers, found that telework in the government had increased from 19% to 41% in 2006, with 43% having started in the past 12 months. Some 32% reported their agency started a telework program in the past year, reflecting a marked acceleration of compliance with 2005 legislation calling on federal agencies to offer all eligible employees the opportunity to telecommute (Thormeyer & Gerin, 2006).

In spite of its increase, telecommuting from home has been found to be an isolating experience in most studies to date (Chapman, Sheehy, Heywood, Dooley, & Collins, 1995; Fitzer, 1997), but not in all such studies (Belanger, 1999). Morgan and Symon (in press) note that e-mail and intranets can be used either to promote or to undermine organizational commitment and goal unity with respect to remote employees. A review of the literature on telework by Ellison (2000) concluded that "telework has not come close to fulfilling early, exuberantly optimistic predictions of its adoption," (p. 259) with employee isolation being the most important of several obstacles. Some studies suggest, however, that remote workers, in contrast to telecommuters, are less isolated and more integrated into the organization (Duxbury & Neufeld, 1999).

The Research Need

One of the major efficiency gains promised by information technology is the flattening of organizational hierarchies, as better information flow and control means that the span of control can be increased, enabling the elimination of many middle-management jobs. Increased span of control can be extended to remote work and even telecommuting, enabling possible further efficiencies. The research focus to date has been primarily on obstacles to remote work and telecommuting, particularly isolation. However research is needed even more to study the underlying premise: Does IT, in fact, increase the span of control not just in principle but in the practice of public management? How widespread is organizational flattening and how do public managers seek to maintain control in the face of flatter organizations where this has come about? How do public managers seek to maintain control where IT-enabled changes lead to remote work and telework?

The Issue of Implementation Success Factors

Researchers have uncovered a large number of factors that underpin successful implementation of information-technology projects. Garson (1995) listed over 50 success factors identified in one research study or another, including product factors (software functionality, appropriate technology, standardization, user friendliness, etc.), support factors (reputation, good marketing, top-management support, placement of the IT unit in the decision-making inner circle, etc.), management factors (strong project manager, clear goals, good evaluation plan, client customization, process flexibility, team management, participative decision making, investment in training, etc.), and environmental factors (competitive pressures, growing resource base, dynamic environment, etc.). Many of these factors, and others, were uncovered in hundreds of interviews by students of Alana Northrop with public-sector IT officials, showing, for instance, the preeminent importance of direct top management support and leadership, and of investment in training (Northrop, in press; see also Coursey & Killingsworth, 2000; Drucker, 1995; Keen, 1986; Martin, 1995), along with factors such as utilizing employee feedback, having a period of pretesting before implementation, having Web page support, and many more. A number of these factors appeared a decade and a half ago in Newcomber and Caudle's (1991) survey seeking to evaluate public-sector information-technology projects.

As Brown (2000) noted, many success factors have to do with the extent to which the IT project has taken into account incentives for all the affected stakeholders to buy into the effort (Organization for Economic Cooperation and Development [OECD], 2001), operates under a clear goal structure (DeSeve, Pesachowitz, and Johnson, 1997) with clear milestones and measurable deliverables (Bowsher, 1994; Flowers, 1996; Lucas, 1975), is monitored with a good quality assurance program (Bowsher, 1994; Davis, Lee, Nickles, Chatterjee, Hartung, & Wu, 1992; Keil, 1995) with performance measures (Cohen, 2000; U.S. Food and Drug Administration [FDA], 2005), employs a team reflecting not only technical but also organizational and project-management competencies (Keider, 1984; Regan & O'Connor, 1994), and uses outsourcing strategically but is not over-dependent on outside contractors (Brown & Brudney, 1998; Lacity, Willcocks, & Feeny, 1996; Martin, 1995).

Neglect of the human dimensions of information technology may be linked to the narrowly technocratic nature of leadership frequently found in IT projects. A Cutter Consortium survey published in *CIO Magazine* found human factors, not lack of expertise or technological experience, to be all of the top five failure factors for information-technology officers. Based on a study of 250 senior IT executives who were asked to describe the attributes of the worst IT manager they had ever known, the five failure factors were as follows:

1. Poor interpersonal skills (58%)
2. Being self-centered (56%)
3. Failure to acknowledge problems (55%)
4. Untrustworthiness (54%)
5. Weak management skills (52%)

Noting that failed IT leadership was associated with lack of empathy, lack of emotional ability, and inability to connect with others, the authors of the survey identified the overall main cause of IT leadership failure as a "lack of emotional intelligence" (Prewitt, 2005).

The Research Need

While it is always useful to do case studies, management opinion surveys, and otherwise gain additional information on implementation success factors, the present-day research need is more to develop middle-range contingency theory relating the factors that are already largely known. For instance, is it true that the more dynamic the environment, the more important participative decision making? Is it true that the more competitive the environment, the more important goal clarification? Is it true that the more user friendly the software, the less the need for training? Under what conditions is a given success factor a prerequisite to successful implementation, not merely one more plus? In general, research is needed on the relationship between pairs of success factors and eventually on the dynamics among multiple success factors. Such research will help advance the state of the art from an enumeration of success factors in IT implementation to a theory of implementation.

Regulatory Issues

Regulating Social Vices

Information technology makes almost everything more accessible, including access to services that have traditionally been considered social vices (e.g., pornography, gambling, matching services for adultery, escort services, sales of recreational substances). While not necessarily illegal, these Internet-based services have evoked debate about regulatory policy. Griffiths and Parke (in press) found that Internet gambling was one of the fastest-growing aspects of information technology, tripling in scope since 1997 (see also Mitka, 2001). Sinclair (2000) predicted some 15 million Internet gamblers by 2004. Moreover, Griffiths and Parke found that the structural characteristics of the Internet make gambling significantly more convenient and enjoyable compared to traditional gambling. For instance, people gamble more using e-cash than they would with real cash (Griffiths, 1999). From the other end, Internet gambling firms can gather data better on a customers' preferences and thus provide tailored access. There is an explicit law against online betting in only three states: Nevada, California, and Louisiana. However, no American citizen has ever been arrested for betting on the Internet (Bell, 2005). The implication of most studies to date is that authorities may anticipate increased potential for gambling and gambling addiction globally, and therefore there is an increased need for Internet gambling regulation, a topic on which there appears to be almost no research to date. Much the same might be said of studies regarding other Internet-supported social vices.

The Research Need

Research is needed to benchmark the existing extent of Internet-enabled social vice, to forecast trends based on projections of growth of Internet access, and to understand which regulations are effective in achieving various goals (e.g., curtailment of use, taxation of use). In addition, legal research is needed to understand the application of paper-era court precedents to digital-era situations, along the lines of research by Biegel (2001).

Other Forms of Regulation

Biegel (2001) notes that in addition to social vices, there are a large number of other regulatory issues having to do with the advent of the Internet. Many of these have to do with the commercial world, including deceptive advertising, dishonest transactions, theft of intellectual property, and copyright violation (Bowie, 2005; Cole & Broucek, 2000), practicing medicine or other professions without a license, and sexual harassment on the job. Governments and corporations generally seek regulation of computer code, which, as in the 1998 Digital Millennium Copyright Act, restricts the freedom and privacy of citizens of the cyberworld, as traced by Lessig (1999) and others.

Yet other areas of regulatory concern include traditional subjects of law such as defamation, invasion of privacy, hate speech, issuing threats, and malicious destruction of (cyber) property. As Biegel (2001) notes, there are essentially three strategies for regulation of these and other IT-related problems: regulations within individual countries, international agreements on regulation, and mandated changes in Internet-related server and other software code. In each substantive problem area, research is needed to assess what combination of these three strategies is needed.

The Research Need

In every specific policy arena one may research whether information technology is changing the dynamics of the policy process. In global environmental and development policies, does IT improve the effectiveness of environmental activists or of their opponents more? Does it promote cooperation or conflict more? Does it change policy outcomes? Similar research questions may be posed for policy arenas such as intellectual property, privacy protection, regional planning, mass transportation, social welfare, adult education, art and culture, Internet voting proposals, and any other policy domain that can be named.

Conclusion

Like the Internet itself, social research on information technology is in its infancy. The would-be student is inundated with volumes of "look at this, look at that" descriptive literature and an equal amount of rose-colored optimistic IT salesmanship literature. Much of what passes for theory—much of systems theory is an example—is not theory at all in the sense of trying to specify the causal dynamics relating variables affecting IT processes, but is instead merely a vocabulary for ordering and describing such processes. Fortunately, research dimensions are emerging based on empirical research on theoretically important concepts. Without vainly aspiring toward a general field theory of information technology, it is still possible to focus on these theoretically important dimensions and thereby advance the state of the art of research on information technology in the public sector and even in general. To achieve the next step, the level of mid-range theory, researchers must avoid the temptation to be mere publicists of technological novelty and possibility or social engineers studying how to co-opt actors into acceptance of technological change. Researchers must instead return to their roots as social scientists, selecting the targets of study based on theoretical importance, operationally defining indicators of concepts, indexing these into latent variables, and modeling the relationship of these variables to one another.

References

Alexander, J. H., & Grubbs, J. W. (1998). Wired government: Information technology, external public organizations, and cyberdemocracy. *Public Administration and Management: An Interactive Journal. 3*(1). Retrieved from http://www.hbg.psu.edu/Faculty/jxr11/alex.html

Aronowitz, S., & DiFazio, W. (1995). *The jobless future.* Minneapolis, MN: University of Minnesota Press.

Barata, K., & Cain, P. (2001). Information, not technology, is essential to accountability: Electronic records and public-sector financial management. *The Information Society, 17*(4), 247-258.

Barry, C. A. (1997). Information skills for an electronic world: Training. Doctoral research students. *Journal of Information Science, 23*(3), 225-238.

Becker, T., & Scarce, R. (1984, August-September). *Teledemocracy emergent: State of the art and science.* Paper presented at the American Political Science Association 1984 Annual Meeting, Washington.

Beckman, D., & Hirsch, D. (1997). New approach to cite-seeing: West/Thompson merger yields all-electronic method of researching cases. *ABA Journal, 83*(1), 85.

Belanger, F. (1999). Workers' propensity to telecommute: An empirical study. *Information & Management, 35*, 139-153.

Bell, R. J. (2005). Online sports betting: The law and you. *About.* Retrieved February 21, 2006, from http://sportsgambling.about.com/od/legalfacts/a/Betting_Laws.htm

Bhatnagar, S. (2004). *From vision to implementation: A practical guide with case studies.* Thousand Oaks, CA: Sage Publications.

Biegel, S. (2001). *Beyond our control? Confronting the limits of our legal system in the age of cyberspace.* Cambridge, MA: The MIT Press.

Bimber, B. (1998). The Internet and political mobilization: Research note on the 1996 election season. *Social Science Computer Review, 16*(4), 391-401.

Blanchard, A., & Horan, T. (2000). Virtual communities and social capital. In G. D. Garson (Ed.), *Social dimensions of information technology: Issues for the new millennium* (pp. 6-22). Hershey, PA: Idea Group Press.

Bonchek, M. (1995). *A grassroots in cyberspace: Using computer networks to facilitate political participation.* Paper presented to the Midwest Political Science Association, Annual Meeting, Chicago.

Bovens, M., & Zouridis, S. (2002). From street-level to system-level bureaucracies: How information and communication technology is transforming administrative discretion and constitutional control. *Public Administration Review, 62*(2), 174-184.

Bowie, N. E. (2005). Digital rights and wrongs: Intellectual property in the information age. *Business and Society Review, 110*(1), 77-96.

Bowman, C. M., Danzig, P. B., Manber, U., & Schwartz, M. F. (1994). Scalable Internet resource discovery: Research problems and approaches. *Communications of the ACM, 37*(8), 98-107.

Bowsher, C. A. (1994). *Improving mission performance through strategic information management and technology: Learning from leading organizations* (GAO/AIMD-94-115).

Brod, C. (1994). *Technostress: The human cost of the computer revolution.* Boston: Addison-Wesley Publishing Company.

Brody, H. (2005). What matters most depends on where you are. *Technology Review, 4*, 43-52.

Brown, A. (1998). Narrative politics and legitimacy in an IT implementation. *Journal of Management Studies, 35*(1), 1-22.

Brown, M. M. (2000). Mitigating the risk of information technology initiatives: Best practices and points of failure for the public sector. In G. D. Garson (Ed.), *Handbook of public information systems* (pp. 153-164). New York: Marcel Dekker.

Brown, M. M., & Brudney, J. L. (1998). A "smarter, better, faster, and cheaper" government? Contracting for geographic information systems. *Public Administration Review, 58*, 335-345.

Burris, B. H. (1993). *Technocracy at work.* Albany, NY: SUNY Press.

Burris, B. H. (1998). Computerization of the workplace. *Annual Review of Sociology, 22*(1), 141-158.

Caudron, S. (1997). The human side of technology launch. *Training and Development, 51*(2), 20-25.

Chapman, A. J., Sheehy, N., Heywood, S., Dooley, S., & Collins, S. C. (1995). The organizational implications of teleworking. In C. L. Cooper & I. T. Robertson (Eds.), *International Review of Industrial and Organizational Psychology.* New York: Wiley.

Coe, A., Paquet, G., & Roy, J. (1999). E-governance and smart communities: A social learning challenge. *Social Science Computer Review, 19*(1), 80-93.

Cohen, S. (2000). The need for strategic information systems planning when contracting-out and privatizing public sector functions. In G. D. Garson (Ed.), *Handbook of public information systems* (pp. 99-112). New York: Marcel Dekker.

Cohill, A. M., & Cavanaugh, A. L. (Eds.). (1997). *Community networks: Lessons from Blacksburg, Virginia.* Boston: Artech House.

Cole, R. J., & Broucek, E. F. (2000). Intellectual property for public managers. In G. D. Garson (Ed.), *Handbook of public information systems* (pp. 215-230). New York: Marcel Dekker.

Coursey, D., & Killingsworth, J. (2000). Managing government Web services in Florida: Issues and lessons. In G. D. Garson (Ed.), *Handbook of public information systems* (pp. 345-362). New York: Marcel Dekker.

Curtis, T. (1988). The information society: A computer-generated caste system? In V. Mosco & J. Wasko (Eds.), *The political economy of information* (pp. 95-107). Madison, WI: University of Wisconsin Press.

Daniels, A. (1997). The billion-dollar privatization gambit. *Governing*, 28-31.

Danziger, J. N., Dutton, W. H., Kling, R., & Kraemer, K. L. (1982).*Computers and politics.* New York: Columbia University Press.

Davis, G. B., Lee, A. L., Nickles, K. R., Chatterjee, S., Hartung, R., & Wu, Y. (1992). SOS: Diagnosis of an information system failure. *Information and Management, 23*, 293-318.

Derber, C., & Schwartz, W. (1991). New mandarins or new proletariat? Professional power at work. *Research in the Sociology of Organizations, 8*, 71-96.

DeSeve, E., Pesachowitz, A. M., & Johnson, L. K. (1997). *Best IT practices in the federal government.* Chief Information Officers Council & Industry Advisory Council.

Dizikes, P. (2005). The scandal sheet. *The Salon.com.* Retrieved April 19, 2006, from http://dir.salon.com/story/news/feature/2005/01/18/scandal/index.html

Drucker, P. (1995). *Managing in a time of great change.* New York: Truman Talley Books/Dutton.

Dudley, K. M. (1994). *The end of the line: Lost jobs, new lives in post-industrial America.* Chicago: University of Chicago Press.

Duxbury, L., & Neufeld, D. (1999). An empirical evaluation of the impacts of telecommuting on intra-organizational communication. *Journal of Engineering and Technology Management, 16*, 1-28.

Ellison, N. (2000). Researching telework: Past concerns and future directions (pp. 255-276). In G. David Garson, (Ed.), *Social dimensions of information technology.* Hershey, PA: Idea Group Publishing.

Elton, K. (2006). State puts out a contract on DOT budget. *Off the Record, 226*, 1. Retrieved April 15, 2006, from http://www.alaskareport.com/kim-elton2.htm

Ferguson, K. (2000). World information flows and the impact of new technology: Is there a need for international communication policy and regulation? In G. D. Garson (Ed.), *Social dimensions of information technology: Issues for the new millennium* (pp. 323-339). Hershey, PA: Idea Group Press.

Fischer, C. S. (1992). *America calling: A social history of the telephone to 1940.* Berkeley, CA: University of California.

Fitzer, M. M. (1997). Managing from afar: Performance and rewards in a telecommuting environment. *Compensation and Benefits Review, 29*(1), 65-73.

Flowers, S. (1996). *Software failure: Management failure: Amazing stories and cautionary tales.* New York: John Wiley & Sons.

Fountain, J. E. (2001). *Building the virtual state: Information technology and institutional change.* Washington, DC: Brookings Institution Press.

Frisch, K. (2005). Alaska opens door to procurement outsourcing. *Government Procurement, 13*(3), 6-8.

Frost, C. (2006). Internet galaxy meets postnational constellation: Prospects for political solidarity after the Internet. *The Information Society, 22*(1), 45-49.

Fuchs, C. (2004). The political system as a self-organizing information system. In R. Trappl (Ed.), *Cybernetics and systems 2004* (pp. 353-358). Vienna: Austrian Society for Cybernetic Studies.

Fulk, J., & deSanctis, G. (1995). Electronic communication and changing organizational forms. *Organizational Science, 6*, 337-349.

Garrison, B. (1997). Computer-assisted reporting. *Editor & Publisher, 130*(25), 40-43.

Garson, G. D. (1995). *Computer technology and social issues.* Hershey, PA: Idea Group Press.

Gasco, M. (2005). Exploring the e-government gap in South America. *International Journal of Public Administration, 28*(7-8), 683-701.

Gattiker, U. E. (2001). *The Internet as a diverse community: Cultural, organizational, and political issues.* Mahwah, NJ: Lawrence Erlbaum.

Griffiths, M. D. (1999). Gambling technologies: Prospects for problem gambling. *Journal of Gambling Studies, 15*, 265-283.

Griffiths, M. D., & Parke, J. (in press). The social impact of Internet gambling. *Social Science Computer Review.*

Grossman, L. K. (1995). *The electronic republic: Reshaping democracy in the information age.* New York: Viking.

Gurak, L. (1997). *Persuasion and privacy in cyberspace.* New Haven, CT: Yale University Press.

Hakken, D. (1993). Computing and social change: New technology and workplace transformation, 1980-1990. *Annual Review of Anthropology, 22*, 107-132.

Hanson, W. (2006). CIO survival guide. *Public CIO, 4*, 26-32.

Harcourt, W. (1999). *Women@Internet: Creating new cultures in cyberspace.* New York: Zed Books, Ltd./St. Martin's Press.

Harris, N. D. (2000). Intergovernmental cooperation in the development and use of information systems. In G. D. Garson (Ed.), *Handbook of public information systems* (pp. 165-178). New York: Marcel Dekker.

Haynes, P. (2005). New development: The demystification of knowledge management. *Public Money and Management, 25*(2), 131-135.

Healey, B. W. (1997). How to use the Internet for legal research. *Trial, 33*(5), 84.

Heckscher, C. (1994). Defining the post-bureaucratic type. In C. Heckscher & A. Donnellon (Eds.), *The post-bureaucratic organization* (pp. 14-62). Thousand Oaks, CA: Sage Publications.

Hedley, R. A. (2000). The information age: Apartheid, cultural imperialism, or global village? In G. D. Garson (Ed.), *Social dimensions of information technology: Issues for the new millennium* (pp. 278-290). Hershey, PA: Idea Group Press.

Heeks, R. (Ed.). (1999). *Reinventing government in the information age: International practice in IT-enabled public sector reform.* London: Routledge.

Heeks, R. (2000). A core-periphery approach to centralization/decentralization issues in public information systems. In G. D. Garson (Ed.), *Handbook of public information systems* (pp. 127-140). New York: Marcel Dekker.

Henderson, C. C., & King, F. D. (1995). The role of public libraries in providing public access to the Internet. In B. Kahin & J. Keller (Eds.), *Public access to the Internet.* Cambridge, MA: MIT Press.

Hiltz, R. S., & Turoff, M. (1978). *The network nation: Human communication via the computer.* Reading, MA: Addison Wesley Advanced Books.

Horan, T. A. (2000). Planning digital places: A new approach to community telecommunications planning and deployment. In G. D. Garson (Ed.), *Handbook of public information systems* (pp. 473-488). New York: Marcel Dekker.

Hubbard, A. (1981). Online research for a state legislature. *Online, 6*(4), 27-41.

Jones, S. (Ed.). (1995). *CyberSociety: Computer-mediated communication and community.* Thousand Oaks, CA: Sage Publications, Inc.

Kawalek, P., & Wastall, D. (2005). Pursuing radical transformation in information age government: Case studies using the SPRINT methodology. *Journal of Global Information Management, 13*(1), 79-101.

Keen, P. (1986). *Competing in time.* Cambridge: Ballinger Publishing Company.

Keider, S. P. (1984). Why system development projects fail. *Journal of Information Systems Management, 1*(3), 33-38.

Keil, M. (1995). Pulling the plug: Software project management and the problem of escalation. *MIS Quarterly*, 421-447.

Kernaghan, K., & Gunraj, J. (2004). Information technology in the public sector: A resource-based approach to informatization. *Administration Publique du Canada, 47*(4), 525-546.

King, C., Feltry, K., & O'Neil, S. B. (1998). The question of participation: Toward authentic public participation in public administration. *Public Administration Review, 58*(4), 353-359.

Kolleck, B. (1993). Computer information and human knowledge: New thinking and old critique. In

M. Leiderman, C. Guzetta, L. Struminger, & M. Monnickendam (Eds.), *Technology in people services: Research, theory, and applications* (pp. 455-464). New York: Haworth Press.

Koppell, J. G. S. (2005). Pathologies of accountability: ICANN and the challenge of Amultiple accountabilities disorder. *Public Administration Review, 65*(1), 94-108.

Kraemer, K. L. (1991). Strategic computing and administrative reform. In C. Dunlop & R. King, (Eds.), *Computerization and controversy* (pp. 167-180). New York: Academic Press.

Lacity, M. C., Willcocks, L. P., & Feeny, D. F. (1996). The value of selective IT sourcing. *Sloan Management Review, 37*(3), 13-25.

Lavoie, J. A. A., & Pychyl, T. A. (2001). Cyber-slacking and the procrastination super highway: A Web-based survey of on-line procrastination, attitudes and emotion. *Social Science Computer Review, 19*(4), 431-444.

Lax, S. (Ed.). (2001). *Access denied in the information age.* New York: Palgrave Publishing/St. Martin's Press.

Lee, E. (1997). *The labour movement and the Internet: The new internationalism.* London: Pluto Press.

Lessig, L. (1999). *Code and other laws of cyberspace.* New York: Basic Books.

Liner, E. B. (1989). *A decade of devolution: Perspectives on state-local relations.* Washington, DC: The Urban Institute Press.

Lively, G. M. (1995, May). *United Nations On-Line Crime and Justice Clearinghouse (UNOJUST).* Paper presented at the Workshop on International Cooperation and Assistance in the Management of the Criminal Justice System: Computerization of Criminal Justice Operations and the Development, Analysis, and Policy Use of Criminal Justice Information, Cairo, Egypt.

Loper, R. (2001). Digital democracy: Civic engagement in the twenty-first century. *National Civic Review, 3.*

Lucas, H. C., Jr. (1975). *Why information systems fail.* Irvington, NY: Columbia University Press.

Lyman, J. (2001). Internet now "ingrained" in American life. *NewsFactor Network.* Retrieved April 10, 2002, from http://www.newsfactor.com/perl/story/15046.html

Lynn, D. B. (2000). Technology launch in government: The human factor. In G. D. Garson (Ed.), *Handbook of public information systems* (pp. 113-126). New York: Marcel Dekker.

Maclay, V. (1989). Selected sources of United States agency decisions. *Government Publications Review, 16*(3), 271-301.

Mantovani, G. (1996). *New communication environments: From everyday to virtual.* Bristol, PA: Taylor & Francis.

Marron, K. (2000, January 20). *Attack of the cyberslackers.* London: *The Globe and Mail*, p. T5.

Martin, J. (1995). *The great transition: Using the seven disciplines of enterprise engineering to align people, technology, and strategy.* New York: American Management Association.

Michaelson, K. L. (1996). Information, community, and access. *Social Science Computer Review, 14*(1), 57-59.

Miller, R. B. (1995). The information society: O brave new world. *Social Science Computer Review, 13*(2), 163-170.

Milward, H. B., & Snyder, L. O. (1996). Electronic government: Linking citizens to public organizations through technology. *Journal of Public Administration Research and Theory, 6*(2), 261-275.

Mitka, M. (2001). Win or lose, Internet gambling stakes are high. *Journal of the American Medical Association, 285*,1005.

Morgan, S. J., & Symon, G. (in press). Computer-mediated communication and remote management: Integration or isolation? *Social Science Computer Review*.

Mosquera, M. (2005). VA will centralize IT budget, personnel authority over next year. *Government Computer News*. Retrieved October 21, 2005, from http://www.gcn.com/vol1_no1/daily-updates/37362-1.html

Murdock, D. (2000). *Digital divide? What digital divide?* Washington, DC: Cato Institute.

National Telecommunications and Information Administration (NTIA). (1999). *Falling through the Net: Defining the digital divide*. Author.

National Telecommunications and Information Administration (NTIA). (2000a). *Falling through the Net: Defining the digital divide, Part II. Access and usage*. Author.

National Telecommunications and Information Administration (NTIA). (2000b). *Falling through the Net: Toward digital inclusion*. Author.

National Telecommunications and Information Administration (NTIA). (2004). *A nation online: Entering the broadband age*. Washington, DC: Author.

Negroponte, N. (1995). *Being digital*. New York: Knopf.

Newcomber, K. E., & Caudle, S. L. (1991). Evaluating public sector information systems. *Public Administration Review, 51*, 377-384.

Nissenbaum, H. (1994). Computing and accountability. *Communications of the ACM, 37*(1), 73-80.

Northrop, A. (in press). Lessons for managing information technology in the public sector. *Social Science Computer Review*.

Novotny, P. (2000). The World Wide Web and local media in the 1996 presidential election. In G. D. Garson (Ed.), *Social dimensions of information technology: Issues for the new millennium* (pp. 64-85). Hershey, PA: Idea Group Press.

O'Dubhchair, K., Scott, J. K., & Johnson, T. G. (2000). Community decision support systems: Managing knowledge for community and economic development. In G. D. Garson (Ed.), *Handbook of public information systems* (pp. 489-500). New York: Marcel Dekker.

Onley, D. S. (2006, February 20). GAO cites need for better management of GIG. *Government Computer News*. Retrieved March 21, 2006, from http://www.gcn.com/print/25_4/38247-1.html

Organization for Economic Cooperation and Development (OECD). (2001). *The hidden threat to e-government: Avoiding large government IT failures* (Public management policy brief). Retrieved October 6, 2005, from http://www.oecd.org/dataoecd/19/12/1901677.pdf

Orr, J. (1997). *Talking about machines*. Ithaca: ILR.

Perelman, M. (1998). *Class warfare in the information age*. New York: St. Martin's Press.

Pew Internet & American Life Project. (2005). *Demographics of Internet users*. Retrieved October 18, 2005, from http://www.pewinternet.org/trends/User_Demo_08.09.05.htm

Pfister, R. (2000). Information in and on Africa: Past, present, and future. In G. D. Garson (Ed.), *Social dimensions of information technology: Issues for the new millennium* (pp. 301-322). Hershey, PA: Idea Group Press.

Pinsonneault, A., & Kraemer, K. L. (1993). The impact of information technology on middle managers. *MIS Quarterly, 17*, 271-292.

Piore, M., & Sabel, C. F. (1984). *The second industrial divide: Possibilities for prosperity*. New York: Basic Books.

Postmes, T., & Brunsting, S. (in press). Collective action in the age of the Internet: Mass communication and online mobilization. *Social Science Computer Review*.

Prewitt, E. (2005, August 1). Why IT leaders fail. *CIO Magazine*. Retrieved February 2, 2006, from http://www.cio.com/archive/080105/tl_manangement.html

Pruijt, H. (in press). Social capital and the equalizing potential of the Internet. *Social Science Computer Review.*

Quigley, K. (2004). The emperor's new computers: Y2K (re)visited. *Public Administration, 82*(4), 801-829.

Quinn, P. C. (1997). Research sites on the World Wide Web. *Trial, 33*(4), 84-92.

Regan, E. A., & O'Connor, B. N. (1994). *End-user information systems.* New York: Macmillan.

Reich, R. (1991). *The work of nations: Preparing ourselves for 21st century capitalism.* NY: Random House.

Rheingold, H. (1994). *The virtual community: Homesteading on the electronic frontier.* Reading, MA: Addison-Wesley.

Reschenthaler, G. B., & Thompson, F. (1996). The information revolution and the new public management. *Journal of Public Administration Research and Theory, 6*(1), 125-143.

Riedel, E., Wagoner, M. J., Dresel, L., Sullivan, J. L., & Borgida, E. (2000). Electronic communities: Assessing equality of access in a rural Minnesota community. In G. D. Garson (Ed.), *Social dimensions of information technology: Issues for the new millennium* (pp. 86-108). Hershey, PA: Idea Group Press.

Rifkin, J. (1995). *The end of work: The decline of the global labor force and the dawn of the post-market era.* New York: Putnam.

Robbin, A. (1992). Social scientists at work on electronic research networks. *Electronic Networking: Research, Applications and Policy, 2*(2), 6-30.

Robertson, P. J., & Seneviratne, S. J. (1995). Outcomes of planned organizational change in the public sector: A meta analytic comparison to the public sector. *Public Administration Review, 55*(6), 547-558.

Robins, K., & Webster, F. (1999). *Times of technoculture: From the information society to the virtual life.* New York: Routledge.

Rocheleau, B. (1999). The political dimensions of information systems in public administration. In G. D. Garson (Ed.), *Information technology and computer applications in public administration: Issues and trends*(pp. 23-40). Hershey, PA: Idea Group Press.

Sackman, H., & Nie, N. (Eds.). (1970). *The information utility and social choice.* Montvale, NJ: AFIPS Press.

Salman, A. (2004). Elusive challenges of e-change management in developing countries. *Business Process Management Journal, 10*(2), 140-157.

Schiller, H. I. (1996). *Information inequality: The deepening social crisis in America.* New York: Routledge.

Schneider, S. (1996). Creating a democratic public sphere through political discussion: A case study of abortion conversation on the Internet. *Social Science Computer Review, 14*(4), 373-393.

Schoenberger, E. (1997). *The cultural crisis of the firm.* London: Blackwell.

Schön, D., Sanyal, B., & Mitchell, W. (Eds.). (1999). *High technology and low-income communities: Prospects for the positive use of advanced information technology.* Cambridge, MA: MIT Press

Schuler, D. (1996). *New community networks: Wired for change.* Reading, MA: Addison Wesley.

Shi, Y., & Scavo, C. (2000). Citizen participation and direct democracy through computer networking. In G. D. Garson (Ed.), *Handbook of public information systems* (pp. 247-264). New York: Marcel Dekker.

Shields, M. A. (1997). Reinventing technology in social theory. *Current Perspectives in Social Theory,*

17, 187-216.

Sinclair, S. (2000). *Wagering on the Internet.* Retrieved from http://www.igamingnews.com/Stata Corporation

Smythe, E., & Smith, P. J. (2006). Legitimacy, transparency, and information technology: The World Trade Organization in an era of contentious trade politics. *Global Governance: A Review of Multilateralism and International Organizations, 12*(1), 31-53.

Sproull, L., & Kiesler, S. (1991). Computers, networks and work. *Scientific American,* 84-91.

Stevens, D. (2006). *Inequality.com: Money, power and the digital divide.* Oxford, UK: Oneworld Publications.

Stowers, G. N. L. (1996). Moving governments online: Implementation and policy issues. *Public Administration Review, 56*(1), 121-125.

T1 (2004). *Internet governance has become a non-issue.* Circle ID. Septermber 23. Retrieved November 21, 2004, from http://www.circeid.com/article/768_0_1_0_C/

Taylor, J. A. (1991). Public administration and the information polity. *Public Administration, 69*(2), 171-90.

Thierer, A. D. (2000). *How free computers are filling the "digital divide."* Washington, DC: Heritage Foundation.

Thormeyer, R. (2005a). OPM study predicts increase in teleworking by federal employees. *Government Computer News.* Retrieved January 22, 2006, from http://www.gcn.com/vol1_no1/daily-updates/37731-1.html

Thormeyer, R. (2005b). USPS to rely on career employees, not contractors, for IT services. *Government Computer News.* Retrieved September 30, 2005, from http://www.gcn.com/vol1_no1/daily-updates/37141-1.html

Thormeyer, R., & Gerin, R. (2006). *Survey finds federal telework increasing.* Retrieved March 8, 2006, from http://www.gcn.com/online/vol1_no1/40054-1.html

Tonn, B., Zambrano, E. P., & Moore, S. (2001). Community networks or networked communities. *Social Science Computer Review, 19*(2), 201-212.

Tsagarousianiu, R., Tambini, D., & Bryan, C. (1998). *Voice and equality: Civic voluntarism in American politics.* Cambridge, MA: Harvard University Press.

Turner, M. T. (2003). Command and control in cyberspace: An analysis of the case studies surrounding electronically networked maneuver control systems of the United States Armed Forces. *The Turner Network.* Retrieved April 16, 2006, from http://www.turnernetwork.com/online/documents/commandcontrol.shtml

United Nations Development Program (UNDP). (1999). *Human development report 1999.* New York: UNDP & Oxford University Press.

U.S. Food and Drug Administration (FDA). (2005). *FDA information technology (IT) strategic plan.* Washington, DC: Author. Retrieved April 11, 2006, from http://www.fda.gov/oc/information-technology/strategicplan.html#1

Vadén, T., & Suoranta, J. (2004). Breaking radical monopolies: Towards political economy of digital literacy. *E-Learning, 1*(2), 283-301.

Vallas, S. (1993). *Power in the workplace: The politics of production at AT&T.* Albany, New York: State University of New York Press.

Vallas, S. (1999). Rethinking post-Fordism: The meanings of workplace flexibility. *Sociological Theory, 17*(1), 68-101.

Vallas, S. (2000). Manufacturing knowledge: Technology, culture, and social inequality at work. In G. D. Garson (Ed.), *Social dimensions of information technology: Issues for the new millennium* (pp. 236-254). Hershey, PA: Idea Group Press.

Villanueva, E. (2003). *Comparative Media Law Journal, 2*, 89-100.

Walsh, D. C. (2006). In QDR, defense focuses on combating cyberthreats. *Government Computer News*. Retrieved February 14, 2006, from http://www.gcn.com/vol1_no1/security/38207-1.html?topic=security

Warren, M. A., & Wechsler, L. F. (1999). Electronic governance on the Internet. In G. D. Garson, (Ed.), *Information technology and computer applications in public administration: Issues and trends* (pp. 118-136). Hershey, PA: Idea Group Press.

Weiskel, T. C. (1991). Environmental information resources and electronic research systems. *Library Hi Tech, 9*(2), 7-19.

Welch, E. W., & Wong, W. (2001). Global information technology pressure and government accountability: The mediating effect of domestic context on Website openness. *Journal of Public Administration Research and Theory, 11*(4), 509-538.

Wells, J., Silk, E., & Torres, D. (1999). Accountability, technology, and external access to information: Implications for IR. *New Directions for Institutional Research, 103*, 23-39.

West, C. K. (2001). *Techno-human mesh: The growing power of information technologies*. Westport, CT: Quorum Books.

West, D. (2005). *Digital government: Technology and public sector performance*. Princeton, NJ: Princeton University Press.

Wilhelm, A. (1999). Virtual sounding boards: How deliberative is online political discussion? In B. N. Hague & B. D. Loader (Eds.), *Digital democracy: Discourse and decision making in the information age*? London: Routledge.

Wilkinson, B. (1983). *Shopfloor politics of new technology*. London: Heinemann.

Wittig, M. A., & Schmitz, J. (1996). Electronic grassroots organizing. *Journal of Social Issues, 52*(1), 53-69.

World Bank. (1999). *Social capital and information technology*. New York: Author. Retrieved from http://www.worldbank.org/poverty/scapital/topic/info1.htm

Zuboff, S. (1988). *In the age of the smart machine: The future of work and power*. New York: Basic Books.

About the Authors

G. David Garson (e-mail: David_Garson@ncsu.edu) is a professor of public administration at North Carolina State University (USA) where he teaches courses on advanced research methodology, geographic information systems, information technology, e-government, and American government. His most recent book is *Public Information Technology and E-Governance: Managing the Virtual State* (2006). For the last 23 years he has also served as editor of the *Social Science Computer Review* and is on the editorial board of four additional journals.

* * * * *

Myria W. Allen (e-mail: myria@uark.edu) is an associate professor at the University of Arkansas, Fayetteville (USA). She is the author of over 20 journal articles as well as numerous conference papers, case studies, and book chapters. Her primary interest areas include management issues related to the information-technology workforce, gender dynamics in the workplace, and organizational communication. She has conducted research or consulted for a variety of for-profit organizations including Tyson Foods, J. B. Hunt Transport, and Unilever.

Deborah J. Armstrong (e-mail: darmstrong@walton.uark.edu) is an assistant professor of information systems in the Sam M. Walton College of Business at the University of Arkansas (USA). Dr. Armstrong's research includes issues at the intersection of IS personnel and mental models involving the human aspects of technology, change, learning, and cognition. Her articles on cognition have appeared in journals such as the *Journal of Management Information Systems* and *Communications of the ACM* among others, and she has co-edited a book *Causal Mapping for Research in Information Technology*.

Jody Baumgartner (e-mail: gsb_jody@yahoo.com) is an assistant professor of political science at East Carolina University (USA). He received his PhD from Miami University and has published works on several subjects, including the presidency, campaigns and elections, and political behavior.

Suzanne Beaumaster (e-mail: beaumast@ulv.edu) is chair of the Doctoral Program in Public Administration (DPA) and associate professor of public administration at the University of La Verne in Southern California (USA). Dr. Beaumaster received her graduate degrees from Virginia Polytechnic Institute and State University (PhD), and her MPA from Northern Kentucky University. She is a specialist in the area of information-technology management in public organizations. Dr. Beaumaster has presented a number of papers at international conferences on IT. Some of the work she is known for includes *Strategic Information Technology Management: Organizational, Political, and Technological Forces in a Public Utility System, The City of Anaheim Technological Initiatives*, and *Local Government IT Implementation Issues: A Challenge for Public Administration*. Dr. Beaumaster has also developed civic networks and computer training workshops for rural communities.

Douglas A. Carr (e-mail: dcarr@netzero.net) is a PhD candidate at the Martin School of Public Policy and Administration at the University of Kentucky. He received his MPA from the University of Kentucky and his BA from Taylor University (Indiana, USA). His research and teaching interests are in the areas of quantitative analysis, public economics, policy analysis, and environmental policy. He has presented papers on environmental policy and the preparation of teachers in environmental science at state and national professional conferences. He has also authored technical reports for state agencies and private organizations.

T. R. Carr (e-mail: tcarr@siue.edu) is a professor and department chair of public administration and policy analysis at Southern Illinois University, Edwardsville (USA). He received his PhD and MPA from the University of Oklahoma and his BA from Minot State University, North Dakota (USA). He teaches primarily in the areas of quantitative methods and local government administration. He is active in academic and professional associations and has held various offices in state and local chapters of the American Society for Public Administration. He was reelected to his third term as mayor of the city of Hazelwood, Missouri, a suburb of St. Louis (2006).

Chris Demchak (e-mail: demchak@u.arizona.edu) holds a Berkeley PhD in political science (technology and organizations emphasis) with a master's degree in energy engineering (Berkeley) and another in economic development (Princeton). With a long-term interest in how organizations evolve in response to security challenges and surprise associated with complex technologies embedded in the networked structure and processes of large-scale technical systems, Dr. Demchak has studied militaries comparatively in their adoption of, and adaptation to, advanced networked systems. An early social-scientist member of the emerging intelligence and security informatics field (2003), Dr. Demchak is in the process of cofounding with colleague Kurt Fenstermacher (Arizona MIS) an interdisciplinary group on security studies, information, and governance called the Eller Security Lab (http://securitylab.arizona.edu) to bring security scholars at all levels together with IT scholars and social scientists in cross-fertilizing research and courses. One near-term goal is to refine and test the BIK (behavior-identity knowledge) model balancing privacy and security needs nationally (http://www.bepress.com/forum/vol2/iss2/art6/). The second is to apply the ATRIUM model of tacit-knowledge capture to both single and multiple-allied security agencies. Dr. Demchak is also cofounder of the Cyberspace Policy Research Group (CyPRG), which completed a 5-year National Science Foundation (NSF) sponsored study of the spread of the Internet into national governments globally and the consequences for accountability and effectiveness (data available online at http://www.cyprg.arizona.edu). Demchak is the author of a book and multiple articles related to these topics. He also teaches courses on public and nonprofit MIS for MPA students, the evolution of security institutions and the state, and international management.

George T. Duncan (e-mail: gd17@andrew.cmu.edu) is a professor of statistics in the H. John Heinz III School of Public Policy and Management and the Department of Statistics at Carnegie Mellon University (USA). He was the Lord Simon visiting professor at the University of Manchester (2005). He served on the faculty of the University of California, Davis (1970-1974), and as a Peace Corps volunteer in the Philippines (1965-1967), teaching at Mindanao State University. Duncan has published more than 60 papers in such journals as *Statistical Science*, *Management Science*, *Journal of the American Statistical Association*, *Econometrica*, and *Psychometrika*. He chaired the Panel on Confidentiality and Data Access of the National Academy of Science (NAS, 1989-1993), resulting in the book *Private Lives and Public Policies: Confidentiality and Accessibility of Government Statistics*. He chaired the American Statistical Association's Committee on Privacy and Confidentiality. He is a fellow of the American Statistical Association, an elected member of the International Statistical Institute, and a fellow of the American Association for the Advancement of Science. He was elected Pittsburgh Statistician of the Year by the American Statistical Association (1996). Duncan has been editor of the theory and methods section of the *Journal of the American Statistical Association*. He received his PhD degree (1970) in statistics from the University of Minnesota.

Kurt D. Fenstermacher (e-mail: KurtF@Eller.Arizona.edu) is an assistant professor in the Departments of Management Information Systems and Computer Science (by courtesy) at the University of Arizona (USA). Dr. Fenstermacher also directs the Eller Security Lab (http://securitylab.eller.arizona.edu), which conducts interdisciplinary research at the intersection of information security and public policy. He researches the construction of just-in-time knowledge-management systems and the development of analytical systems that preserve privacy.

Carl Grafton (e-mail: cgrafton@mail.aum.edu) is a professor of political science and public administration at Auburn University (Montgomery, USA). He has written numerous articles and book chapters on computer applications in public administration and political science, and publications on public budgeting, political ideology and public policy, and other topics. His teaching areas include public and nonprofit budgeting, quantitative decision making, and science and technology policy. He currently serves as book-review editor of the *Social Science Computer Review*.

Charles C. Hinnant (e-mail: Chris.Hinnant@hinnantassociates.org) is an academic fellow with the IT team at the U.S. Government Accountability Office in Washington, D.C. His research interests include social and organizational informatics, digital government, information management and policy, public management, social-science research methods, and applied statistics. He is particularly interested in how public organizations employ ICT to alter organizational processes and structures and how the use of ICT ultimately impacts institutional governance mechanisms. His research has appeared in journals such as *Administration and Society*, *Journal of Public Administration Research and Theory*, and *IEEE Transactions on Engineering Management*. He earned his BS and MPA at North Carolina State University and his PhD in public administration from the Maxwell School at Syracuse University.

Stephen H. Holden (e-mail: holden@umbc.edu) is an assistant professor in the Department of Information Systems at the University of Maryland, Baltimore County (UMBC). His research interests include electronic government, the management of public-sector information technology, and information policy. He has published in *Administration and Society*, *IEEE Internet Computing*, *Government Information Quarterly*, *The Information Society*, *International Journal of Public Administration*, and *Public Performance and Management Review*. Holden was also a coauthor of two reports by an NAS study team examining the privacy impacts of authentication and is a member of a NAS team studying the Social Security Administration's electronic-government strategy. He holds a PhD (public administration and public affairs) from Virginia Polytechnic and State University, and an MPA and BA (public management) from the University of Maine.

Alana Northrop (e-mail: anorthrop@fullerton.edu) is a professor of political science at California State University, Fullerton (USA). She received her PhD from the University of Chicago. Dr. Northrop has worked in the field of information systems for 30 years, being on the original URBIS study team that first surveyed U.S. cities' use of computers. She has continued to study government use of computing both in the United States and cross-nationally, and recently has written on city Web-site features and evaluation methods. In addition, she has published research on municipal reform, bureaucratic effectiveness, quantitative methods, and the initiative process.

John O'Looney (e-mail: olooney@cviog.uga.edu), a senior public-service associate at the Carl Vinson Institute of Government, University of Georgia (USA), has worked with government and nonprofit organizations for over 25 years to help improve their management, enhance their performance, and evaluate their impacts through the effective use of technology. He is responsible for consultation and technical assistance to local and state government agencies in the areas of information systems and electronic government, and has developed computer applications and decision-support systems for public-sector budgeting, emergency management, human-services planning and program evaluation, energy education, and land-use planning.

Ali O. Ozturk (e-mail: aoozturk@chass.ncsu.edu) is the assistant director of Leadership in the Public Sector, a distance education degree-completion program at the College of Humanities and Social Sciences, North Carolina State University (USA).

Anne Permaloff (e-mail: apermalo@mail.aum.edu) is a professor of political science and public administration at Auburn University (Montgomery, USA). Public-policy analysis, applied research and program evaluation, and public and nonprofit budgeting are among the courses she teaches. Her publications include numerous articles on computer applications in public administration and political science as well as works related to Alabama and state government, public budgeting, and other topics.

Charles Prysby (e-mail: prysby@uncg.edu) is a professor of political science at the University of North Carolina (UNC) at Greensboro (USA). He received a BS degree in political science from Illinois Institute of Technology (1966) and a PhD in political science from Michigan State University (1973). His main areas of research are in American elections, voting behavior, and political parties. He is co-author of the SETUPS (*Supplementary Empirical Teaching Unit in Political Science*) series of computer-based instructional packages on voting behavior in U.S. elections.

Nicole D. Prysby (e-mail: nprysby@earthlink.net) is an attorney with interests in the area of employment law. She received her BA with highest honors (economics and history) from the University of North Carolina at Chapel Hill (1992) and her JD with honors from the University of North Carolina at Chapel Hill School of Law (1995). She is a contributing author to several annual publications on the employment and human-resource law area, including the *State by State Guide to Human Resource Law*, *Multistate Guide to Benefits Law*, *Multistate Payroll Guide*, and *Equal Employment Opportunity Compliance Guide* (all published by Aspen/Panel). She and Charles Prysby are the co-authors of three previous articles dealing with legal aspects of e-mail in the workplace. She currently works as a consultant in Charlottesville, Virginia.

Christopher G. Reddick (e-mail: Chris.Reddick@utsa.edu) is an assistant professor of public administration at the University of Texas at San Antonio (USA). Dr. Reddick's research interests are in e-government, public budgeting, and employee health benefits. Some of his publications can be found in *Public Budgeting & Finance*, *Government Information Quarterly*, and the *e-Service Journal*.

Margaret F. Reid (e-mail: mreid@uark.edu) is a professor of political science at the University of Arkansas, Fayetteville. Her research focuses on gendered workplaces, complexities involving the implementation of multiactor policy partnerships, and sustainable community development (domestic and international). Her book *Glass Walls and Glass Ceilings: Women's Representation in State and Municipal Bureaucracies*, co-authored with colleagues Kerr and Miller, came out in 2003. She has been author or co-author of research that has been published in, among others, *Sex Roles*, *Women & Politics*, *Administration & Society*, *Public Administration Review*, *Urban Affairs Review*, *State and Local Government Review*, and in numerous edited works.

Cynthia K. Riemenschneider (e-mail: criemen@walton.uark.edu) is an associate professor of information systems in the Sam M. Walton College of Business at the University of Arkansas (USA). Her publications have appeared in *Information Systems Research*, *IEEE Transactions on Software Engineering*, *Journal of Management Information Systems*, *Sex Roles: A Journal of Research*, and others. She currently conducts research on IT workforce issues, women in IT, IT adoption, and IT and small businesses.

Bruce Rocheleau (e-mail: brochele@niu.edu) is a professor of political science at Northern Illinois University (USA). He has been researching information management in the public sector for 25 years. He has written extensively on the subject in various articles and chapters on topics ranging from expert systems to e-payment rates. Recently, he completed a book titled *Public Management Information Systems* and is currently working on a book titled *Case Studies in Digital Government*.

Stephen F. Roehrig (e-mail: roehrig@andrew.cmu.edu) is a teaching professor in information systems and public policy at the H. John Heinz III School of Public Policy and Management at Carnegie Mellon University (USA). He has worked as a mathematician for the U.S. Navy (1981-1983) and the U.S. Coast Guard (1983-1991). He has published papers in the *Proceedings of the American Mathematical Society*, *Management Science*, *Decision Support Systems*, and the *Annals of Operations Research*. He has consulted with the U.S. Census Bureau, the U.S. Bureau of Labor Statistics, and the U.S. Energy Information Administration. He has a PhD degree (1991) in decision sciences from the Wharton School, University of Pennsylvania.

Carmine Scavo (e-mail: scavoc@ecu.edu) is an associate professor and director of the Master of Public Administration Program at East Carolina University in Greenville, North Carolina (USA). He received his PhD in political science from the University of Michigan (USA). He has been coauthor of the American national election study SETUPS published by the American Political Science Association (since 1984). His published work has appeared in *Public Administration Review*, *Urban Affairs Review*, *Journal of Urban Affairs*, *Social Science Quarterly*, *Politics and Policy*, *Journal of Public Affairs Education*, and in chapters in edited volumes. For a number of years, he has taught graduate courses on public information technology, public-policy analysis, intergovernmental relations, and urban management.

Shannon Howle Schelin (e-mail: schelin@sog.unc.edu), PhD, is the director of the Center for Public Technology at the University of North Carolina's School of Government (USA). She is also a faculty member and teaches several courses in the MPA program, including research methods, program evaluation, and public-policy analysis. Dr. Schelin designed and implemented the first local government chief information officers certification program in the nation and successfully graduated the first class (November, 2005). Shannon received her PhD in public administration from North Carolina State University (2003) and was named Departmental Teaching Assistant of the Year for 2 consecutive years. She received her MPA from UNC Charlotte (2000) and her BA from UNC at Chapel Hill (1997). She has numerous publications on public information technology, including *Humanizing IT: Advice from the Experts* with G. David Garson, and "Training for Digital Government" in *Digital Government.*

Stuart W. Shulman (e-mail: Shulman@pitt.edu, Web site: http://shulman.ucsur.pitt.edu) is an assistant professor with a joint appointment in the School of Information Sciences and the Graduate School of Public and International Affairs at the University of Pittsburgh. He is a senior research associate in the University of Pittsburgh's Center for Social and Urban Research (UCSUR) and in the E-Democracy Centre of the Université de Genève, European University Institute, and Oxford Internet Institute. Dr. Shulman is the founder and director of UCSUR's Qualitative Data Analysis Program (QDAP). Dr. Shulman has been the principal investigator and project director on related National Science Foundation-funded research projects focusing on electronic rule making, language technologies, digital citizenship, and service-learning efforts in the United States. He has been the organizer and chair for federal-agency-level electronic rule-making workshops held at the Council for Excellence in Government (2001), the National Defense University (2002), the National Science Foundation (2003), and the George Washington University (2004). For the last 3 years, Dr. Shulman has served on the program committee for the NSF's National Conference on Digital Government Research, and is the senior contributing editor for the *International Journal of E-Government* and an editorial board member for the *International Journal of Electronic Democracy*, and was the 2004 and 2005 president of the American Political Science Association's organized section on information technology and politics. He may be contacted at.

James E. Swiss (e-mail: swiss@social.chass.ncsu.edu) is an associate professor in the public-administration program of the School for Public and International Affairs, North Carolina State University (USA).

Ellen Storey Vasu (e-mail: ellen_vasu@ncsu.edu) is a professor and head of the Department of Curriculum and Instruction at North Carolina State University. She has also served as an adjunct professor, and teaches methodology at the doctoral level in the College of Education.

Michael L. Vasu (e-mail: vasu@chass.ncsu.edu) is an assistant dean of information technology at North Carolina State University (USA). He has twice served as president of the Southern Association for Public Opinion Research (SAPOR). He is the author of numerous books and articles on computing, research methods, and organizational behavior. Professor Vasu and Professor Raymond Taylor shared the second-place cash award in the 1998 Franz Edleman Award for Achievements in Operations Research and Management Sciences.

Index

Symbols

A

B

C

D

E

F

G

H

I

J

L

M

N

Q

R

S

T

U

V

W

Y

CPSIA information can be obtained
at www.ICGtesting.com
Printed in the USA
BVHW020245210723
667580BV00003B/30